Large Marine Ecosystems of the World

Trends in Exploitation, Protection, and Research

Large Marine Ecosystems – Volume 12

Series Editor: Kenneth Sherman
Director, Narragansett Laboratory and Office of
Marine Ecosystem Studies
NOAA-NMFS, Narragansett, Rhode Island, USA
And Adjunct Professor of Oceanography
Graduate School of Oceanography, University of Rhode Island
Narragansett, Rhode Island, USA

On the cover

The global map of average primary productivity and the boundaries of the 64 Large Marine Ecosystems (LMEs) of the world, available at www.edc.uri.edu/lme. The annual productivity estimates are based on SeaWIFS satellite data collected between September 1998 and August 1999, and the model developed by M. Behrenfeld and P.G. Falkowski (Limnol.Oceangr. 42(1): 1997, 1-20). The color-enhanced image (provided by Rutgers University) depicts a shaded gradient of primary productivity from a high of 450 gCm^2yr^{-1} in red to less than 45 gCm^2yr^{-1} in purple.

Large Marine Ecosystems of the World

Trends in Exploitation, Protection, and Research

Edited by

Gotthilf Hempel
Science Advisor, Senate of Bremen, Germany
Professor emeritus, Bremen and Kiel Universities
Director, Alfred-Wegener Institute for Polar and Marine Research, Bremerhaven until 1992
Director, Center for Tropical Marine Ecology, Bremen until 2000
Director, Institute for Baltic Research, Rostock until 1997

Kenneth Sherman
Director, Narragansett Laboratory and Office of Marine Ecosystem Studies
NOAA-NMFS
Narragansett, Rhode Island, USA
and
Adjunct Professor of Oceanography,
Graduate School of Oceanography
University of Rhode Island, Narragansett, Rhode Island, USA

2003

ELSEVIER

Amsterdam - Boston - Heidelberg - London - New York - Oxford - Paris
San Diego - San Francisco - Singapore - Sydney - Tokyo

ELSEVIER B.V.
Sara Burgerhartstraat 25
P.O. Box 211, 1000 AE Amsterdam, The Netherlands

First edition 2003

Library of Congress Cataloging in Publication Data
A catalog record from the Library of Congress has been applied for.

British Library Cataloguing in Publication Data
A catalogue record from the British Library has been applied for.

ISBN: 0 444 51027 3

♾ The paper used in this publication meets the requirements of ANSI/NISO Z39.48-1992 (Permanence of Paper).
Printed in The Netherlands.

Series Editor's Introduction

The world's coastal ocean waters continue to be degraded from unsustainable fishing practices, habitat degradation, eutrophication, toxic pollution, aerosol contamination, and emerging diseases. Against this background is a growing recognition among world leaders that positive actions are required on the part of governments and civil society to redress global environmental and resource degradation with actions to recover depleted fish populations, restore degraded habitats and reduce coastal pollution. No single international organization has been empowered to monitor and assess the changing states of coastal ecosystems on a global scale, and to reconcile the needs of individual nations to those of the community of nations for taking appropriate mitigation and management actions. However, the World Summit on Sustainable Development convened in Johannesburg in 2002 in recognition of the importance for coastal nations to move more expeditiously toward sustainable development and use of ocean resources, declared that countries should move to introduce ecosystem-based assessment and management practices by 2010, and by 2015, restore the world's depleted fish stocks to maximum levels of sustainable yields. At present, 126 developing countries are moving toward these targets in joint international projects supported, in part, by financial grants by the Global Environment Facility in partnership with scientific and technical assistance from UN partner agencies (e.g. UNIDO, UNEP, UNDP, IOC, FAO), donor countries and institutions and non-governmental organizations including the IUCN (World Conservation Union). Many of these projects are linked to ecosystem-based efforts underway in Europe and North America in a concerted effort to overcome the North-South digital divide.

The volumes in the new Elsevier Science series on Large Marine Ecosystems are bringing forward the results of ecosystem-based studies for marine scientists, educators, students and resource managers. The volumes are focused on LMEs and their productivity, fish and fisheries, pollution and ecosystem health, socioeconomics and governance. This volume in the new series, "Large Marine Ecosystems of the World: Trends in exploitation, protection and research," encompasses a broad spectrum of assessments of changing states of LMEs from around the globe with a focus on the physical and biological forces driving the observed changes. The volume provides a comprehensive synthesis of information on the present state of marine fisheries, other resources and environments. The contributions have been peer reviewed. They have been selected for inclusion in the series as a contribution toward a continuing assessment and evaluation of the changing conditions within the World's LMEs as scientists and ocean stewardship agencies move ahead toward ecosystem improvement targets endorsed by world leaders at the Johannesburg Summit. Production of this volume was supported in part by financial assistance of the IUCN.

Kenneth Sherman, Series Editor
Narragansett, Rhode Island

Acknowledgements

The editors are indebted to the willingness of the contributors to take time from their normal schedules to prepare the expert syntheses and reviews that collectively serve to move us forward toward ecosystem-based assessment and management of the world's Large Marine Ecosystems. We are pleased to acknowledge the interest and financial support of the International Union for the Conservation of Nature (World Conservation Union), the U.S. National Oceanic and Atmospheric Administration and the U.S. National Marine Fisheries Service and the Senate of Bremen, Germany.

This volume would not have been possible without the capable cooperation of many people who gave unselfishly of their time and effort. We are indebted to Dr. Sally Adams, North Scituate, Rhode Island for her extraordinary dedication, care and expertise in technically editing and preparing the volume in camera-ready format for publication. Also, we extend our thanks to Donna Busch for her meticulous review of each of the chapters to ensure that the volume was free of any inadvertent scientific or technical omissions or misrepresentations. Thanks go also to Ms. Franca Hinrichsen for editorial assistance. Further, we extend thanks to Ms. Mara Vos-Sarmiento and Ms. Els Bosma of Elsevier Science for their care in the final production of this volume.

The Editors

Editors' Introduction

Gotthilf Hempel and Kenneth Sherman

During the ten years between the UN Conference on Environment and Development (UNCED) in Rio in 1992 and the World Summit on Sustainable Development (WSSD) in Johannesburg in 2002, the movement toward a science based Large Marine Ecosystem approach to marine resources assessment and management has been extended from earlier studies of the North Sea (Hempel 1978, Daan et al. 1996) and the US Northeast Shelf ecosystem (Sherman et al. 1996) to a global series of projects with substantial financial support provided to 126 countries in Asia, Africa, Latin America and eastern Europe by the Global Environment Facility.

The ecosystem-based movement is now well underway on the recovery of depleted fish stocks, restoration of degraded habitats, and reduction of pollution. The Global Environment Facility (GEF) has been replenished with pledges of $3.0 billion in grant funding for developing nations requiring financial assistance as they move toward ecosystem-based assessment and management practices. Integration of ongoing research on ecological processes (eg GLOBEC in China, Germany and the US; the BENEFIT Program of Angola, Namibia and South Africa) is encouraged by scientists participating in ecosystem based resource projects.

The early results are encouraging. As described in Chapter 5, significant recovery of depleted demersal fish stocks is underway in the US Northeast Shelf ecosystem, while countries bordering on the North Sea are undertaking management measures to restore its depleted fish stocks (Hempel and Pauly 2002[1], ICES 2003[2]). In other parts of the world, the GEF is assisting countries in adopting assessment and management practices to recover depleted fish stocks of LMEs in sub-Saharan Africa, southeast Asia, China, Latin America, and countries bordering the Baltic Sea.

LME projects have become important vehicles for international partnership and mutual assistance for scientific technical and administrative capacity building. In addition to the GEF, funding by World Bank, as well as the UNDP, UNIDO, IUCN and UNEP, is supporting national assistance programmes. Experience gained by highly developed marine institutes is being shared to assist scientists

---

[1] Hempel, G. and D. Pauly. 2002. Fisheries and Fisheries Science and their search for sustainability. Chapter 5 in: J.G. Field, G. Hempel and C.P. Summerhayes, eds. Oceans 2020: Science for future needs. Island Press, Washington, D.C.

[2] ICES. 2003. Environmental Status of the European Seas. International Council for the Exploration of the Sea, Copenhagen 2003. 75p

in less developed countries in tailoring research, and monitoring and management methodologies to needs and capabilities within their LMEs (Duda and Sherman 2002[3], Fortes and Hempel 2002[4]).

The LME approach bridges the gap between research and monitoring, as strategies and methods of monitoring have to be based on research in the area under consideration, while large, well-designed long term data-sets are indispensable for advanced system modeling and prediction. The interdisciplinary dialogue produces new lines of scientific thinking. The LME projects have greatly promoted our knowledge of the coastal and marine ecosystems in tropical and subtropical regions. Scientifically compelling are comparisons between various ecosystems which are studied simultaneously by largely the same methods. Many insights can be obtained from the data on different trophic levels of the various systems. In combination with the intra-LME analysis of time series, inter-LME comparisons are more likely to provide more causal clues than the study of decadal changes in individual LMEs alone. Scientific complementarity can also be realized where LME assessments can be linked to ongoing, more basic research activities.

In the present volume the authors present new information on global trends in LME exploitation, protection and research. These studies are focused on 8 polar and boreal LMEs, 4 LMEs in upwelling areas, and 3 tropical LMEs. The penultimate chapter provides a global hierarchical system for merging open-ocean biogeochemical biomes and provinces with the more coastal 64 LMEs in a GIS-based approach. This hierarchy allows for scaling of biological, physical and chemical indicators of changing ecosystem states from LMEs to biogeochemical provinces and biomes. A synopsis that highlights and connects the principal findings by the authors is given in the concluding chapter of the volume.

---

[3] Duda, A.M. and K. Sherman. 2002. A new imperative for improving management of large marine ecosystems. Ocean & Coastal Management 45 (2002) 797-833.

[4] Fortes, M.D. and G. Hempel. 2002. Capacity building. In Field, J.G., G. Hempel and C.P. Summerhayes, eds. Oceans 2020: Science, trends, and the challenge of sustainability. Island Press: Washington, DC (USA). 283-307.

Contributors

Nickolas A. Bond
University of Washington, JISAO
Seattle, Washington
USA

Richard D. Brodeur
NOAA/National Marine Fisheries
Service, NWFSC
Newport, Oregon
USA

Jon Brodie
James Cook University
Townsville, Queensland
AUSTRALIA

Peter Celone
NOAA, NMFS,
Narragansett Laboratory
Narragansett, Rhode Island
USA

Villy Christensen
Fisheries Centre
University of British Columbia
Vancouver, BC
CANADA

Philippe Cury
IRD Research Associate
Marine and Coastal Management
Rogge Bay, Cape Town
SOUTH AFRICA

Georgi M. Daskalov
Institute of Fisheries
Varna, BULGARIA

Vladimir J. Denisov
Murmansk Marine Biological Inst.
Murmansk, RUSSIA

Sergei K. Dzhenuk
Murmansk Marine Biological Inst.
Murmansk, RUSSIA

Werner Ekau
Center for Tropical Marine Ecology
Bremen, GERMANY

Rainer Froese
Center for Tropical Marine Ecology
Bremen, GERMANY

Gotthilf Hempel
c/o Center for Tropical Marine Ecology
Bremen, GERMANY

Gerd Hubold
Director, Institute of Sea Fisheries
Hamburg, GERMANY

Bengt-Owe Jansson
Professor, Department of Systems
Ecology
Stockholm University
SWEDEN

Joseph Kane
USDOC/NOAA/NMFS
Narragansett Laboratory
Narragansett, RI
USA

Bastiaan A. Knoppers
Departamento de Geoquimíca
Universidade Federal Fluminense
Rio de Janeiro, BRAZIL

Patricia A. Livingston
NOAA/National Marine Fisheries
Service, AFSC
Seattle, Washington, USA

Alan Longhurst
Place de l'Eglise
Carjac, Lot,
FRANCE

Daniel Lluch-Belda
Centro Interdisciplinario de Ciencias
Marinas (CICIMAR-IPN)
La Paz, Baja California
Sur 23000,
MEXICO

Daniel B. Lluch-Cota
Centro de Investigaciones Biologicas del
Noroests (CIBNOR)
La Paz, Baja California
Sur 23000,
MEXICO

Salvador E. Lluch-Cota
Centro de Investigaciones Biologicas del
Noroests (CIBNOR)
La Paz, Baja California
Sur 23000,
MEXICO

Claudia Wosnitza-Mendo
Instituto del Mar del Peru
Callao, PERU

Jaime Mendo
Universidad Nacional Agraria La Molina
Facultad de Pesqueria
Lima, PERU

Jeffrey M. Napp
NOAA/National Marine Fisheries
Service, AFSC
Seattle, Washington
USA

Gennady G. Matishov
Murmansk Marine Biological Institute,
KSC RAS
Murmansk,
RUSSIA

John O'Reilly
USDOC, NOAA, NMFS
Narragansett Laboratory
Narragansett, RI
USA

Michael O'Toole
Benguela Current Large Marine
Ecosystem Programme
Ausspannplatz, Windhoek, Namibia
AFRICA

Trevor Platt
Bedford Institute of Oceanography
Fisheries and Oceans Canada
Ocean Sciences Division
Dartmouth, Nova Scotia
CANADA

Daniel Pauly
Fisheries Centre
University of British Columbia
Vancouver, BC,
CANADA

Claude Roy
Centre IRD de Bretagne
Plouzane,
FRANCE

Shubha Sathyendranath
Bedford Institute of Oceanography
Dartmouth, Nova Scotia
CANADA

James D. Schumacher
Two Crow Environmental, Inc.
Friday Harbor, Washington
USA

Vere Shannon
Department of Oceanography
University of Cape Town
SOUTH AFRICA

Kenneth Sherman
USDOC, NOAA, NMFS
Narragansett Laboratory
Narragansett, Rhode Island
USA

Phyllis J. Stabeno
NOAA/Pacific Marine Environmental
Laboratory,
Seattle, Washington
USA

Qisheng Tang
Yellow Sea Fisheries Research Institute
Qingdao,
P.R. CHINA

Reg Watson
Fisheries Centre
University of British Columbia
Vancouver, BC
CANADA

Matthias Wolff
Center for Tropical Marine Ecology
Bremen,
GERMANY

Kees C.T. Zwanenburg
Marine Fish Division
Bedford Institute of Oceanography
Dartmouth, Nova Scotia
CANADA

Contents

I
Polar and Boreal LMEs

Large Marine Ecosystems of the World
G. Hempel and K. Sherman (Editors)

1

The Antarctic Weddell Sea

Gerd Hubold

GEOGRAPHY

The Weddell Sea is a geographically and hydrographically well-defined sub-system in the Atlantic Sector of the Southern Ocean (Figure 1-1). To the west and north-west it is bordered by the Antarctic Peninsula, the South Orkney Islands and the Scotia Ridge at about 60°S/60-30°W. The 60°S latitude can be taken as northern boundary of the Weddell Sea, and in the south, the continental coast is at 70 – 78°30` S. The underwater ridge of the Maud Rise, approximately located at the 0° meridian, forms a natural margin to the deep basin in the East. In this definition, the area under consideration is about 4,2 million km². The Weddell circulation, however, can be traced as far as 30°E in the Lazarev Sea (Deacon 1937, Gordon *et al.* 1981). Thus, the Weddell Sea LME may be defined as an area of at least 5,7 million km², representing 16 percent of the 36 million km² Southern Ocean.

Most of the Weddell Sea belongs to the Southern Ocean circumpolar deepwater system. Less than 20 percent of the area is shallower than 2000 m. The shelf itself is between 200 and 500 m deep, and in the eastern part less than 100 km wide. Only in the south and west of 20°W, does the shelf widen to more than 500 km. Deep trenches cut the shelf, the largest and deepest (> 1000 m) being the Filchner Trench in the southernmost part of the Weddell Sea. This deep cut separates the eastern from the western shelf.

Large parts of the coast are characterised by the vertical barrier of the continental ice shelf. The ice shelves are grounded at bottom depths of about 200 – 300 m or float over greater depths, thus leading to an almost complete lack of shallow water habitats in the Weddell Sea.

The seaward extension of the ice barrier is not constant. As an example, in the south, the Filchner-Rönne Ice Shelf grows by 1 – 1,8 km per year (Robin *et al.* 1983; Lange and Kohnen 1985). As a compensation, from time to time, large icebergs break off the barrier and drift to the west and north (e.g. in October 1998, a berg of 2900 km² carrying the German Filchner station). The variability of the ice barrier and the grounding of icebergs on shallower parts of the shelf strongly affect benthic and to a lesser extent pelagic coastal habitats in the Weddell Sea.

HYDROGRAPHY

The southern part of the Weddell Sea is under the influence of the continental East Wind Drift (Figure 1-1). Relatively constant katabatic winds from the Continent force the Coastal Current (CC) to flow westward parallel to the coast. The CC enters

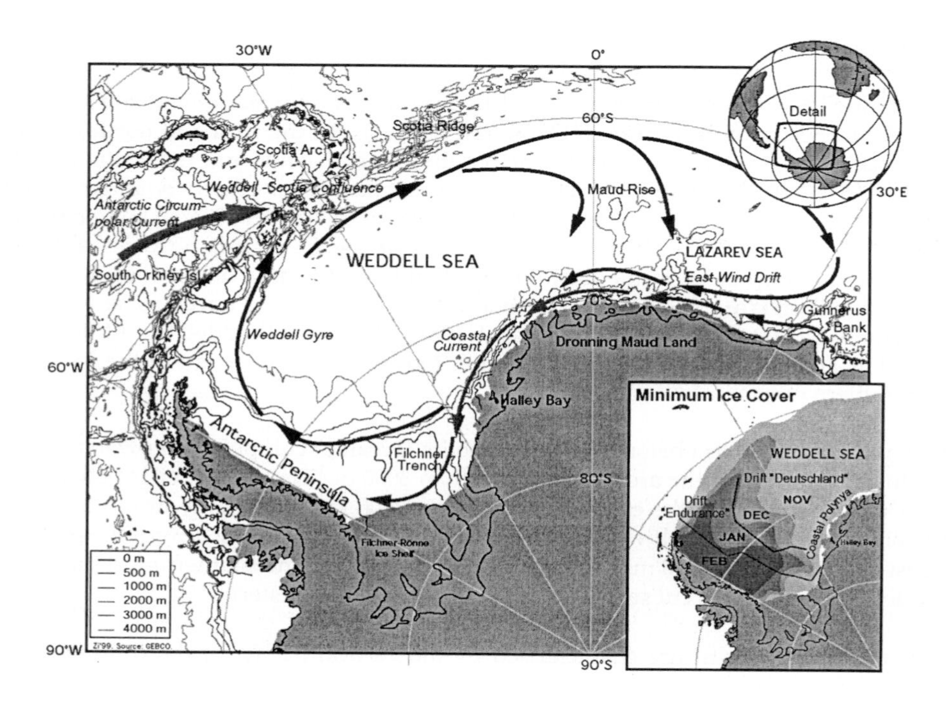

Figure 1-1. Map of the Weddell Sea currents

the Weddell Sea from the Lazarev Sea and follows the Dronning Maud Land coast to the southwest, then turns west near Halley Bay, and later north to the tip of the Antarctic Peninsula, where it forms the southern branch of the Weddell - Scotia - Confluence. As part of the west wind driven Antarctic circumpolar circulation, the Weddell Sea Water then takes an easterly direction, until at 30° E the current turns south again, thus closing the circulation pattern of the Weddell Gyre (Deacon 1937, 1976; Gordon *et al.* 1981).

Mean current velocity in the Coastal Current is 5 – 10 cm s^{-1} (Hellmer and Bersch 1985), in the eastern Weddell Sea it may be up to 30 – 40 cm s^{-1} (Carmack and Foster 1977), then slowing gradually down on its way to the west, where it can be traced as "v-shaped" flow of 6 – 9 cm s^{-1} between shelf water and oceanic water in the southern Weddell Sea (Gill 1973). There is a counter current at the continental slope between 1000 and 2000 m depth with a speed of 5 – 8 cm s^{-1} (Kottmeier and Fahrbach 1989).

Westward surface transport in the CC as estimated from drifting buoys and sea ice drift is significant. It takes about 6 months for a buoy to drift from 2° E to 32° W along the coast. In three more months, a position at 70°S/50°W is reached, and in another three to four months the tip of the Antarctic Peninsula can be reached (Hubold 1992).

According to the large-scale water circulation, the Weddell Gyre as a whole may be considered a retention structure for long-lived pelagic organisms or populations with an ocean-wide distribution (some copepods, krill, fishes). Pelagic species of the coastal zone will rather have to adapt their life cycles to the coastal current and its counter current in the southern and eastern Weddell Sea.

TEMPERATURE, SALINITY, OXYGEN

Surface water temperatures in the Weddell Gyre range from +2°C in the North to ⁻1.8°C in the South. During summer, a surface layer of 30 – 80 m thickness may have a +6 to +1°C (Hufford and Seabrooke 1970; Wegner 1982). Typical salinity is 34 – 35 PSU; melting sea ice in summer can lower surface salinity locally to 30 (Fukuchi *et al.* 1985).

The Warm Deep Water of 0 to +0.7°C and 34.7 PSU fills the main part of the Weddell Sea below the surface layer. There is an oxygen minimum of 4 – 4.7 ml l^{-1} at 400 - 600 m depth. At the deep sea floor, temperatures of the Antarctic Bottom Water are between –0.2 and –0.8°C (Fahrbach 1993).

Over the eastern shelf, water temperatures range from –1.8°C at the bottom to +0.4°C at the surface. Here, salinity is 33.9 – 34.4 PSU (Seabrooke *et al.* 1971; Carmack and Foster 1975). Eastern Shelf Water (ESW) is less dense (σ = 27.7) than Warm Deep Water (WDW) (σ = 27.8). Over the outer shelf, ESW is therefore bordered by the WDW which hits the shelf at depths of about 500 m, and bottom temperatures rise to + 0.3 to + 0.7°C (Bullister *et al.* 1985, Fahrbach *et al.* 1987). Due to a strong vertical convection on the shelf, the ESW is well mixed and saturated with oxygen from surface to depth (6.6 to 9 ml l^{-1}).

Western Shelf Water (WSW) and Ice Shelf Water (ISW) are specific cold water masses in the south of the Weddell Sea formed by the contact with the ice shelves and due to surface cooling in latent heat coastal polynyas.

Temperature, salinity and oxygen in the Weddell Sea show an overall low variability both regionally and temporally, in the course of the year and over the years. Variations are basically limited to surface layers and are in the range of no more than 4-5°C and 4 salinity units within the entire extension of the gyre system. In the South, variability decreases to less than 2°C over the seasonal cycle. By horizontal advection of water masses, the oxygen content of shelf- and bottom water is high at all depths, even under the permanent sea ice cover. It is likely, that such stable environmental conditions have prevailed since Antarctica became cold and isolated, i.e. about 20 million years ago.

SEA ICE AND LIGHT

The principal factors of variability in the Weddell Sea are the light cycle and the seasonal dynamics of sea ice. The seasonal light cycle in the Antarctic is less pronounced than in the high Arctic due to the relatively low latitudes of the circumpolar ocean between 60 and 80°S. In the southern parts, the sun disappears completely below the horizon for a few days or even weeks in winter. Nevertheless the system undergoes strong seasonal changes, because sea ice formation in autumn rapidly impedes penetration of light into the water column.

The Weddell Sea is situated in the seasonal and permanent pack ice zones *sensu* Hempel (1985 a, b). Sea ice is at maximum extent in late winter (November), when it covers an area from the continent to 60°S (small inlay in Figure 1-1). Due to strong winds exerting a diverging force on the ice, the Weddell Sea ice cover is not completely closed even in winter. There are numerous leads and polynyas allowing for gas- and heat exchange and, to a certain degree, for the penetration of light. The coastal polynyas, which are maintained by strong katabatic winds, are found at any time, yet in varying extent and location along the ice shelf coast. Emperor penguins depend on the availability of polynyas in the vicinity of their breeding sites on the continental fast ice during winter and spring. A unique large offshore oceanic polynya, which was convection- rather than wind-driven, appeared in the years 1974 to 1976 between 65° and 70° S to the west of the Maud Rise (Gordon and Comiso 1988).

In summer (February), most of the Weddell Sea is ice-free. Only an area of approximately 800 000 km^2 of strongly ridged sea ice remains in the south and west. In some years, however, the entire area west of 25° W can remain ice-covered (Strübing 1982).

The sea ice hosts an ecosystem of its own consisting of protozoa, ice algae, crustaceans (amphipods, krill) fishes, birds and seals. Sea ice is both a habitat and a transport medium. The sea ice community influences the physical conditions of the ice itself, e.g. by faster melting due to the dark colour of the enclosed organisms (Littlepage 1965). Melting ice stabilizes the water column in spring. By seeding of algae from the ice, algal blooms appear typically close to retreating ice edges.

Sea ice formation in the southern coastal polynya areas and a predictable drift pattern of the newly formed ice towards the west and north may play a key role in the life cycles of symbatic or temporarily ice-associated organisms such as Antarctic krill (*Euphausia superba*). Drift velocities calculated from the path of the ice-locked expeditions of the "Deutschland" (1911-1912) and "Endurance" (1914-1916) (inlay in Figure 1-1) are 15-20 cm s^{-1}, and thus in the same order of magnitude or higher than the underlying water current, and in a more northerly direction (Hubold 1992).

Considering Antarctica´s geological history, the opening of the Drake Passage about 20 million years ago has set the stage for the development of today's circulation system and for the abiotic characteristics of the Weddell Sea LME. Its organisms and communities represent several million years of evolution under relatively stable conditions. Separation from all other continents by deep oceanic basins, and the strong circumpolar current system with the Antarctic convergence zone as a zoogeographical boundary, enabled the development of the present highly adapted and highly endemic Antarctic flora and fauna.

PHYTOPLANKTON AND PRIMARY PRODUCTION

Low water temperatures limit algal growth in adapted Antarctic species also (Neori and Holm-Hansen 1982). South of 66°S, phytoplankton growth is largely confined to the months December to March (El-Sayed 1987). Algae growing in sea ice can extend their growth season beyond this period. In the southern Weddell Sea, daily production in summer can reach 1.6 g C m^{-2} d^{-1} (mean 0.4-0.7 g C m^{-2} d^{-1}, El-Sayed and Taguchi 1981, von Bröckel 1985). The northern oceanic parts are less productive with 0.1 – 0.2 g C m^{-2} d^{-1} (El-Sayed and Taguchi 1981; Heywood and Whitaker 1984) but, at retreating ice edges, production can be considerably enhanced even in the oceanic realm to 0.3 – 1.0 g C m^{-2} d^{-1} (von Bodungen *et al.* 1986).

During the short summer phase, a succession of species was observed in the southern Weddell Sea, dominated by Phaeocystis sp. in January and diatoms in February (Nöthig

1988). Total annual production (100 days) on the eastern shelf was estimated as 30 – 40 g C m^{-2} and 20 – 30 g C m^{-2} in the northern oceanic parts (Jennings *et al.* 1984). Ice algae produce an additional 10 %. Annual primary production in the Weddell Sea is thus comparable to northern boreal seas, yet it is highly pulsed by a very short summer season.

ZOOPLANKTON

The zooplankton community of the Weddell Gyre is composed mainly of copepods, euphausiids, salps, amphipods, chaetognaths and molluscs (Boysen-Ennen and Piatkowski 1988). Larval fishes are rare, except for the southern shelf zones, where they may constitute a large fraction of the macro-zooplankton (Hubold *et al.* 1988).

Maximum zooplankton standing stock in the Southern Ocean is found between 50 and 55°S (Foxton 1956), i.e. to the north of the Weddell Gyre. Estimated zooplankton biomass (combined micro-, meso- and macro-zooplankton; wet weight m^{-2}) in the Weddell Sea decreases from 42 g m^{-2} north of 60°S to 16 g m^{-2} at 66 - 73°S and is higher again near the coast between 70 - 74°S (28 g m^{-2}). On the southern coast between 75 and 78°S a lower value of 11 g m^{-2} is found (Hubold 1992). In spite of the notorious local mass aggregations of krill (*E. superba*) mostly in the north, the more important fraction in the zooplankton, which accounts for 75 - 80 percent of the biomass, is much smaller than 15 mm size.

Accordingly to the biomass distribution, zooplankton production decreases from north to south from 162 g m^{-1} yr^{-1} to 42 g m^{-1} yr^{-1} , and 90 percent of secondary production is due to the smaller <15 mm zooplankton fraction. The production is confined to a narrow seasonal window in summer with a maximum in December and January in the south.

A significant part of the annual new phytoplankton production sinks to the seafloor. In the southern Weddell Sea, it was estimated that from a pelagic primary production of 11-13 g C m^{-2}, 2.4 g C m^{-2} sedimented mostly as faecal pellets of zooplankton (von Bodungen *et al.* 1986, 1988). In shelf areas, this organic matter reaches the sea floor and can sustain diverse benthic communities. Due to slow bacterial degradation rates in the water column (Smith *et al.* 1989), the organic matter can be preserved over long periods and is advected over long distances serving as allochthoneus feed in remote parts of the Weddell Sea (e.g. under ice shelves and in permanent pack ice areas). Due to the sedimentation and advection processes, deep-water benthic communities of the Weddell Sea can be largely uncoupled from direct regional and seasonal surface phytoplankton signals (Hubold 1992).

KRILL

Antarctic krill *Euphausia superba,* has a main distribution area in the northwestern part of the Weddell Gyre, but is comparatively rare in the south, where the smaller ice krill, *E. crystallorophias* occurs. Krill lives in large swarms and may be taken as analogous to pelagic mass-fishes of other oceans, which are not present in the Antarctic seas (Hempel 1981). Several specialised adaptations enable krill to live both pelagically and sea ice-associated (cryopelagic or epontic) and make this animal the most "successful" Antarctic species with a biomass estimated between 300 and 500 million tons (Hempel 1985b). These estimates are based on the international BIOMASS experiments FIBEX and SIBEX in the seventies and eighties. Since then, a comprehensive repetition of krill biomass determination has not been attempted. Recent smaller scale surveys, however, show that krill biomass undergoes strong year-to-year fluctuations (Siegel and Loeb 1995). These investigations also showed positive correlations of krill recruitment with seasonal sea ice cover in the northern Weddell Sea, whereas in years of less ice, the gelatinous plankton, e.g. salps predominate. The same authors stated an overall decrease of krill biomass in the Atlantic sector over recent years. These changes cannot be attributed to direct human factors such as fishing, but are rather due to responses of the species to regional climatic variability. A comprehensive review on krill is found in Everson (2000).

FISHES

Three major groups can be distinguished in the Weddell Sea fish fauna: The oceanic community of widely distributed mesopelagic species, a northern group of island related notothenioid coastal fishes, and a southern continental shelf group of endemic notothenioid taxa (Andriashev 1965). At present, more than 83 fish species (69 notothenioid, and 14 non-notothenioid species) representing 8 orders, 14 families and 43 genera are known to occur in the Weddell Sea (Hubold 1992).

Life cycles of mesopelagic Antarctic fishes are poorly known. Abundant species such as *Electrona antarctica* (Myctophidae), paralepidids and bathylagids occur over the entire Southern Ocean. Single populations may be associated with the Weddell LME circulation, yet this remains purely speculative. The mesopelagic fishes are strictly confined to the WDW and avoid both cold surface water and cold shelf water even when these water masses are found over great depths e.g. in the Filchner trench.

The northern island fish fauna is composed mostly of notothenioid families Nototheniidae, Channichthyidae, and Artedidraconidae. Most abundant species around the South Orkney and South Shetland Islands are *Notothenia gibberifrons,* N. *squamifrons, Chaenocephalus aceratus, Pseudochaenichthys georgianus,* N. *rossii, Chionodraco rastrospinosus*, *Champsocephalus gunnari*, and *Nototheniops larseni*, which account for approximately 98 percent of total fish biomass (Kock 1992, Kock, pers. comm.)

Over the outer island shelves and oceanic seamounts, *Dissostichus eleginoides*, one of the two species of this largest Antarctic fish genus is found in commercial concentrations.

The southern Weddell Sea shelf ichthyofauna was systematically studied during several German "Polarstern" expeditions in the 1980s. Of more than 200,000 fishes collected, 99 percent were of the suborder Notothenioidei, more than 90 percent belonged to one species, *Pleuragramma antarcticum*. Other important species were *Chionodraco myersi*, *Pagetopsis maculatus*, *Trematomus scotti*, T. *lepidorhinus*, T. *eulepidotus*, and *Dolloidraco longedorsalis*. On the outer shelf, Macrourids *Macrouras holotrachys* occurred (Hubold 1992). *D. mawsoni*, the southern species of the *Dissostichus* genus, which is locally abundant in the Ross Sea, has been found only in low numbers in the Weddell Sea.

In spite of the numerical dominance of the pelagic *Pleuragramma antarcticum*, the fish community of the southern Weddell Sea is surprisingly diverse. For the demersal ichthyofauna, α-diversity indices (from large bottom trawl hauls) of H = 1.8 compare with lower values of H = 1.0 – 1.5 near the Antarctic Peninsula, and H = 1.7 in the Ross Sea. For comparison, diversity values on the Greenland shelf were calculated as H = 1.2; and H = 1.1 is a typical diversity value in the boreal North Sea (Hubold 1992).

Demersal fish biomass on the eastern Weddell Sea shelf was determined as between 260 and 3,700 kg km^{-2} (mean 900 kg km^{-2}). Towards the south, biomass decreases to 300 kg km^{-2}. Total demersal fish biomass of the southern and eastern Weddell Sea shelf is probably not more than 100,000 t (Hubold 1992). In contrast to demersal fish, the abundance of pelagic fishes increases from north to south. Pelagic fish biomass was found to be in the order of 1 t km^{-2} in the south and 0.1 t km^{-2} on the eastern shelf. A total of 200,000 – 500,000 t of pelagic fish biomass – almost exclusively *P. antarcticum* - was estimated.

Slow growth and low fecundity of Antarctic fishes is well documented. From north to south, growth parameters tend to decrease. Typical growth performances (Pauly 1979) of southern Weddell Sea fishes range between P = 0.7 and 2, because fishes have both low growth efficiency (k) <u>and</u> small maximum sizes (Hubold 1992). Fecundity values of notothenioid Weddell Sea fishes are low and range between 10^2 to 10^4 eggs per female (7 – 200 eggs g^{-1} body weight). Possibly, spawning takes place only every second year. Low growth performance combined with low fecundity and biomass results in a very low total fish production in the order of 0,5 t km^{-2} yr^{-1} for the Weddell Sea shelf area.

FOOD WEB

In the northern parts of the Weddell Sea, krill (*Euphausia superba*) is the dominant species and is the staple food for most of the higher predators (fish, squid, birds, mammals). Towards the south, however, the relative importance of finfish in the diet of top predators increases. In spite of the low fish biomass and production, a number of top predators depend completely on fish rather than krill. Weddell Seals (*Leptonychotes weddellii*) and Ross Seals (*Ommatophoca rossii*) are fish and squid predators, whereas the most abundant species, the Crab-eater Seal (*Lobodon carcinophagus*) feeds primarily on krill. Emperor Penguins (*Aptenodytes forsteri*) rely on fish (mostly *Pleuragramma antarcticum)* and squid. Adelie penguins (*Pygoscelis adeliae*) and snow petrels (*Pagodroma nivea*) feed on krill in the north, and switch to fish in the south.

Summer guests in the southern Weddell Sea such as minke whales (*Balaenoptera acutorostrata*) feed on krill and fish. Only the northern Weddell Sea is visited by the big baleen whales during austral summer, which feed on krill in the Weddell-Scotia Confluence zone.

HUMAN IMPACT

Whaling and sealing were massive human impacts only in the northern Weddell Sea and around the sub Antarctic islands. Whaling was most intensive in the 1930s and ceased after the 1986/87 season.

Commercial fishing of krill as reported by the FAO statistics started in the early 1970s in the Bransfield Strait and around Elephant Island with catches of 100,000 t. Annual catches increased to 350,000 – 425,000 t in 1985/86 - 1992 and decreased after 1992 to approximately 100,000 t. Area 48 of FAO (Atlantic, Antarctic) includes the Weddell Sea, Antarctic Peninsula, Scotia Arc, South Sandwich Islands and South Georgia.

A significant fishery for fin-fish in the Atlantic Antarctic sector started in 1969/70 outside the Weddell Sea, when around South Georgia island 400,000 t of marbled notothenia (*Notothenia rossii marmorata*) were caught. Catches declined to almost zero in 1972. In 1978 and 1983, 146,000 t and 140,000 t of the white-blooded ice-fish *Champsocephalus gunnari* were caught, respectively.

Since 1987, and increasingly since 1990, the large pelagic notothenioid fish *Dissostichus eleginoides* is fished in the Southern Ocean, including the northern parts of the Weddell Sea. In the season 1995/96, 4857 t were reported to the Commission for the Conservation of Antarctic Marine, Living Resources (CCAMLR) and FAO from the long-line fishery around the sub Antarctic islands in area 48. An additional, possibly large quantity is caught illegally, and may present a serious threat to the population of these large and slow growing Antarctic fishes. Oceanic myctophid species *Electrona carlsbergi* was caught in the northern Weddell Sea between 1987 and 1992 with a maximum quantity of 78,500 t in 1990/91 for industrial (reduction) purposes. Later this fishery was not continued.

The southern parts of the Weddell Sea have never been subjected to significant human impact, neither by whaling/seal hunting nor fishing. A USSR test fishery in the south-eastern Weddell and Lazarev Seas in the early 1980s yielded a total of 110 t of *P. antarcticum* and was not further developed. Since that time, no commercial activities have been reported from the Weddell Sea, so that this LME may be still considered a widely pristine ecosystem, as far as direct human impact is concerned.

CHANGES IN THE SYSTEM

Indirect impacts due to anthropogenic environmental change may well be relevant in the Weddell Sea. Being situated at high polar latitudes, the Antarctic ozone depletion area directly affects the Weddell Sea. Increased UV radiation may negatively impact on surface phytoplankton productivity, because the pelagic algae are already stressed by ambient UV radiation levels (El-Sayed *et al.* 1990). Under an extreme ozone hole, a carbon loss of 6.4 percent in the upper 20 m water column was estimated because of enhanced UV-B radiation. However, taking into account the magnitude and variations of ozone depletion, losses in primary production over the whole Southern Ocean were estimated to be less than 0.15 percent for the entire year (Helbling *et al.* 1994). More relevant, however, could be the known differential sensitivity of different taxa to UV radiation. For example, flagellates (the most sensitive) and diatoms (the least sensitive) organisms might shift species composition and size distribution of phytoplankton in ozone depletion areas towards diatom dominated communities.

The Weddell Sea seasonal production cycle is strongly determined by ice formation in autumn and ice melting in spring and summer. The resulting convection in winter and the stabilizsation of a thin surface layer during summer are important parameters in the planktonic life cycles, e.g. of copepods and larval fish. Krill (E. *superba*) seems especially closely and positively related to the extent of seasonal sea ice. Climatic warming, resulting in less sea ice may impact on this sensitive mechanism and could lead to decreased krill biomass. On the other hand, pelagic species such as salps may profit from a warmer, less ice covered sea, and overall primary production might be enhanced due to earlier and more consistent stabilisation of the water column.

In contrast to most of the Large Marine Ecosystems, direct human impact is insignificant in the Weddell Sea to date. Due to its remoteness, low overall production, and lack of native coastal populations, pollution and over exploitation have not been significant so far. Thus, the Weddell Sea LME may have a fair chance to remain in a widely natural condition in the 21st century as it did for the last 20-30 million years.

REFERENCES

Andriashev, A. P. 1965. A general review of the Antarctic fish fauna. In: Oye P. van, J. van Mieghem, eds. Biogeography and Ecology in Antarctica. Junk Publ., The Hague, Monographiae Biologicae 15: 491-550

Bodungen, B. von, V.S. Smetacek, M.M. Tilzer, B. Zeitzschel. 1986. Primary production and sedimentation during spring in the Antarctic Peninsula region. Deep Sea Res. 33: 177-194. ANT 880930.3

Bodungen, B. von, Nöthig E M, Q. Sui. 1988. New production of phytoplankton and sedimentation during summer 1985 in the southeastern Weddell Sea. Comp. Biochem. Physiol. 90B (3):475-487

Boysen-Ennen E. and U. Piatkowski. 1988. Meso- and Macrozooplankton Communities in the Weddell Sea. Polar Biol. 9:17 – 35

Bröckel, K. von. 1985. Primary production data from the southeastern Weddell Sea. Polar Biol. 4: 75 - 80

Bullister J., H. Hellmer, G. Krause, G. Rohardt, P. Schlosser, H. Witte. 1985. Physical Oceanography. In: Hempel G., ed. Die Expedition ANTARKTIS III mit FS "Polarstern" 1984/85. Ber. z. Polarforschung 25:99-103

Carmack E.C. and T.D. Foster. 1975. Circulation and distribution of oceanographic properties near the Filchner Ice Shelf. Deep Sea Res. 22: 77-90

Carmack E.C. and T.D. Foster. 1977. Water masses and circulation in the Weddell Sea. In: Dunbar M.J., ed. Polar Oceans. Proc. SCOR/SCAR polar ocean conf., Montreal 1974: Arctic Inst. of North America. 151 – 165

Deacon G. E. R. 1937. The Hydrology of the Southern Ocean. Discovery Reports. 15:124p

Deacon G.E.R. 1976. The Cyclonic Circulation in the Weddell Sea. Deep Sea Res. 23 (1):125-126

El-Sayed S. Z. and S. Taguchi. 1981. Primary production and standing crop of phytoplankton along the, ice-edge in the Weddell Sea. Deep Sea Res. Vol. 28A, (9):1017 – 1032

El-Sayed S. Z. 1987. Biological productivity of Antarctic waters: present paradoxes and emerging paradigms. In: El-Sayed SZ, Tomo AP, eds. Antarctic Aquatic Biology. Proceedings of the Regional Symposium on Recent Advances in Antarctic Aquatic Biology. San Carlos de Bariloche, Argentina 6.-10. June, 1983. BIOMASS Scientific Series No 7: 1 - 22

El-Sayed S.Z., F.C. Stephens, R.R. Bidigare, M.E. Ondrusek. 1990. Effect of ultraviolet radiation on Antarctic marine phytoplankton. In: Kerry and Hempel, eds. Antarctic Ecosystems. Ecological Change and Conservation. Berlin, Springer Verlag: 379-385

Everson I, ed. 2000. Krill biology and fisheries. Blackwell, Oxford, 372p

Fahrbach E., H. Klindt, D. Muus, G. Rohardt, P. Salameh. 1987. Physical Oceanography. In: Schnack-Schiel S., ed. Die Winter-Expedition mit FS "Polarstern" in die Antarktis (ANT V/1-3). Ber z Polarforsch. 39: 156 – 169

Fahrbach E. 1993. Zirkulation und Wassermassenbildung im Weddellmeer. Die Geowissenschaften, Nr. 7:246-252

Foxton P. 1956. The distribution of the standing crop of zooplankton in the Southern Ocean. Discovery Rep. 28: 191-236

Fukuchi M., A. Tanimura, H. Ohtsuka. 1985. Marine Biological and Oceanographical Investigations in Lützow-Holm Bay, Antarctica. In: W.R. Siegfried, P.R. Condy, R.M. Laws, eds. Antarctic nutrient cycles and food webs. Proc. 4th SCAR Symp. on Antarctic Biology. Springer Verlag Berlin, Heidelberg, N.York, 52 - 5

Gill A.E. 1973. Circulation and Bottom Water Production in the Weddell Sea. Deep Sea Res. 20: 111-140

Gordon A. L., D.G. Martinson, H.W. Taylor. 1981. The wind driven circulation in the Weddell-Enderby Basin. Deep Sea Res. 28 A: 151 - 163

Gordon A.L., J.C. Comiso. 1988. Polynjas im Südpolarmeer. Spektrum der Wissenschaften, Aug. 1988:92-99

Helbling E.W., V. Villafane, O. Holm-Hansen. 1994. Effects of ultraviolet radiation on antarctic marine phytoplankton photosynthesis with particular attention to the influence of mixing. In: Ultraviolet radiation in Antarctica: Measurements and Biological effects. Antarctic research series, Vol 62. Am. Geophys. Union, Washington DC. 207-227

Hellmer H. and M. Bersch. 1985. The Southern Ocean. Ber. Polar Forsch. Vol. 26: 115p

Hempel G. 1981. Das antarktische Ökosystem und seine fischereiliche Nutzung. Jahrb. der Wittheit zu Bremen, Band XXV: 55 – 68

Hempel G. 1985a. Antarctic Marine Food Webs. In: Siegfried WR, Condy PR, Laws RM (eds) Antarctic nutrient cycles and food webs. Proc. 4th SCAR Symp. on Antarctic Biology. Springer Verlag Berlin, Heidelberg, N.York, 266-270

Hempel G. 1985b. On the Biology of Polar Seas, Particularly the Southern Ocean. In: Marine Biology of Polar Regions and Effects of Stress on Marine Organisms. J.S. Gray, M.E. Christiansen, eds. Wiley and Sons. 3 - 33

Heywood R.B., Whitaker T.M. 1984. The Antarctic marine flora. In: R.M. Laws, ed. Antarctic Ecology. Academic Press, London. 373-420

Hubold G., I. Hempel, M. Meyer. 1988. Zooplankton communities in the southern Weddell Sea (Antarctica). Polar Biol. 8:225 -233

Hubold G. 1992. Zur Ökologie der Fische im Weddellmeer. Ecology of Weddell Sea fishes. Ber z Polarforsch. Reports on Polar Research 103:1-157

Hufford G.L., J.M. Seabrooke. 1970. Oceanography of the Weddell Sea in 1969 (WSOE). US Coast Guard Oceanographic Report No 31 CG 373-31, 32p

Jennings J.C., L.J. Gordon, D.M. Nelson. 1984. Nutrient depletion indicates high primary productivity in the Weddell Sea. Nature 309: 51 – 53

Kock K.H. 1992. Antarctic fish and fisheries. Cambridge University Press, Cambridge, New York. 375p

Kottmeier C.H., E. Fahrbach. 1989. Wechselwirkung zwischen Wasser, Eis und Luft in der antarktischen Küstenzone. Promet 19, 1/2:15 – 22

Lange M.A. and H. Kohnen. 1985. Ice front fluctuations in the eastern and southern Weddell Sea. Annals of Glaciology 6

Littlepage I.L. 1965. Additional Oceanographic Studies in McMurdo Sound, Antarctica. In: Biology of the Antarctic Seas II. Am Geophys. Union, Antarct. Res. Ser. 5: 1-37

Neori A., O. Holm-Hansen. 1982. Effect of temperature on rate of photosynthesis in Antarctic phytoplankton. Polar Biology 1:33-38

Nöthig E.M. 1988. Untersuchungen zur Ökologie des Phytoplanktons im südöstlichen Weddellmeer (Antarktis) im Januar/Februar 1985. Berichte zur Polarforschung, 53:1-118

Pauly D. 1979. Gill size and temperature as governing factors in fish growth. A generalization of Von Bertalanffy's growth formula. Ber Inst Meeresk Kiel, no 63: 156p

Robin G.Q., C.S.M. Doake, H. Kohnen, R.D. Crabtree, S.R. Jordan, D. Möller. 1983. Regime of the Filchner Ronne Ice Shelves, Antarctica. Nature, Vol 302 No 5909:582-586

Seabrooke I.M., G.L. Hufford, R.B. Elder. 1971. Formation of Antarctic Bottom Water in the Weddell Sea. J. Geophys. Res. 76: 2164-2178

Siegel V., V. Loeb. 1995. Recruitment of Antarctic Krill (*Euphausia superba*) and possible causes for its variability. Mar. Ecology Progr. series. 123(1):45-56

Smith G.A, J.D. Davis, A.M. Muscat, R.L. Moe, D.C. White. 1989. Lipid Composition and Metabolic Activities of Benthic Near-Shore Microbial Communities of Arthur Harbor, Antarctic Peninsula: Comparisons with McMurdo Sound. Polar Biol. 9 (8):517 - 524

Strübing K. 1982. Die Zugänglichkeit von Forschungsstationen am Rande der Weddellsee in Abhängigkeit von den Meereisverhältnissen in: Proceedings of the Intermaritec 12, Hamburg, 29-30 Sept. 82 IMT 82-210:431-442

Wegner G. 1982. Two Temperature Sections Through the Southern Atlantic and Southeastern Weddell Sea. Meeresforschung 29 (4): 239-243

Large Marine Ecosystems of the World
G. Hempel and K. Sherman (Editors)

2

Climate Change in the Southeastern Bering Sea and Some Consequences for its Biota

J. D. Schumacher, N.A. Bond, R.D. Brodeur, P.A. Livingston, J.M. Napp and P.J. Stabeno

INTRODUCTION

Dramatic changes are occurring throughout the Arctic (Weller *et al.* 1997). The systematic modification in wintertime climate of Alaska and the Bering Sea that occurs in 1976-1977 illustrates the magnitude and nature of these changes. Among the effects documented are a step-like increase of nearly 2° C in air temperature (Bowling 1995), an ~5 percent decrease in sea ice extent (Niebauer 1998), and decreases in sea-ice thickness (Wadhams 1995). Many local residents around the Bering Sea have also noted changes in ice thickness and strength (Gibson and Schullinger 1998). Over longer time scales, the extent of glaciers has decreased markedly. Permafrost temperatures measured in bore-holes in northern Alaska are 2 to 4°C warmer than they were 50 to 100 years ago (Lachenbruch and Marshall 1986). Discontinuous permafrost has warmed considerably and is thawing in some locations (Osterkamp 1994). For the Bering Sea region (Figure 2-1), the warming may reflect a natural multi-decadal cycle (regime shift) superimposed on a warming trend due to the greenhouse effect. Given our limited understanding of the Earth climate system, the latter mechanism provides compelling reasons for legitimate public and management concern over future changes resulting from increased concentrations of greenhouse gases (AGU Report 1999).

Substantial natural and human induced variability occurs in the ecosystem of the eastern Bering Sea which includes some of the most productive fisheries on Earth (NRC Report 1996). Besides lucrative crab, halibut and salmon fisheries, most of the world catch of walleye pollock (*Theragra chalcogramma)* occurs here and these resources represent 2 to 5 percent of the world's fishery production and ~50 percent of the U.S. fishery production (NRC Report 1996). The Bering Sea is home to at least

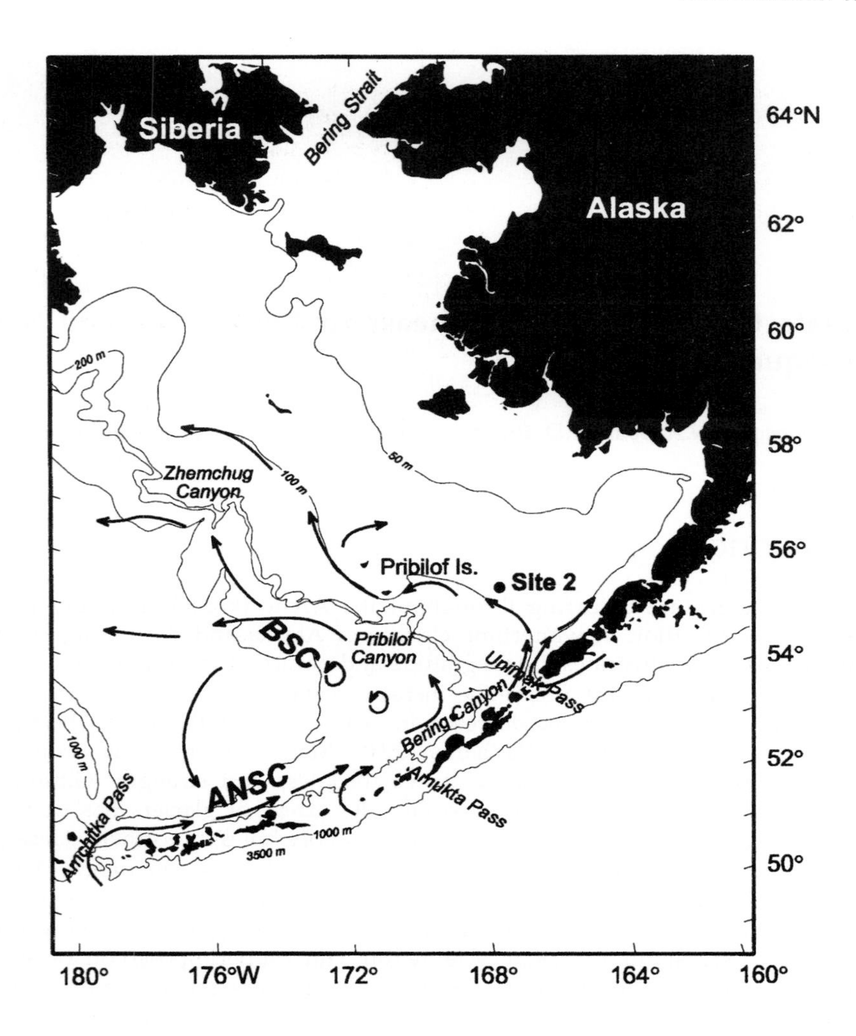

Figure 2-1. Eastern Bering Sea showing geographic names and the location of the Fisheries-Oceanography Coordinated Investigations (FOCI) moored biophysical platform (Site #2). Also shown is a schematic of the mean circulation. BSC= Bering Slope Current; ANSC-Aleutian North Slope Current.

450 species of fish, crustaceans and mollusks; 50 species of seabirds; and 25 species of marine mammals. Walleye pollock is a nodal species in the food web (NRC Report 1996) with juveniles providing the dominant prey of fishes, seabirds, and marine mammals (Springer and Byrd 1989, Livingston 1993, Brodeur *et al.* 1996). In the past, climate variations altered species composition and at present may be reducing carrying capacity (Kruse 1998, Napp and Hunt 2001). Such changes impact sustained ecosystem health and should dictate management of human activities and utilization of this rich ecosystem.

Changes forecast for the Bering Sea due to warming (US GLOBEC Report 1996, Weller *et al.* 1997, Schumacher and Alexander 1999) include decreases in storms (mixing energy), the supply of nutrients and sea ice extent/thickness and an increase in sea surface temperature. Physical conditions observed during 1997 fit these predictions well. During 1998, some conditions continued to be abnormal. It is early to tell whether these conditions will persist or transition back to those characteristic of the historical record. Nevertheless, it is clear that recent changes in the physical environment had immediate and profound consequences on biota. The complex mechanisms that link physical and biological elements preclude precise prediction of ecosystem changes, but it is likely that major evolutions will occur. Below, we update previous work on biophysical processes (Incze and Schumacher 1986) and use recent conditions to suggest mechanisms and biotic responses. We conclude with a brief discussion of management concerns and a scenario for potential future ecosystem change.

THE BIOPHYSICAL ENVIRONMENT

Recent results germane to biophysical mechanisms in the southeastern Bering Sea include: refinement of our understanding of the Bering Slope and Aleutian North Slope Current (Stabeno *et al.* 1999, Reed and Stabeno 1999) and the potential importance of these to fish stocks (Reed 1995), discovery of eddies over the outer shelf (Reed 1998) and slope (Schumacher and Stabeno 1994) and suggestions of their biological importance, discovery of a mean flow across the shelf (Reed and Stabeno 1996, Reed 1998), observations from moored instruments of the ice-associated phytoplankton bloom without water column stratification and the role advection has on stratification (Stabeno *et al.* 1998), determination of the critical nature of timing of sea ice melt and wind mixing to bloom dynamics (Stabeno *et al.* 2001), clarification of the relationship between inner front dynamics and prolonged production (Hunt *et al.* 1999), the potential influence that wind drift, currents and cannibalism have on recruitment of pollock (Wespestad *et al.* 2000) and the importance of warm season climate (Overland *et al.* 2001).

The southeastern Bering Sea consists of oceanic and shelf regimes. Within the latter regime, three distinct domains exist, characterized by contrasts in water column structure, currents and biota (Iverson *et al.* 1979a, Coachman 1986, Schumacher and Stabeno 1998). These are the coastal (<50 m deep, weak stratification), middle shelf (50 to 100 m deep, strong stratification), and outer shelf (100 to 200 m deep, mixed upper and lower layers separated by slowly increasing density). The domains provide unique habitats for biota. For example, the zooplankton community in the two shallower domains is comprised primarily of small copepods and euphausiids whereas, in the outer shelf domain and oceanic region, large copepods dominate (Cooney and Coyle 1982, Vidal and Smith 1986).

The Oceanic Regime is influenced by the Alaskan Stream flowing through Amchitka and Amukta Pass, producing the Aleutian North Slope Current (ANSC) (Reed and Stabeno 1999). The ANSC provides the main source of the Bering Slope Current (BSC) which exists either as an ill-defined, variable flow interspersed with eddies and meanders, or as a more regular, northwestward flowing current (Stabeno *et al.* 1999). Shelf/slope exchange likely differs depending upon which mode is dominant. The importance of these currents to dissolved or planktonic material is threefold: they can provide transport from oceanic (including an important spawning region for pollock) to shelf waters, their inherent eddies may temporarily provide habitat that favors survival of larvae (Schumacher and Stabeno 1994) and temperature characteristics that are potentially important to fish stocks (Reed 1995). Eddies are also common in waters just seaward of the shelf break (Schumacher and Reed 1992) and exist even in the region between 100 and 122 m (Reed 1998). Transport of high concentrations of pollock larvae onto the shelf by eddies occurs (Schumacher and Stabeno 1994).

The amount of sea ice cover depends on storm tracks (Schumacher and Stabeno 1998) and varies by > 40 percent about the mean (Niebauer 1998). Ice advection and melting play a critical role in fluxes of heat and salt and generation of both baroclinic flow and the cold pool located over the middle shelf domain (Schumacher and Stabeno 1998, Wyllie-Echeverria and Wooster 1998). The positive buoyancy from melting ice initiates both baroclinic transport along the marginal ice zone and stratification. Cooling and mixing associated with ice advance help to condition the entire water column over the middle shelf domain (Stabeno *et al.* 1998). With seasonal heating, the lower layer becomes insulated and temperatures often remain below 2.0°C (Reed 1995). The area of this cold pool varies by ~2.0×10^5 km^2 between maximum and minimum extent. A spring bloom of phytoplankton is associated with the sea ice and accounts for 10 to 65 percent of the total annual primary production (Niebauer *et al.* 1995). Ice and the cold pool both influence distributions of higher trophic level biota (Ohtani and Azumaya 1995, Wyllie-Echeverria and Wooster 1998, Brodeur *et al.* 1999b).

RECENT ANOMALIES

Recently, the Bering Sea exhibited a host of noteworthy physical and biological conditions (Kruse 1998, Vance *et al.* 1998, Tynan 1998, Baduini *et al.* 2001, Napp and Hunt 2001, Hunt *et al.* 1999, Stabeno *et al.* 2001): the first recorded major coccolithophorid blooms (1997, 1998 and 1999), large die-off of shearwaters (1997), salmon returns far below predictions (1997 and 1998), the unusual presence of whales over the middle shelf (1997 and 1998), unusually warm summer sea surface temperatures (1997 and 1998), and a decrease in the onshore transport of slope water (1997). It is instructive to consider these events in the context of the historical records.

Changes in the regional environment are related to the state of the entire North Pacific/Arctic climate system. Much of the variance herein is accounted for by a few modes. Particularly systematic effects appear to be linked to the El Niño-Southern Oscillation (ENSO) on 2 to 7 year time scales and associated with the Pacific Decadal Oscillation (PDO) (Mantua *et al.* 1997) on decadal to multi-decadal time scales. In brief, both ENSO and the PDO impact the Bering Sea during winter through their positive correlation with the strength of the Aleutian Low (Figure 2-2).

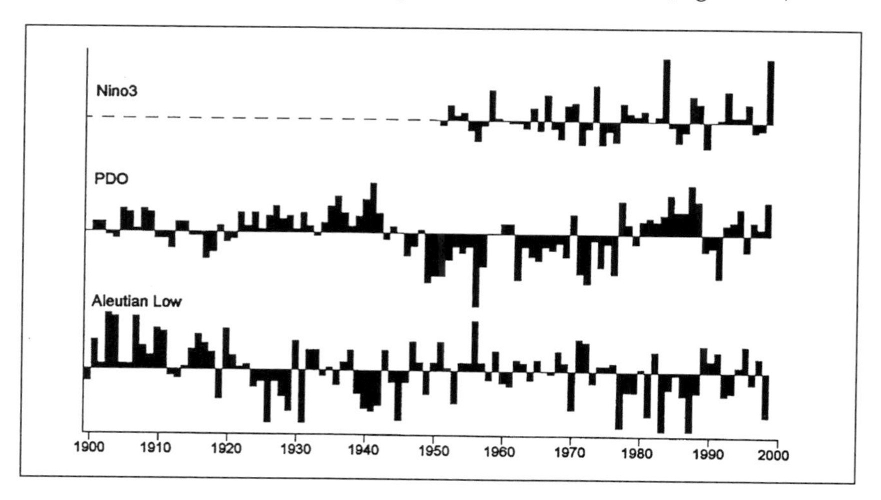

Figure 2-2. Atmospheric indices for the North Pacific and Bering Sea: an index of ENSO (the NINO3 index), the PDO (after Mantua *et al.* 1997) and the Aleutian Low (from Stabeno *et al.* 2001).

The Aleutian Low is important to the Bering Sea through its impact on surface winds (advection and mixing of the upper ocean and production/advection of ice) and heat flux (mixing and ice formation). The time series of both the ENSO and PDO indices show a marked change between 1976 and 1977, the well known "regime shift" (eg., Trenbberth and Hurrell 1995, Hare and Mantua 2000). While the anomalies were less extreme than during the 1970s, the Aleutian Low has undergone significant recent variations between being weaker than normal in 1995, stronger than normal from 1996 through 1998 and then weaker than normal in 1999. As noted below, this contributed to substantial differences in timing and persistence of ice cover.

Sea ice characteristics represent an integrated measure of winter atmospheric forcing. An index of sea ice developed for Site #2 (Figure 2-3) shows that the most extensive ice years coincided with a strong negative PDO (Figure 2-2). Sea ice arrived as early as January and remained as late as mid May. Between 1979 and 1981, ice was largely absent. Beginning in the early 1990s, ice again became more common, although not to the extent observed in the early 1970s. Even though ENSO and the PDO are the best understood and potentially predictable components of the climate

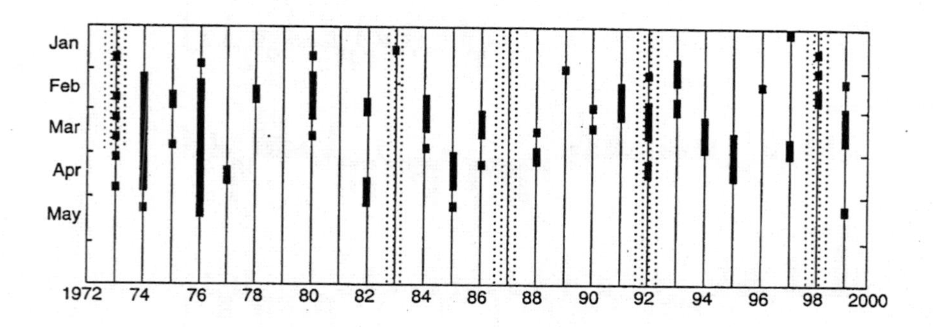

Figure 2-3. The persistence of ice cover at the position of Site #2 is indicated by the dark bars. The shaded areas indicate periods when an El Niño was occurring on the equator (after Stabeno *et al.* 2001).

variability for the Bering Sea in winter, they do not actually account for much of the interannual fluctuation in sea ice cover: ENSO accounts for only ~7 percent of the variance (Niebauer 1998). To illustrate the variability in sea ice extent over the

Bering Sea shelf, we divided the ice observations into three subsets according to generally agreed upon short-term climate regimes (Figure 2-4): the 1972 –1976 cold

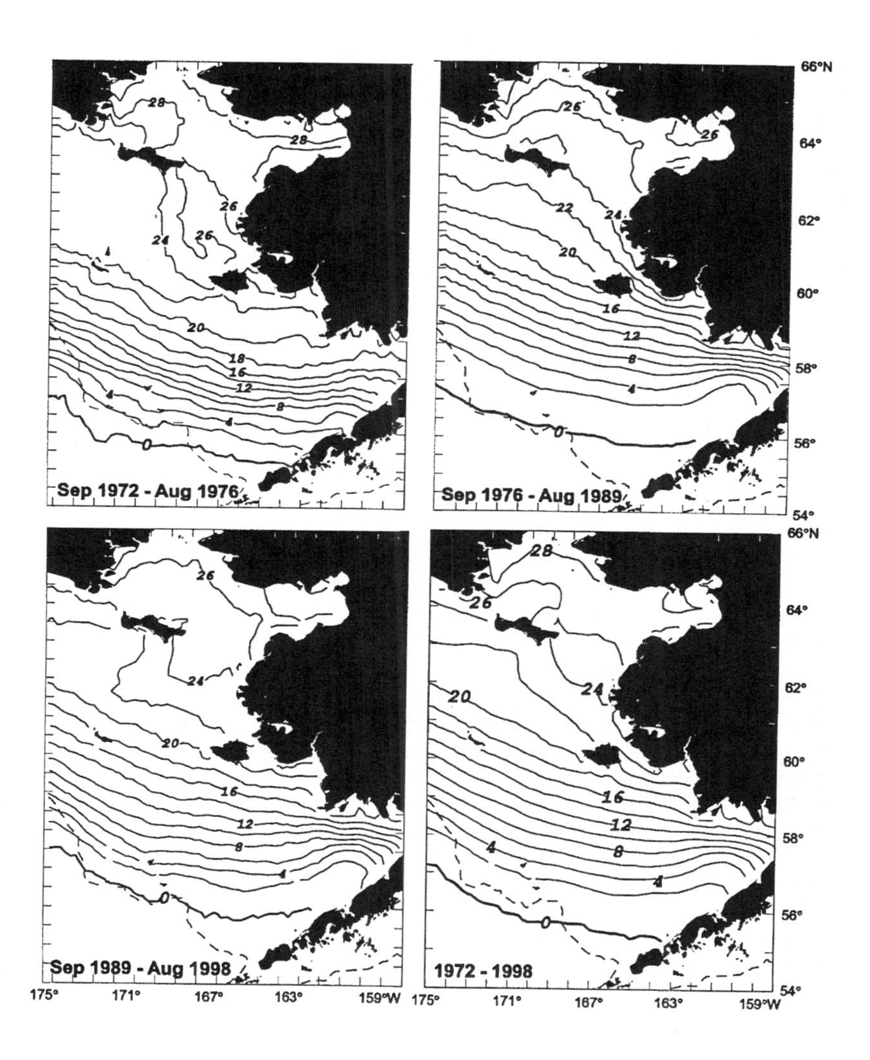

Figure 2-4. Contours of the number of weeks that sea ice was present over the Bering Sea shelf shown as the average ice coverage during: a) 1972-1976, b) 1977-1988, c) 1989-1998, and d) 1972-1998 (from Stabeno *et al.* 2001).

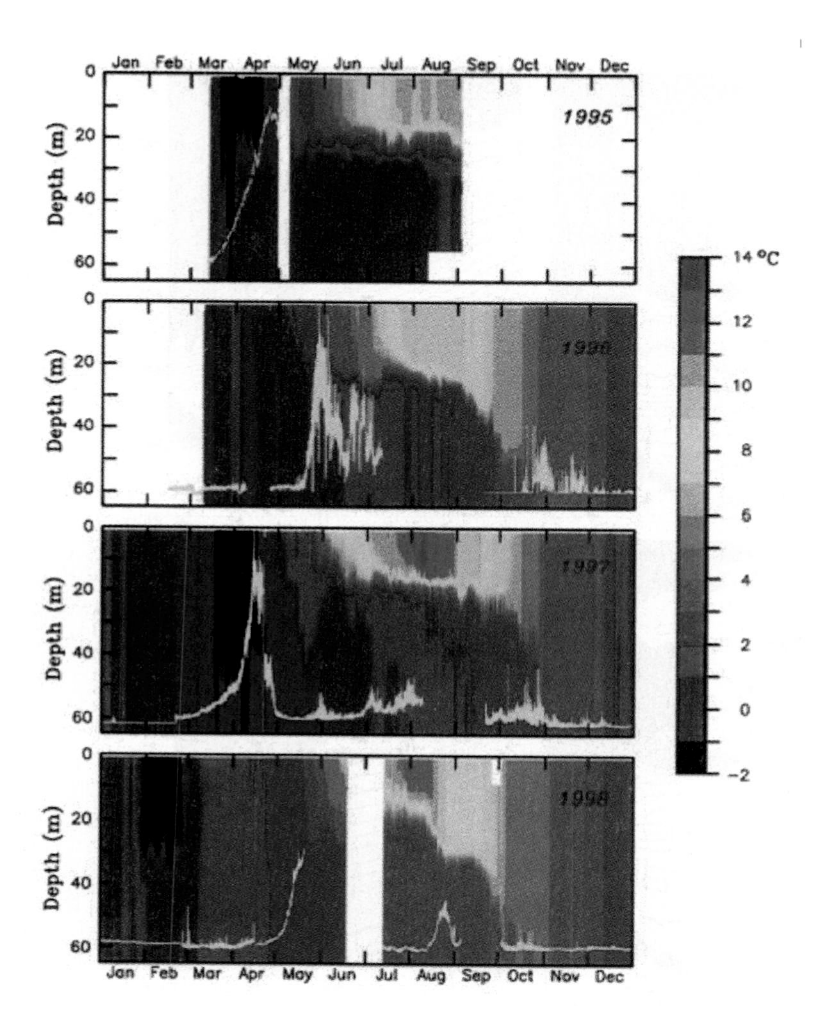

Figure 2-5. Time-series of water temperature and chlorophyll/fluorescence (llm) collected at Site #2. The coldest temperatures indicate the presence of ice (from Stabeno *et al.* 2001.

period, the 1977–1988 warm period, and the 1989–1998 weaker cold period, to characterize the temporal variability in the spatial pattern of sea ice (Stabeno *et al.* 1999). During the first period, ice covered the shelf out to and over the slope and remained around St. Paul Island for more than a month. During the later years of this period, ice did not extend as far seaward and its residence time was typically 2 to 4 weeks less than during the cold period. The differences between the two latter regimes are more subtle, but still evident. Along 59°N, there were 2 to 4 weeks more ice during the 1989 –1998 period than during 1977–1988.

Until recently, studies of the causes and implications of climate variations for the North Pacific and Bering Sea have focused almost exclusively on the winter season. Interest and awareness is growing, however, in warm season climate variations. While these may not be as large as those during winter, they can stand out above background atmospheric conditions (Trenberth *et al.* 1998) and can impact the upper ocean and its biota. For example, the unusually warm SSTs in the eastern Bering Sea (summer 1997) were ascribed mainly to atmospheric anomalies associated with the concurrent strong El Niño (Overland *et al.* 2001). This type of research is in its infancy, but we now have new tools, in particular ~40-year long data sets (e.g. the NCEP/NCAR [National Centers for Environmental Prediction/ National Center for Atmospheric Research] atmospheric reanalysis) for estimating aspects of atmospheric forcing (e.g. radiative effects) that were formerly unavailable. Preliminary results using these data indicate that the eastern shelf is experiencing a trend toward sunnier summers. The consequence is that about 20 watts/m^2 of additional heating has occurred during recent years as compared with 30 to 40 years ago. This is a substantial change and may be attributed to a positive feedback between sea surface temperature and cloud cover. We expect rapid progress in identifying and understanding the variability in the warm season climate of the Bering Sea.

Temperature records from 1995 into 1999 (Figure 2-5) illustrate the seasonal cycle typical for the southeastern middle shelf. In January, the water column is well mixed. The coldest temperatures typically occur in February or early March with the arrival of sea ice. This condition persists until buoyancy is introduced to the water column either through ice melt or solar heating. Generally, this stratification develops during April. The water column exhibits a well defined two-layer structure throughout the summer consisting of a 15-25 m wind mixed layer and a 35-40 m tidally mixed bottom layer. Deepening of the mixed layer by strong winds and heat loss begins as early as mid August, and by early November the water column is again unstratified.

During 1997 through 1999, variations in winds resulted in dramatic changes in the structure and function of the ecosystem. The timing and duration of the transition from winter to spring and summer conditions dictated whether there was an ice associated bloom (1997) or prolonged primary production by diatoms (1998). In 1997, moderate winds resulted in average ice cover that persisted into April; in 1998 weak winds with a more southwesterly component resulted in minimal ice cover of brief duration; whereas in 1999, northerly winds resulted in less extensive ice cover than occurred in 1997, but it remained over the southeastern shelf into May.

In 1997, ice was present when adequate light existed to support an ice-related bloom. In 1998, however, ice departed prior to the existence of sufficient light levels and as a consequence a bloom occurred in May/June when stratification of the water column occurred. The strength and stability of the inner front was also modified by the prevailing wind conditions. Weak winds in spring and summer 1997 resulted in a broad, diffuse front with undetectable nitrate to 60 m depth and few nutrients entering the upper mixed layer. In contrast, the strong storms of spring and the lack of an ice-edge bloom in 1998 resulted in a slow draw down of nutrients. During 1999, the stability of the front permitted events of vertical mixing/upwelling to supply nutrients to the upper mixed layer throughout summer, supporting prolonged production. Interannual variation in the shelf production regimes was reflected in the body condition of migrant short-tail shearwaters (*Puffinus tenuirostris*) which starved in fall 1997, were emaciated but survived in fall 1998, and were of healthy body mass in 1999 (Badiuini *et al.* 2001, Hunt *et al.* 1999).

BIOLOGICAL TRENDS

The relative importance of bottom-up and top-down processes in controlling production in marine ecosystems continues to be debated (Micheli 1999). In addition to losses to natural predators, human harvests of living marine resources in the eastern Bering Sea have averaged over 1.6 million metric tons in the period from 1979 to 1998 (Figure 2-6), with the majority of the catch consisting of pollock. Harvest rates of pollock and other resources have been relatively conservative over the last twenty-five years, with exploitation rates (catch/mature biomass) of 20 percent or less (Livingston *et al.* 1999). Despite these conservative exploitation rates, a variety of species in diverse trophic groups has shown either long-term increases or decreases in abundance while others have shown cyclic fluctuations in abundance over the last two decades (Figure 2-7).

Total biomass in the inshore benthic infauna consumer trophic guild is higher than two decades ago. Within this guild, rock sole and predatory starfish biomass have increased, yellowfin sole and Alaska plaice have decreased, and crabs have shown two periods of fluctuations in abundance, with recent trends indicating several crab species are now at very low abundance and three stocks have been placed in the

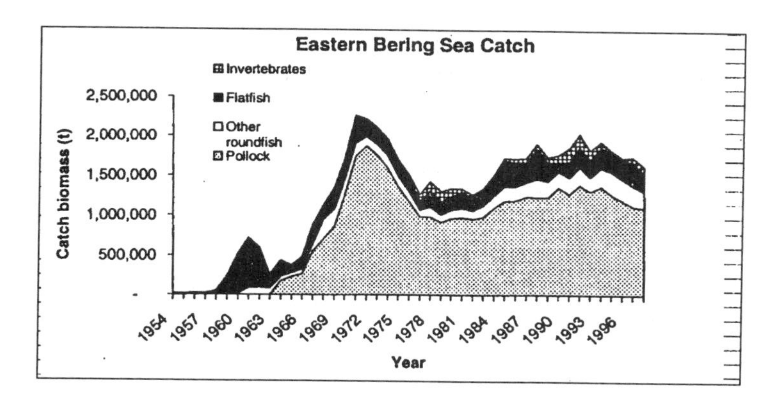

Figure 2-6. Catch biomass of pollock, other roundfish, flatfish, and invertebrates in the eastern Bering Sea from 1954 to 1998.

overfished category. Offshore pelagic fish consumers are dominated in biomass by walleye pollock, which have undergone at least two periods of fluctuation and are now at a lower abundance level than in the 1980s. Several lines of evidence suggest continuation of the long term pattern of increasing arrowtooth flounder abundance and decreasing Greenland turbot abundance. Arrowtooth flounder abundance may now be leveling off. Northern fur seal and piscivorous bird populations declined in the late 1970s and early 1980s but have also leveled off. Crab/fish consumer's biomass is generally declining although skate and halibut biomass within that group is higher in the 1990s than in the 1980s.

Several lines of evidence suggest that the overall production or allocation of resources has changed over the last few decades. The fact of smaller than average adult salmon returning to western Alaska has been linked to changes in ocean conditions that influence growth and survival (Kruse 1998). Decreases in the numbers of seabirds breeding on the Pribilof Islands since the mid 1970s point to a recent change in the carrying capacity of the middle shelf region (Hunt and Byrd 1999). Zooplankton biomass in this region appears to have declined in the early 1990s relative to the mid 1980s (Figure 2-8). Marked increases in jellyfish and rebounding populations of planktivorous marine mammals (Tynan 1998) may put further pressure on food resources (zooplankton) used by larval and juvenile fishes (Brodeur *et al.* 1999a). As noted by Napp and Hunt (1999), if food to apex predators becomes limiting, then climate induced perturbations that affect production and availability of zooplankton may have an even greater effect on the structure of trophic webs than was observed in 1997.

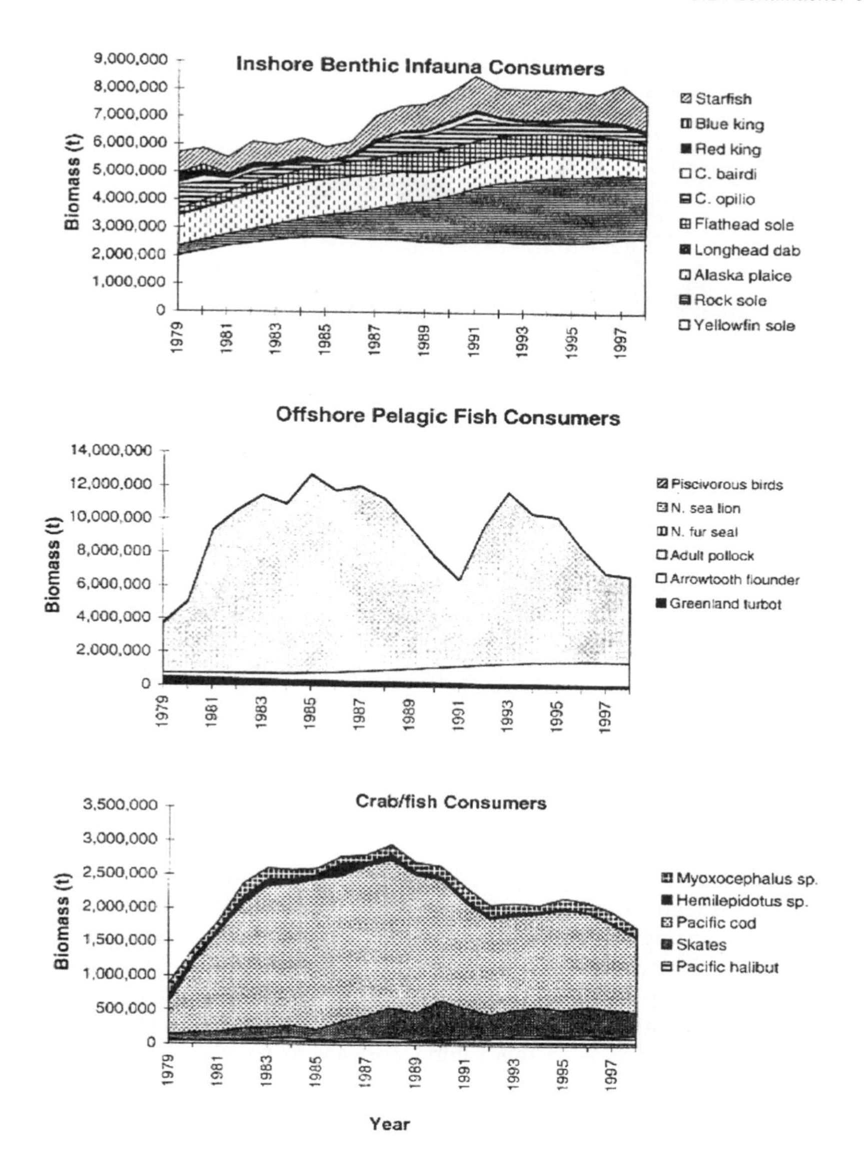

Figure 2-7. Biomass trends of three major trophic guilds in the eastern Bering Sea from 1979 to 1998.

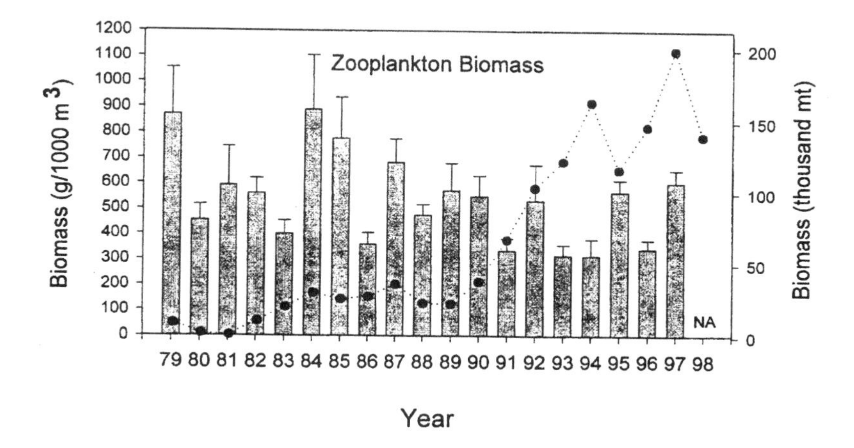

Figure 2-8. Biomass of zooplankton in the Eastern Bering Sea. The bars represent mesozooplankton collected by NORPAC nets (see Sugimoto and Tadokoro 1997) and the circle-and-dash lines represent the biomass of large medusae (in thousand metric tons) collected in standard trawl surveys (modified from Brodeur *et al.* 1999a). All error bars are errors of the mean.

There are indications that the population changes of fish, marine mammals and marine birds in the last twenty years were caused, at least in part, by environmentally driven recruitment changes (Hollowed *et al.* 1998, Kruse 1998, Rosenkranz *et al.* 1998, Springer 1998, Zheng and Kruse 1998). Different species may respond to environmental changes that occur at either interannual or decadal scales. The biomass trends seen in eastern Bering Sea species may be due to decadal or interannual forcing of their prey. Understanding the responses of species to climate forcing is progressing rapidly in this ecosystem. Predicting future climate states and developing fishery management strategies that take changing climate into account must now make similar advancements.

BIOTIC RESPONSES

Identifying and understanding mechanisms that transfer climate change via the ocean to biota (Figure 2-9) is essential if we are to understand ecosystem dynamics (Francis *et al.* 1998). Fluctuations in the physical environment can impact the ecosystem through both changes in the nutrient-phytoplankton-zooplankton sequence (i.e., bottom-up control) and/or by altering habitat resulting in changes in abundance and/or composition of higher trophic level animals (i.e. top-down control). Sugimoto and Tadokoro (1997) hypothesize that for the eastern Bering Sea, top-down control may be responsible for year-to-year fluctuation of zoo- and phytoplankton biomass, while bottom-up control is the mechanism responsible for longer period (decadal) variations.

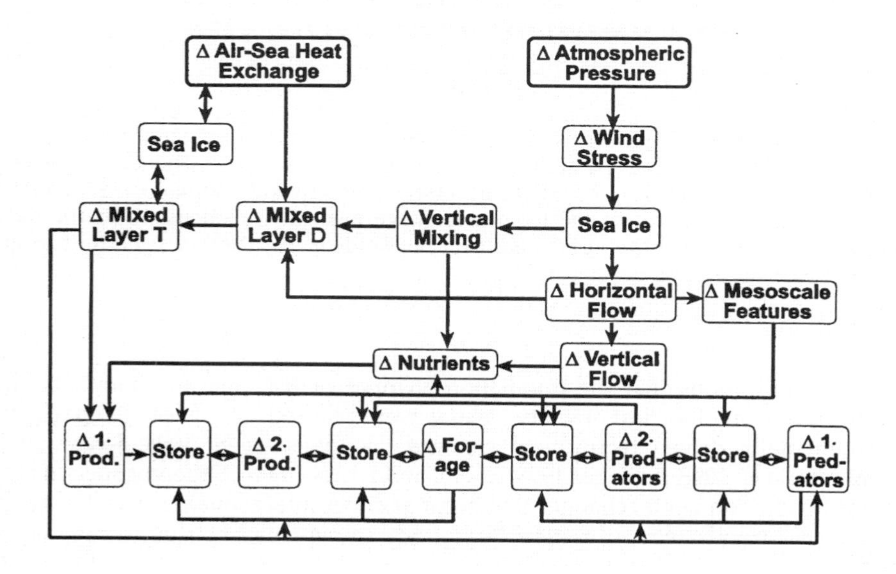

Figure 2-9. Pathways of influence of climate changes in the biological environment (after Francis *et al.* 1998). We have added ice which couples atmospheric phenomenon to the ocean thereby effecting biota in a bottom-up mode. The presence of ice also directly influences distributions of marine mammals exerting an aspect of top-down control. Note that the presence of a coccolithophorid bloom also has direct influence on light penetration, hence on primary production and success of visual feeders.

Sea ice is a feature not present in more temperate ecosystems. Sea ice and its interannual variation has marked ramifications for both the physical and biological environment. As the climate over the Bering Sea warmed, changes occurred in sea ice, the ocean itself and biota. Distributions of marine mammals (Tynan and DeMaster 1997) and fish (Wyllie-Echeverria and Wooster 1998, Brodeur *et al.* 1999b) as well as survival of age-1 pollock (Ohtani and Azumaya 1995) respond to the extent of sea ice itself and to its associated cold pool. The substantial increase in jellyfish biomass over the eastern shelf of the Bering Sea may be linked to climate change through ice cover (Brodeur *et al.* 1999a). The extent, timing, and persistence of ice cover can dramatically alter time/space characteristics of primary production (Niebauer *et al.* 1995, Stabeno *et al.* 1998), and secondary production as food for larval fishes (Napp *et al.* 2000).

Recent results support the belief that interannual and decadal changes in the environment (e.g., using temperature and wind-drive transport) play a significant role in standing stock variability. For example, water temperature has been implicated as an important regulating factor of salmon production in Alaskan waters (Downton and Miller 1998, Kruse 1998, Welch *et al.* 1998). In the southeastern Bering Sea, wind-drive advection of surface waters containing planktonic stages of pollock (Wespestad *et al.* 2000) and Tanner crabs (Rosenkranz *et al.* 1998) accounts for some of the observed fluctuations in year-class strength. In these studies the mechanism which links advection to year-class strength is predation. The advection models use wind-drift of the planktonic stages either to or away from regions where strong predation pressure exists. In the case of marine mammals and seabirds, climate effects appear to be mediated through the food web, although in some cases the links may be direct (Springer 1998).

Coccolithophorid blooms have been observed in the eastern Bering Sea since 1997 under different environmental conditions (Hunt *et al.* 1999). Why this happened and what the implications are for the future is not known. Perhaps coccolithophores will replace the small flagellates that normally dominate in summer. It is likely that favoring another trophic level between primary producer and consumer (i.e., microzooplankton (Nejstgarrd *et al.* 1997) will affect ecosystem dynamics (Napp and Hunt 2001). Attenuation and scattering of light by whole cells and detached liths alters submarine light fields (Voss *et al.* 1998), influencing competition among phytoplankton species and affecting the quality and quantity of light for subsurface visual predators (e.g. diving seabirds). Dense concentrations of coccolithophores also alter regional biogeochemical cycles making large positive contributions to calcite and dimethylsulfide production (Matrai and Keller 1993, Brown and Podesta 1997).

Changes in the ecosystem occurred which emphasize the importance of biological interactions. During the regime shift of the mid-1970s, the physical system experienced significant changes. Atmospheric conditions which had favored strong winds from the north, extreme ice cover and an extensive cold pool became those

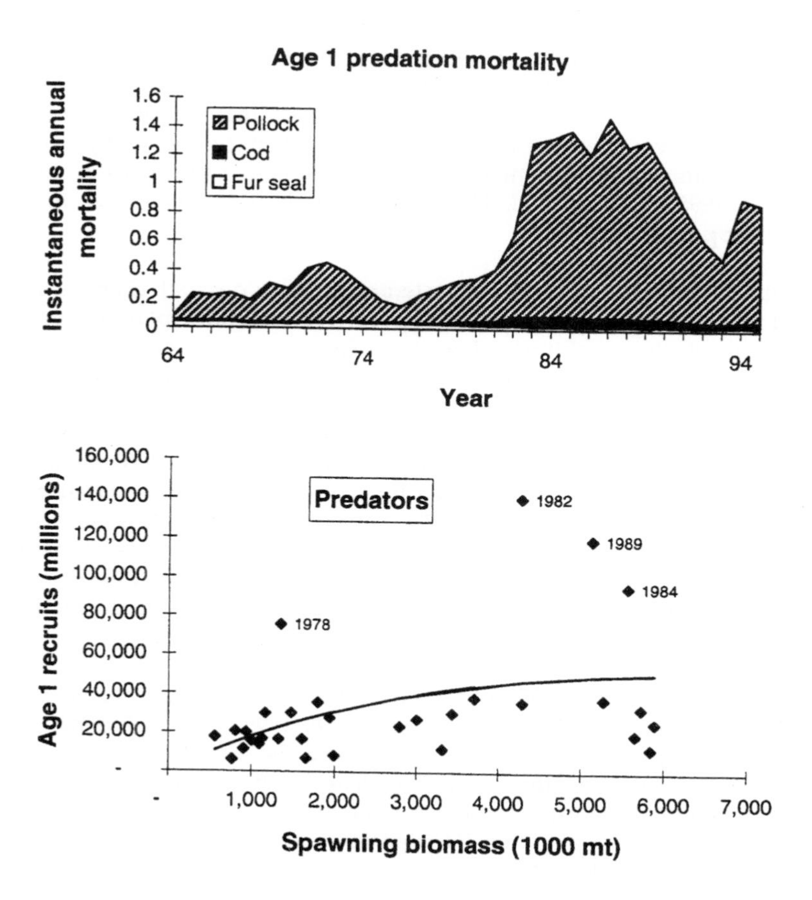

Figure 2-10. Predation mortality of age-1 walleye pollock from 1963 to 1995 estimated by the eastern Bering Sea pollock population model with predation by adult pollock, Pacific cod and northern fur seals (top). Estimates of age-1 recruitment of pollock recruits versus pollock spawning biomass from the pollock population model with predators and the fitted Ricker stock recruitment curves (bottom) (from Livingston and Methot 1998.

experienced significant changes. Atmospheric conditions which had favored strong winds from the north, extreme ice cover and an extensive cold pool became those which generated minimal ice cover and magnitude/extent of the cold pool. During winter and spring of 1977-1978, a set of environmental conditions occurred that promoted development of the extremely large 1978 year class of pollock. The role of cannibalism in influencing population dynamics resulting from the strong year class has been examined (Livingston and Methot 1998). Age-1 pollock mortality due to cannibalism increased in the mid-1980s due to the influence of the large 1978 year class attaining the adult, more cannibalistic size (Figure 2-10). Predation models suggest that the recruitment of subsequent year classes was reduced by cannibalism which played a large role in reducing the number of juvenile pollock that eventually recruited to the fishery (Livingston and Methot 1998). Model results also indicate the possibility of two different spawner-recruit relationships, one for average to cold years and one for warmer years. The authors hypothesize differential survival of pollock in cold versus warm years may initially be due to differences in weather mediated transport of surface waters that separates age-1 pollock from cannibalistic adults (Wespestad *et al.* 2000). These results highlight the importance of studying biotic interactions in order to pinpoint which life history stage of a population is most influenced by abiotic and biotic factors. Because of the interplay between predation and climate factors, it is possible that a year-class as large as the 1978 year-class will not be observed again until there is a combination of a relatively low adult population and winter/spring conditions favoring larval survival.

MANAGEMENT CONCERNS

Many issues surround management of human use of shell fish and fin fish stocks in the Bering Sea. Activities associated with these industries can impact the ecosystem through many pathways, including: overfishing, altering trophic pathways, waste from by-catch, habitat destruction, benthic disturbance and marine debris. These threats to maintaining a healthy ecosystem are compounded by variations in climate that can cascade through the ecosystem. In this region where indigenous peoples abound, issues such as subsistence harvest jurisdiction, and the lack of involvement of local communities in resource management decisions add crucial concerns. These human factors play out on Earth's natural variability which itself has already been impacted by greenhouse gasses (AGU 1999). Part of the solution to this complex challenge is to attain an understanding of the mechanisms which dictate ecosystem vitality and health. We have provided only a few examples of changes in ecosystem dynamics of the eastern Bering Sea and have suggested that these are mainly due to climate change. We must continue discovering how mechanisms causing natural fluctuations function in order to have a reliable basis for managing human impact on various populations (Livingston *et al.* 1999). To make further advances in our understanding of how this ecosystem functions requires a greater effort monitoring and conducting process-oriented investigations of both physical and biological parameters. This will allow identification of the important fluctuations and elucidation of the mechanisms by which changes in physical phenomena are transferred to biota. In addition, this would permit the development of more complete knowledge of the life histories of the central species in the ecosystem and clarify the role of biological mechanisms such as predation. Long-term monitoring programs appear to be the most direct way to establish causal mechanisms responsible for inducing ecosystem change.

THE NEXT DECADE

Prediction through mechanistic understanding is the goal of many applied sciences (e.g. Schumacher and Kendall 1995). Using our increasing knowledge of processes important to the functioning of the Bering Sea ecosystem, we speculate here about what might happen if the majority of years within the next decade resemble environmental conditions observed during spring and summer of 1997.

From a mechanistic view, the reduction of on-shelf transport during 1997 stands out as a fundamental process, regulating production on the shelf. This transport is important for supplying inorganic nutrients, heat and salt (thereby affecting stratification) to the shelf. Assuming that, in the next decade, there is a decrease in shelf flux of nutrients, together with weaker stratification, a reduced influence of sea ice and warmer water temperatures, we envision the following changes. Annual primary production will decline and the spring phytoplankton bloom (in the absence of ice) will also be of lower magnitude but longer duration. This will favor planktonic rather than benthic production. In addition, those zooplankton species that are temperature, rather than food limited, may initially have higher rates of production under a warming scenario. Predation by invertebrates, mammals, and seabirds also has an important role in structuring marine ecosystems. In recent decades, marine mammal populations have been protected from harvest. As their population sizes increase, we expect to see increased inter- and intra-specific competition for food. In 1999, sightings of dead beach-cast Pacific gray whales increased (Rugh *et al.* 1999) with scientists hypothesizing that inadequate food resources in their traditional summer feeding grounds (northern Bering and Chukchi seas) was the cause. Under limiting food resources, prey switching (increasing diet breadth) can be expected. Recent examples include hypothesized use of sea otters by killer whales, *Orcinus orca,* (Estes *et al.* 1998) and diet breadth expansion in planktivorous short-tailed shearwaters (*Puffinus tenuirostris*) when adult euphausiids were not available (Baduini *et al.* 2001). Increased competition for planktonic prey such as euphausiids may also explain the recent decline in size at maturity of Bristol Bay sockeye salmon (Kruse 1998). A decrease in the presence of sea ice in the southeastern Bering Sea and under ice phytoplankton blooms may reverse the recent dramatic increase in scyphomedusae (jellyfish) which prey on zooplankton.

We emphasize that these speculations form, in part, a test of our knowledge of structuring mechanisms of the Bering Sea ecosystem. They are based on observations of its previous states. It is possible that multiple stable points exist for this resource-rich ecosystem. Formulation of hypotheses and collection of observations designed to test them is an essential part of increasing and refining our knowledge.

ACKNOWLEDGMENTS

We acknowledge and thank all those scientists, technicians and support personnel who have collected, analyzed and interpreted observations that form the foundation of this Chapter. In particular, we thank the leaders of the Alaska Fisheries Science Center and the Pacific Marine Environmental Laboratory for their continued support of research toward understanding the ecosystem of the eastern Bering Sea. A Review by A.W. Kendall, Jr. greatly improved the manuscript. We thank R.L. Whitney for technical editing and K. Birtchfield for graphics. The first author thanks Creator for Grandmother Bering Sea and sends a voice for respectful use of her gifts. The research presented herein was partially supported by National Oceanic and Atmospheric Administration's (NOAA's) Coastal Ocean Program through Southeast Bering Sea Carrying Capacity and by National Marine Fisheries Service (NMFS) and Oceanic and Atmospheric Research (OAR) through Fisheries Oceanography Coordinated Investigations (FOCI). This is PMEL Contribution 2155 and JISAO Contribution 730 and Southeast Bering Sea Carrying Capacity contribution S379.

REFERENCES

AGU (American Geophysical Union Report). 1999. EOS, Trans. Am. Geophys. Union. 80(5) 49.

Baduini, C.L., K.D. Hyrenbach, K.O. Coyle, A. Pinchuk, V. Mendenhall, and G.L. Hunt, Jr. 2001. Mass mortality of short-tailed shearwaters in the southeastern Bering Sea during summer 1997. Fish. Oceanogr. 10:117-130

Bowling, S.A. 1995. Geophysical Institute, University of Alaska Fairbanks. Personal communications (see **http://www.besis.uaf.edu/**)

Brodeur, R.D., P.A. Livingston, T.R. Loughlin, and A.B. Hollowed, eds. 1996. Ecology of Juvenile Walleye Pollock, *Theragra chalcogramma*. NOAA Tech. Rep. NMFS 126,227p

Brodeur, R.D., C.E. Mills, J.E. Overland, G.E. Walters, and J.D. Schumacher. 1999a. Evidence for a substantial increase in gelatinous zoooplankton in the Bering Sea, with possible links to climate change. Fish. Oceanogr. 8:296-306.

Brodeur, R.D., M.T. Wilson, G.E. Walters, and I.V. Melnikov. 1999b. Forage fishes in the Bering Sea: Distribution, species associations, and biomass trends. In: T. R. Loughlin and K. Ohtani, eds. Dynamics of the Bering Sea. University of Alaska Sea Grant, AK-SG-99-03:509-536

Brown, C.W. and G. P. Podesta. 1997. Remote sensing of coccolithophore blooms in the western South Atlantic Ocean. Remote Sens. Environ. 60:83-91

Coachman L.K. 1986. Circulation, water masses, and fluxes on the southeastern Bering Sea shelf. Cont. Shelf. Res. 5:23-108

Cooney, R.T. and K.O. Coyle. 1982. Trophic implications of cross-shelf copepod distributions in the southeastern Bering Sea. Mar. Biol. 70:187-196

Downton, M.W. and K.A. Miller. 1998. Relationships between Alaskan salmon catch and North Pacific climate on interannual and interdecadal time scales. Can. J. Fish. Aquat. Sci. 55:2255-2265

Estes, J.A., M.T. Tinker, T.M. Williams and D.F. Doak. 1998. Killer whale predation on sea otters linking oceanic and nearshore ecosystems. Science 282:473-476

Francis, R.C., S.R. Hare, A.B. Hollowed and W.S. Wooster. 1998. Effects of interdecadal climate variability on the oceanic ecosystems of the NE Pacific. Fish. Oceanogr. 7:1-21

Gibson, M.A. and S.B. Schullinger. 1998. In: Answer from the ice edge. The consequences of climate change on life in the Bering and Chukchi seas. 1998. Arctic Network and Greenpeace, USA, Washington DC. 32p

Hollowed, A.B., S.R. Hare and W.S. Wooster. 1998. Pacific-Basin climate variability and patterns of northeast Pacific marine fish production. In: G. Holloway, P. Muller and D. Henderson, eds. Proceedings of the 10th Aha Huliko'a Hawaiian Winter Workshop on Biotic Impacts of Extratropical Climate Variability in the Pacific, January 20-26, 1998. NOAA Award No. NA67RJ0154, SOEST Special Publication.

Hare, S.R. and N.J. Mantua. 2000. Empirical indicators of climate variability and ecosystem response since 1965. Prog. in Oceanogr. 47:103-145

Hunt, G.L. Jr. and G.V. Byrd Jr. 1999. Marine bird populations and carrying capacity of the eastern Bering Sea. In: T.R. Laughlin and K. Ohtani, eds. Dynamics of the Bering Sea. Alaska Sea Grant Press Pub. AK-SG-99-03. 509-536

Hunt, G.L. Jr., C.L. Baduini, R.D. Brodeur, K.O. Coyle, N.B. Kachel, J.M. Napp, S.A. Salo, J.D. Schumacher, P.J. Stabeno, D.A. Stockwell, T.E. Whitledge, S.I. Zeeman. 1999. The Bering Sea in 1998: The second consecutive year of weather forced anomalies. EOS, Trans. Am. Geophys. Union 80 (47):561-566

Incze, L.S. and J.D. Schumacher. 1986. Variability of the environment and selected fisheries resources of the Eastern Bering Sea ecosystem. In: Sherman, K and L.M. Alexander, eds. Variability and Management of Large Marine Ecosystems. AAAS Selected Symposium 99, 109-143

Iverson, R.L., L.K. Coachman, R.T. Cooney, T.S. English, J.J. Goering, G.L. Hunt, Jr., M.C. Maccauley, C.P. McRoy, W.S. Reeburgh, T.E. Whitledge. 1979. Ecological significance of fronts in the southeastern Bering Sea. In: R.J. Livingston, ed. Ecological Processes in Coastal and Marine Systems. Plenum Press, New York. 437-466

Kruse, G.H. 1998. Salmon run failures in 1997-1998: A link to anomalous ocean conditions? Alaska Fish. Res. Bull. 5:55-63

Lachenbruch, A.H. and B.V. Marshall. 1986. Changing climate: geothermal evidence from permafrost in the Alaskan Arctic. Science 234:689-696

Livingston, P.A. 1993. The importance of predation by groundfish, marine mammals, and birds on walleye pollock *Theragra chalcogramma* and Pacific herring *Clupea pallasi* in the eastern Bering Sea. Mar. Ecol. Prog. Ser. 102:205-215

Livingston, P.A. and R.D. Methot. 1998. Incorporation of predation into a population assessment model of eastern Bering Sea walleye pollock. In Fishery Stock Assessment Models, Alaska Sea Grant College Program. AK-SG-98-01: 16p

Livingston, P.A., L.L. Low and R.J. Marasco. 1999. Eastern Bering Sea ecosystem trends. In: Q. Tang and K. Sherman, eds. Large Marine Ecosystems of the Pacific Rim: Assessment, Sustainability and Management. Blackwell Science, Boston. 140-162.

Mantua, N.J., S.R. Hare, Y. Zhang, J.M. Wallace and R.C. Francis. 1997. A Pacific interdecadal oscillation with impacts on salmon production. Bull. Am. Meteorol. Soc. 78:1069-1079

Matrai, P.A. and M.D. Keller. 1993. Dimethylsulfide in a large-scale coccolithophore bloom in the Gulf of Maine. Cont. Shelf Res. 13:831-843

Micheli, F. 1999. Eutrophication, fisheries, and consumer-resource dynamics in marine pelagic ecosystems. Science. 285:1396-1398

Napp, J.M. and G.L. Hunt, Jr. 2001. Anomalous conditions in the southeastern Bering Sea, 1997: Linkages among climate, weather, ocean and biology. Fish. Oceanogr. 10:61-68

Napp, J.M., A.W. Kendall, Jr. and J.D. Schumacher. 2000. A synthesis of biological and physical processes affecting the feeding environment of larval wall eye pollock (*Theragra chalcogramma*) in the Eastern Bering Sea. Fish. Oceanogr. 9:147-162

Nejstgarrd, J.C., I. Gismervik and P.T. Solberg. 1997. Feeding and reproduction by *Calanus finmarchicus* and microzooplannkton grazing during mesocosm blooms of diatoms and the coccolithophore *Emiliania huxleyi*. Mar. Ecol. Prog. Ser. 147:197-217

Niebauer, H.J. 1998. Variability in the Bering Sea ice cover as affected by a regime shift in the North Pacific in the period 1947-1996. J. Geophys. Res. 103(27):717-727, 737

Niebauer, J.J., V. Alexander and S.M. Henrichs. 1995. A time-series study of the spring bloom at the Bering Sea ice edge: Physical processes, chlorophyll and nutrient chemistry. Con. Shelf Res. 15:1859-1878

NRC (National Research Council). 1996. The Bering Sea Ecosystem. National Academy Press, Washington, D.C. 324p

Ohtani, K. and T. Azumaya. 1995. Influence of interannual changes in ocean conditions on the abundance of walleye pollock (*Theragra chalcogramma*) in the eastern Bering Sea. In: R.J. Beamish, ed. Climate Change and Northern Fish Populations. Can. Spec. Publ. Fish. Aquat. Sci. 121:87-95

Osterkamp, T.E. 1994. Evidence for warming and thawing of discontinuous permafrost in Alaska. EOS, Trans. Am. Geophys. Union. 75, 85

Overland, J.E., N.A. Bond and J. Miletta. 2001. North Pacific atmospheric and SST anomalies in 1997: Links to ENSO? Fisheries Oceanography. 10(1):69-80

Reed, R.K. 1995. On the variable subsurface environment of fish stocks in the Bering Sea. Fish. Oceanogr. 4:317-323

Reed, R.K. 1998. Confirmation of a convoluted flow over the Southeastern Bering Sea Shelf. Cont. Shelf Res. 18:99-103

Reed, R.K. and P.J. Stabeno. 1999. The Aleutian North Slope Current. In T.R. Loughlin and K. Ohtani, eds. Dynamics of The Bering Sea. University of Alaska Sea Grant, AK-SG-99 99-03, 177-192

Rosenkranz, G.E., A.V. Tyler, G.H. Kruse and H.J. Niebauer. 1998. Relationships between wind and year class strength of tanner crabs in the southeastern Bering Sea. Alaska Fish. Res. Bull. 5:18-24

Rugh, D.J., M.M. Muto, S.E. Moore and D.P. DeMaster. 1999. Status review of the eastern north Pacific stock of gray whales. NOAA Tech. Memo. NMFS-AFSC-103. 96p

Schumacher, J.D. and A. W. Kendall, Jr. 1995. An example of fisheries oceanography: walleye pollock in Alaskan waters. Rev. Geophys. Suppl. 33:1153-1163

Schumacher, J.D. and R.K. Reed. 1992. Characteristics of currents over the continental slope of the eastern Bering Sea. J. Geophys. Res. 97:9423-9433

Schumacher, J.D. and P.J. Stabeno. 1994. Ubiquitous eddies of the eastern Bering Sea and their coincidence with concentrations of larval pollock. Fish. Oceanogr. 3:182-190

Schumacher, J.D. and P.J. Stabeno. 1998. Continental shelf of the Bering Sea. Chapter 27 In A.R. Robinson and K.H. Brink, eds. The Sea, Vol.11. 789-822 John Wiley and Sons.

Schumacher, J.D. and V. Alexander. 1999. Variability and role of the physical environment in the Bering Sea ecosystem. In: T.R. Loughlin and K. Ohtani, eds. Dynamics of The Bering Sea. University of Alaska Sea Grant, AK-SG-99-03. 147-160

Springer, A.M. 1998. Is it all climate change? Why marine bird and mammal populations fluctuate in the North Pacific. In: G. Halloway, P. Muller and D.

Henderson, eds. Biotic Impacts of Extratropical Climate Change in the Pacific. Univ. Hawaii. 121-125

Springer, A.M. and G.V. Byrd. 1989. Seabird dependence on walleye pollock in the southeastern Bering Sea. In: Proceedings of the International Symposium on the Biology and Management of Walleye Pollock. Alaska Sea Grant AK-SG-89-01. 667-677

Stabeno, P.J., J.D. Schumacher, R.F. Davis and J.M. Napp. 1998. Under-ice observations of water column temperature, salinity and spring phytoplankton dynamics: Eastern Bering Sea shelf. 1995. J. Mar. Res. 56:239-255

Stabeno, P.J., J.D. Schumacher and K. Ohtani. 1999. Overview of the Bering Sea. 1999. In Loughlin, T.R. and K. Ohtani eds. Dynamics of The Bering Sea. University of Alaska Sea Grant, AK-SG-99-03. 1-28

Stabeno, P.J., N.A. Bond, N.B. Kachel, S.A. Salo and J.D. Schumacher. 2001. On the temporal variability of the physical environment over the southeastern Bering sea. Fish. Oceanogr. 10(1):81-98

Sugimoto, T. and K. Tadokoro. 1997. Interannual-interdecadal variations in zooplankton biomass, chlorophyll concentration and physical environment of the subarctic Pacific and Bering Sea. Fish. Oceanogr. 6:74-93

Trenberth, K.E. and J.W. Hurrell. 1995. Decadal coupled atmospheric-ocean variations in the North Pacific Ocean. In: R.J. Beamish, ed. Climate Change and Northern Fish Populations. Can. Spec. Publ. Fish. Aquat. Sci. 121:15-24

Trenberth, K.E., G.W. Branstator, D. Karoly, A. Kumar, N.C. Lau, and C. Ropelewski. 1998. Progress during TOGA in understanding and modeling global teleconnection associated with tropical sea surface temperatures. J. Geophys. Res. 103:14,291-14,324

Tynan, C.T. 1998. Coherence between whale distributions, chlorophyll concentration, and oceanographic conditions on the southeast Bering Sea shelf during a coccolithophore bloom, July–August, 1997

Tynan, C.T. and D.P. DeMaster. 1997. Observations and predictions of Arctic climate change: potential effects on marine mammals. Arctic 50:308-322

U.S. GLOBEC. 1996. Report on Climate Change and Carrying Capacity of the North Pacific Ecosystem. Scientific Steering Committee Coordination Office, Dept. Integrative Biology, Univ. Calif., Berkeley, CA, U.S. GLOBEC Rep. 15. 95p

Vance, T.C., C.T. Baier, R.D. Brodeur, K.O. Coyle, M.B. Decker, G.L. Hunt Jr., H.M. Napp, J.D. Schumacher, P.J. Stabeno, D. Stockwell, C.T. Tynan, T.E. Whitledge, T. Wyllie-Echeverria and S. Zeeman. 1998. Aquamarine waters recorded for first time in eastern Bering Sea. EOS, Trans. Am. Geophys. Union. 79:122-126

Vidal, J. and S.L. Smith. 1986. Biomass, growth and development of population of herbivorous zooplankton in the southeastern Bering Sea. Deep-Sea Res. 33:523-566

Voss, K.J., W.M. Balch and K.A. Kilpatrick. 1998. Scattering and attenuation properties of *Emiliania huxleyi* cells and their detached coccoliths. Limnol. Oceanogr. 43:870-876

Wadhams, P. 1995. Arctic sea ice extent and thickness. Phil. Trans. R. Soc. London A. 352:301-319

Welch, D.W., Y. Ishida and K. Nagasawa. 1998. Thermal limits and ocean migrations of sockeye salmon (*Oncorhynchus nerka*): long-term consequences of global warming. Can. J. Fish. Aquat. Sci. 55:937-948

Weller, G., A. Lynch, T. Osterkamp and G. Wendler. 1997. Climate trends and scenarios. In: G. Weller and P.A. Anderson, eds. Implications of Global Change for Alaska and the Bering Sea Region: Proceedings of a Workshop, University of Alaska Fairbanks, 3 – 6 June, 1997

Wespestad, V.G., L.W. Fritz, W.J. Ingraham, Jr. and B.A. Megrey. 2000. On relationships between cannibalism, climate variability, physical transport and recruitment success of Bering Sea walleye pollock, *Theragra chalcogramma. ICES* J. Mar. Sci. 56:272

Wyllie-Echeverria, T. and W.S. Wooster. 1998. Year-to-year variations in Bering Sea ice cover and some consequences for fish distributions. Fish. Oceanogr. 7:159-170

Zheng, J. and G.H. Kruse. 1998. Stock recruitment relationships for Bristol Bay Tanner crab. Alaska Fish. Res. Bull. 5:116-130

Large Marine Ecosystems of the World
G. Hempel and K. Sherman (Editors)

3

Contemporary State and Factors of Stability of the Barents Sea Large Marine Ecosystem

G.G. Matishov, V.V. Denisov, S.L. Dzhenyuk

INTRODUCTION

The Barents Sea Large Marine Ecosystem is at the highest latitude among the LMEs of the World Ocean. It is situated within the European part of the Arctic shelf, completely to the north of the Northern Polar Circle (Figure 3-1). The area of the Barents Sea is 1424 km^2, the volume of the water is 316 thousand km^3, the average depth is 222 m (Atlas 1980).

The Barents Sea is well known as one of the most productive ocean areas. An intensive fishery has been conducted here from the beginning of the twentieth century by Russia (the Soviet Union from 1922-1991), Norway and some other European countries. Together with significant stocks of marketable fish (cod, herring, capelin, plaice and many others) it is rich in such resources as invertebrates (shrimps, mussels, crustaceans etc.), sea algae, many species of sea mammals and birds. The coastal zone of the Kola Peninsula is a prospective area for the artificial breeding of valuable fish species and for the organization of a recreational fishery.

For a relatively long period, the state of the Barents Sea ecosystem was determined only by the natural factors. However, the rapid growth of the fishing industry in the second half of the twentieth century became the most important factor influencing the quantitative and qualitative composition of the marine biota. The overfishing of certain fish species led to a population dominance in the fishery by less valuable and profitable species. In general, the exhaustion of the Barents Sea caused the expansion of the Soviet fishery fleet in 1970-80s to the distant ocean areas of the North-East and Southern Atlantic, the Southern Pacific, and the Antarctic. Only in the last decade, due to the changing geopolitical situation and to the strengthening of protective measures, has the role of the Barents Sea as the leading fishery basin been restored to a great extent.

At the same time, the development of other branches of the economy, such as the oil and gas industries, has begun on the Barents Sea shelf. Two large offshore projects have already started and several more oil and gas structures are prospective for future exploitation. New ports and industrial enterprises are projected for the Barents Sea coast for the transportation and processing of oil hydrocarbons from deposits of Pechora Basin and the Yamal Peninsula. The environmental impact of the oil and gas industries will be among the most important factors affecting the state of the Barents Sea ecosystem in the near future.

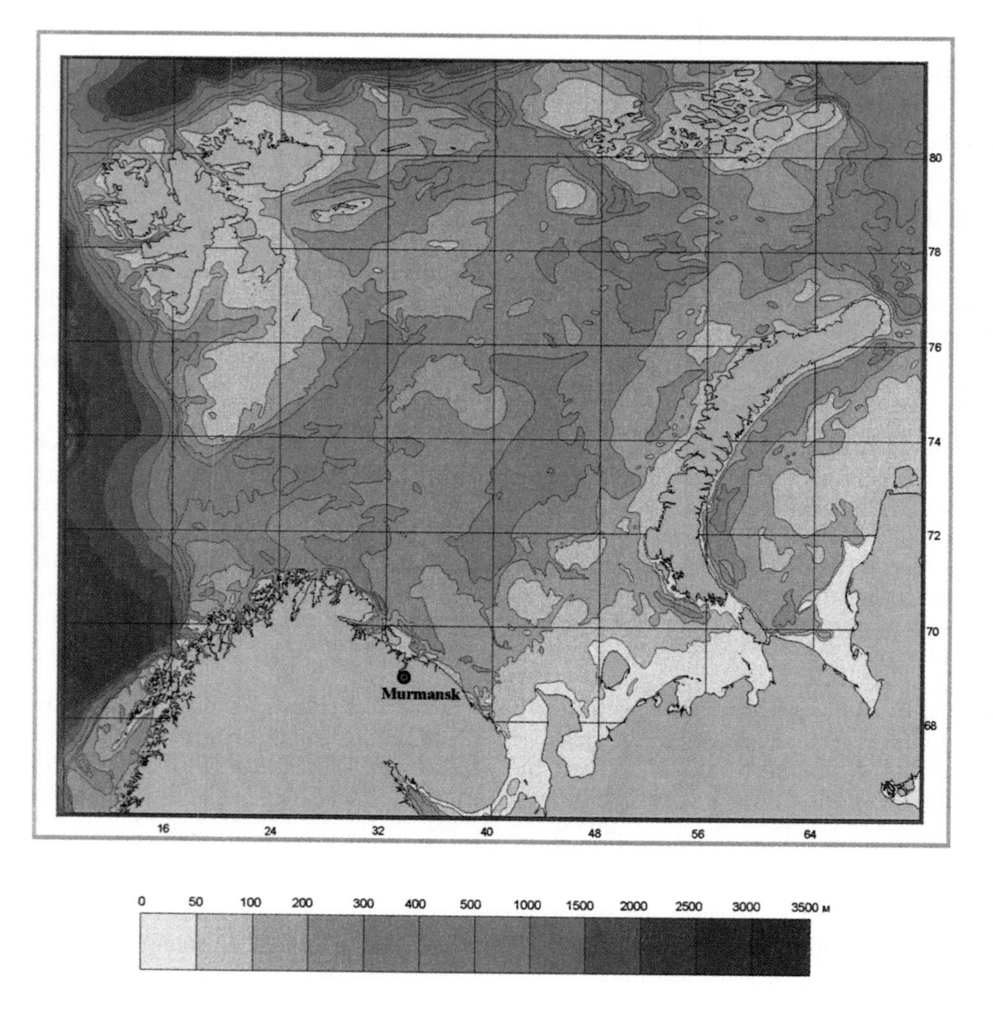

Figure 3-1. Bathymetric map of the Barents Sea. Degrees of longitude and latitude are given on the bottom and side of the figure respectively.

BIOLOGICAL PRODUCTIVITY OF THE BARENTS SEA ECOSYSTEM

Factors of Bioproductivity

The present state of the Barents Sea ecosystem is determined by its evolutional history and by the variability of numerous environmental factors. The Barents Sea is a young post-glacial basin. At the maximum of Quaternary glaciation development, the greater part of it was occupied by continental and shelf glaciers (Figure 3-2) (Matishov and Pavlova 1990). The advection of warm oceanic waters was strictly limited because of ice barriers and the decrease of sea level. After the deglaciation, which occurred about 10,000 years ago, the free connection between the Atlantic and Arctic basins through the Barents Sea shelf was restored, and the contemporary environmental state became established in a relatively short time.

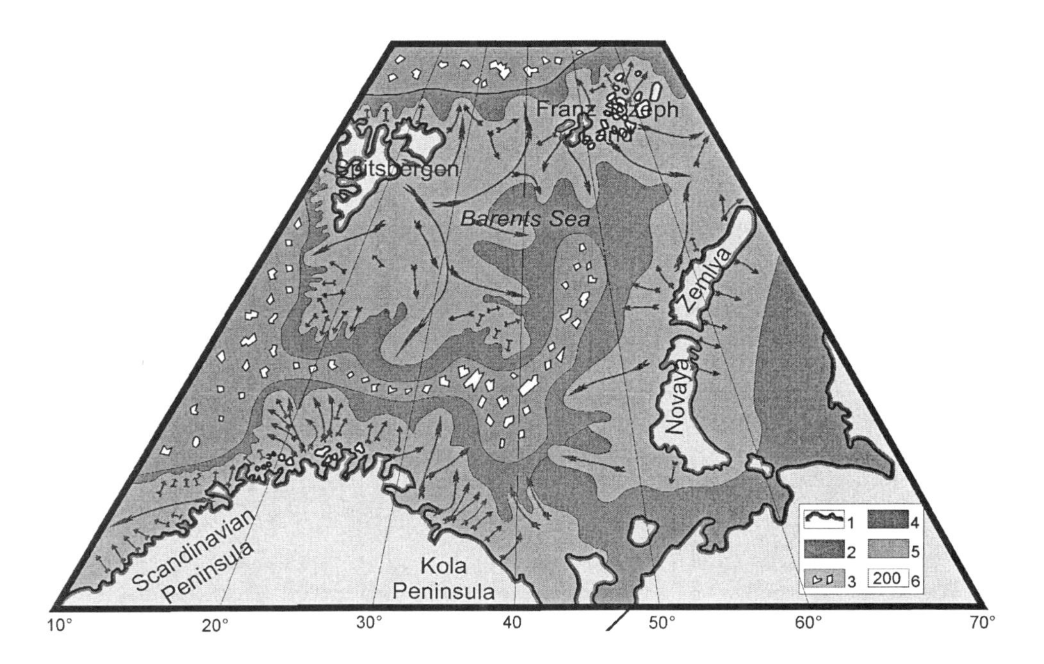

Figure 3-2 Probable development of the continental glaciation on the Barents Sea shelf during the Last Pleistocene (18 to 20 thousand years ago): 1) contemporary coast line; 2) floating ice shelves, i.e. shelf glaciers; 3) pack ice and drifting icebergs; 4) sea bottom permafrost; 5) continental (or lying on the bottom) ice and trajectories of its flow.

The most prominent feature of the Barents Sea hydrography, which has numerous climatic, biological and economic consequences, is the stable interaction between the Atlantic and Arctic water masses divided by the polar hydrological front. This interaction is most distinctly expressed in the western part of the sea, to the south of Spitsbergen. The surface temperature of the Atlantic water mass changes in the annual cycle from 8-10 °C in summer to 3-4 °C in winter and spring. Its salinity is less variable and close to values typical of the North Atlantic (34-34.5 ‰). Transformed Arctic waters, also named the Barents Sea water mass, are warmed on the surface in summer to not more than 6 °C and are cooled to the freezing point by the end of winter. They are less saline (from 34 to 32 ‰ on the surface) because of low evaporation and ice melting. In the coastal zone of the Kola Peninsula and in the shallow south-east part of the Barents Sea (often named the Pechora Sea) the local water masses, subject to the freshening from river runoff, are distinguished.

The presence of Atlantic waters is noted also in the deep layers (100-300 m) of the northern part of the Barents Sea where they belong to the subsurface water mass of the Arctic Ocean transferred from the Atlantic. They are slightly warmer (to 1.5 °C) than the upper water layer, and the stable density distribution is maintained because of their higher salinity.

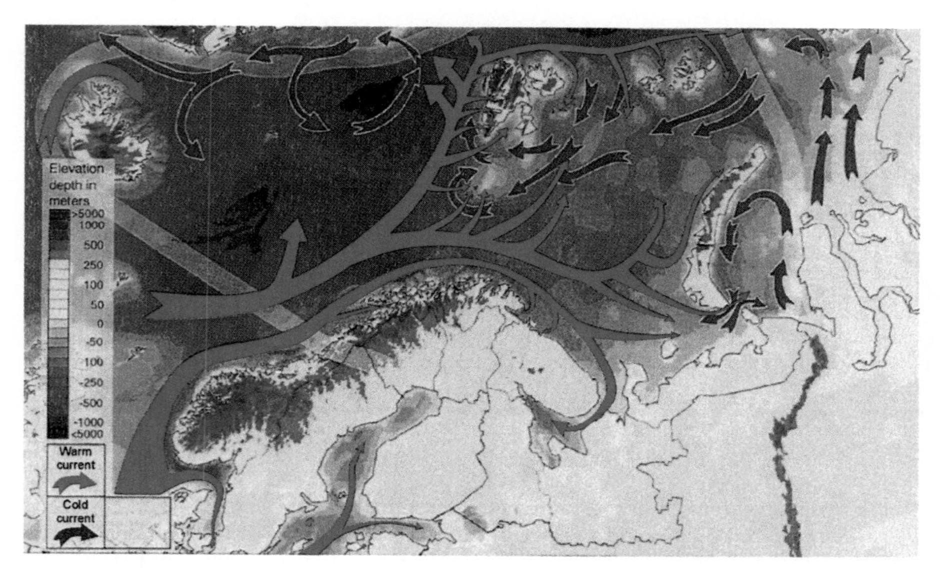

Figure 3-3. Advection of warm Atlantic water of the Gulf Stream system.

Many biological processes in the Barents Sea are governed by the seasonal phenomena of thermocline formation and convective mixing (Hydrometeorology 1990). In summer the warming from the surface, together with mixing by waves and currents, leads to the formation of the upper homogeneous layer divided from the deep cold water by the density difference. Its lower boundary reaches 20-30 m depth. During the long period of cooling in autumn and winter, intensive vertical mixing (autumn-winter convection) takes place, and homogeneous density distribution is established from surface to bottom. This process enriches deep layers with oxygen and contributes to the high biological productivity of the pelagic zone.

Barents Sea water is characterized by a stable system of currents that carries Atlantic waters along the southern coast up to the Pechora Sea and the Novaya Zemlya. This eastward transfer is the most important factor determining the drift of plankton organisms, including fish larvae and therefore the migratory pathways of marketable fishes. Opposing current directions prevail in the northern part of the sea (Figure 3-3). The period of total replacement of the Barents Sea water mass takes, on average, 5 years. However, the water movements in the real time-scale are determined mainly by tidal currents, which form closed trajectories of water particles, and irregular wind currents connected with deep and fast-moving cyclones that are typical for the Barents Sea.

The environmental conditions and bioproductivity of the Barents Sea depend to a great extent on the state of the ice cover. Drifting and fixed ice masses are inhabited by several species of sea mammals. These ice masses have a significant influence on the dynamics and seasonal cycles of plankton communities that, in turn, affect representatives of higher trophic levels. The ice-edge zones are distinguished by increased primary production in spring and early summer when the blooming of plankton starts as the ice retreats. The position of the ice edge during the period of its farthest advancement is close to the polar hydrological front, though it may deviate because of the wind variability.

The state of this marine ecosystem depends also on the structure and dynamics of the ice cover (forms and sizes of floes, fissures and open-water areas, drift velocity and direction, number and sizes of ice ridges).

All environmental factors that relate to biological productivity are subject to spatial and temporal variability of different scales. The climate fluctuations, with a duration on the order of several decades, result in steady anomalies of water temperature and salinity (Figure 3-4). The most distinct period of warming was noted in 1920-30s. It was part of the well-known "Arctic warming" which manifested itself also in an increase of air temperature, the retreat of Arctic sea ice, and the penetration of Atlantic fauna to the east and north. Among the long-term changes of salinity "The great salinity anomaly" at the end of the twentieth century is characterized by Belkin *et al.* (1997). The freshening,

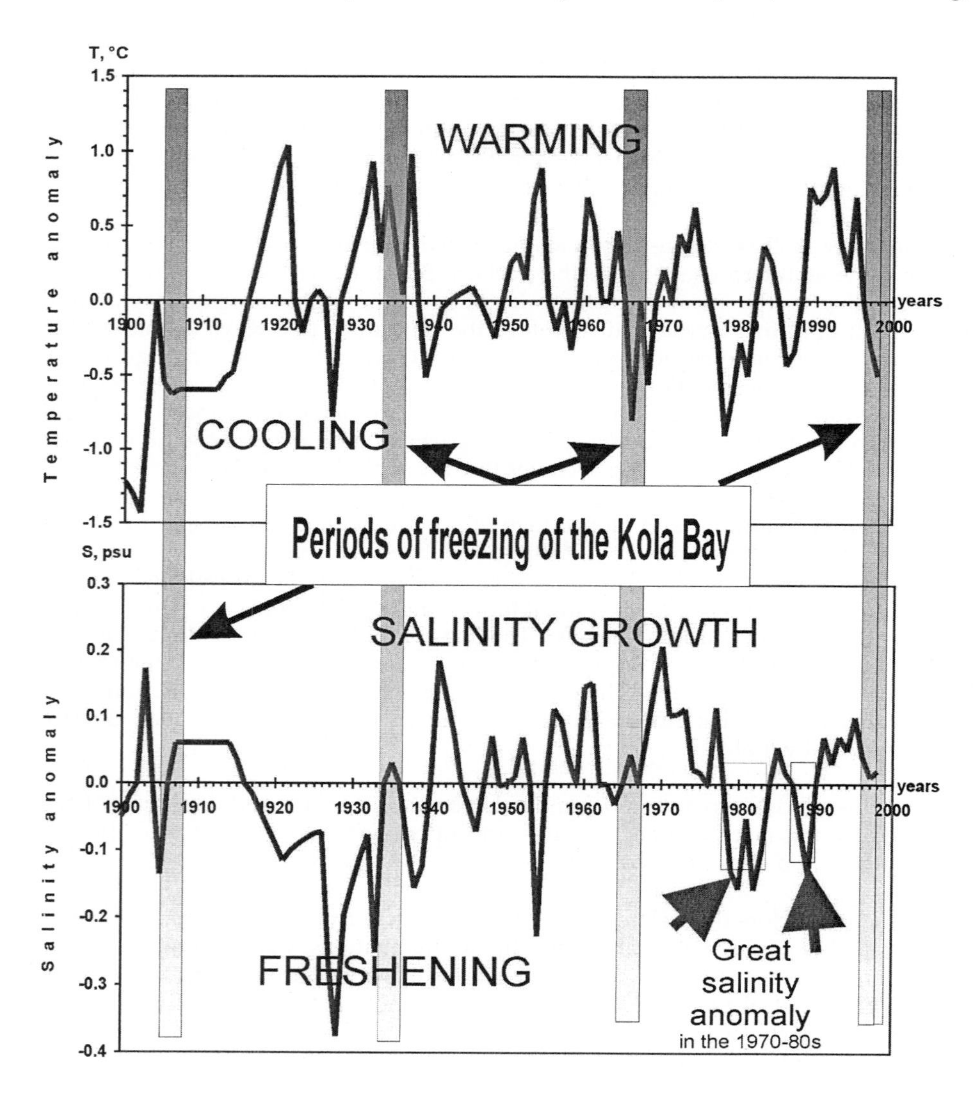

Figure 3-4. Long-term changes of temperature and salinity in the Barents Sea, 0 to 50 m

which was observed in the North Atlantic in the mid-1970s, extended to the Barents Sea in the beginning of 1980s. Such anomalies affect the rate of convection and ice formation, so they have an indirect influence on ecosystem processes.

The Barents Sea water and air temperature changes of the last decades, contrary to climatic data from other regions, show no definite warming trend. The relatively stable temperature can be explained by patterns of atmospheric circulation in the West Arctic hindering the longitudal transfer of air masses. However, there is a high probability of an expansion of the warming, already observed in adjacent regions, to the Barents Sea. Warming may reduce sea ice area and thickness, increase water temperature in all seasons and, possibly, displace the polar front to the north-east.

The depletion of the ozone layer is another important global scale process, one that can increase ultraviolet radiation received by organisms during periods of low ozone content in the upper atmosphere. "Ozone holes" are observed mainly in continental areas of high latitudes. So far there has been no proof of this phenomenon in the Barents Sea region. At least, in 1998-1999 the ozone content over the Kola Peninsula was not less than, and in some months even exceeded, the climatic averages (Nikulin *et al.* 2000). The spatial and temporal variability of the ice cover is an important environmental factor responsible for anomalies in the distribution and seasonal cycles of marine organisms. The annual average values of ice-covered area in the Barents Sea varies from 22 to 52 percent of its total area. A complete absence of ice within the sea is possible from August to October in mild years, but in severe years the maximum ice cover in September reaches 35 percent. In April, when the ice cover is the greatest in the annual cycle, its extent can vary from 50 to 92 percent. The south-west area of the Barents Sea, traditionally referred to as "unfreezing" varies from 180 to 700 thousand km^2, depending on the ice conditions of the current year.

The environmental parameters' variations of synoptic scale are determined by the trajectories and frequency of appearance of the Barents Sea cyclones. The cyclonic activity is the greatest in autumn and winter. These seasons are favourable for development of stormy waves because vast areas are free of ice. The wave heights, typical of the Barents Sea storms, are 5-7 m. Maximum individual wave heights, observed by wave recorders, reached 11-12 m. The centennial calculated maximums are estimated at 20-23 m on the western boundary of the Barents Sea and 10-12 m in the Pechora Sea, where the wave activity is limited by ice and shallow depths.

The rotation of stormy and calm periods has an influence on the trophic and reproductive behaviour of sea mammals, and on the passive and active migrations of pelagic organisms. The rotation also has indirect ecosystem consequences because the fishing operations, especially those of small coastal fleets, strongly depend on the sea state.

General Features of Pelagic Ecosystems

The Barents Sea LME is of the shelf type, with a prevalence of pelagic trophic chains (Zenkevich 1963). The input of detritus plays no significant role because the river runoff is small in comparison with the sea water mass and is poor in organic matter. Not less than 80 percent of the fish biomass is sustained by the consumption of plankton.

Three types of pelagic habitats belonging to corresponding water masses are distinguished in the Barents Sea: Atlantic, Arctic and coastal. Peculiarities of life activity, in each of the water masses, are determined mainly by abiotic environmental factors: seasonal changes of the temperature and light regimes, advection or ice cover. The common feature of all areas in the Barents Sea large marine ecosystem is the short period of phytoplankton blooming. This phenomenon starts in April in the south-west part of the sea and advances to the north-east during late spring and summer, finishing in September in the Arctic waters.

The main trophic levels of the Barents Sea ecosystem are the same as everywhere in the ocean: primary producers (phytoplankton); primary consumers (microzooplankton); secondary consumers (macro- and mesozooplankton represented correspondingly by Euphausiacea and Copepoda) with mixed feeding on plant and animal microorganisms; tertiary consumers (planktivorous fishes); consumers of higher orders (piscivorous fishes, sea birds and mammals). However, the individual trophic links may differ significantly in various environmental conditions.

The Atlantic portion of the ecosystem is the most productive and therefore it has been studied relatively well (Figure 3-5). It is characterized by a consortive type of organization (Matishov *et al.* 1994), i.e. containing one central species or group of species on which the rest of the biota depends. These dominating groups on different trophic levels are pelagic crustaceans (*Calanus finmarchicus* and Euphausiacea), herring or capelin (in the north-eastern part of the sea, polar cod), and cod. The planktivorous fishes (capelin) are the connecting link between all pelagic components and so their state is representative of the Barents Sea ecosystem as a whole.

The structure and dynamics of the Arctic ecosystem are shown on Figure 3-6. The life activity here is closely connected with the ice cover. Spring biological processes begin during the ice melting. They play an important role in trophic chains. Numerous zooplanktivorous bird species transfer organic substances from the breeding zone to the coastal areas.

The coastal waters are characterized by close interaction of pelagic and benthic communities. The spring processes begin here earlier than in offshore areas and develop more smoothly, without sharp changes in plankton biomass (Kamshilov *et al.* 1958).

This is explained by the abundance of larvae of benthic organisms in the plankton composition. The coastal waters are the spawning zone of planktivorous fishes (capelin, herring).

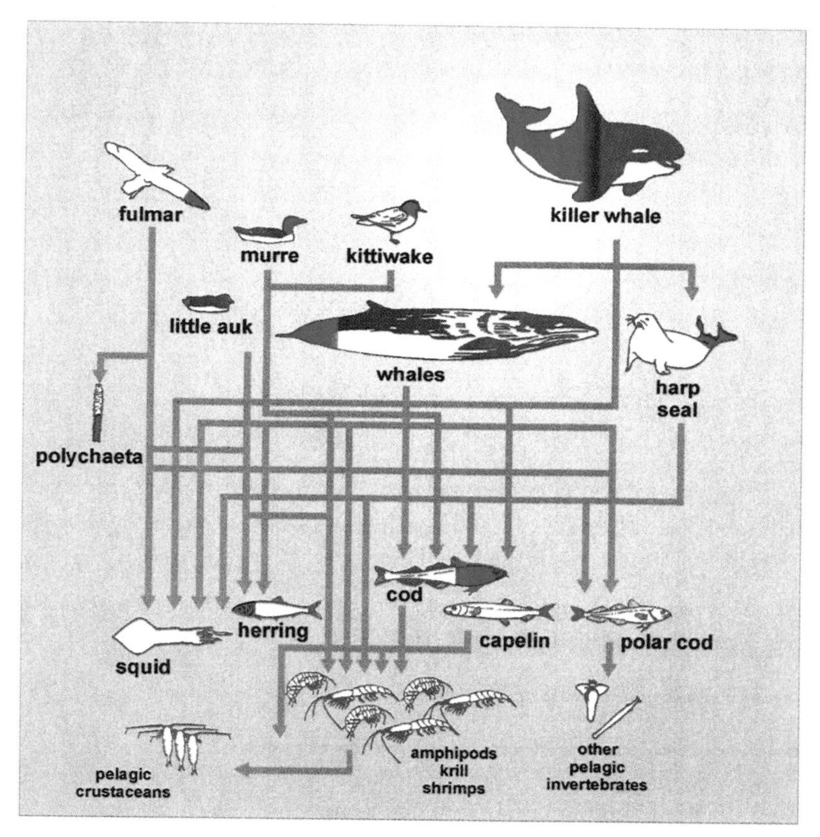

Figure 3-5. The Barents Sea ecosystem (Atlantic waters) (the pattern principles from The State of the European Arctic, 1997).

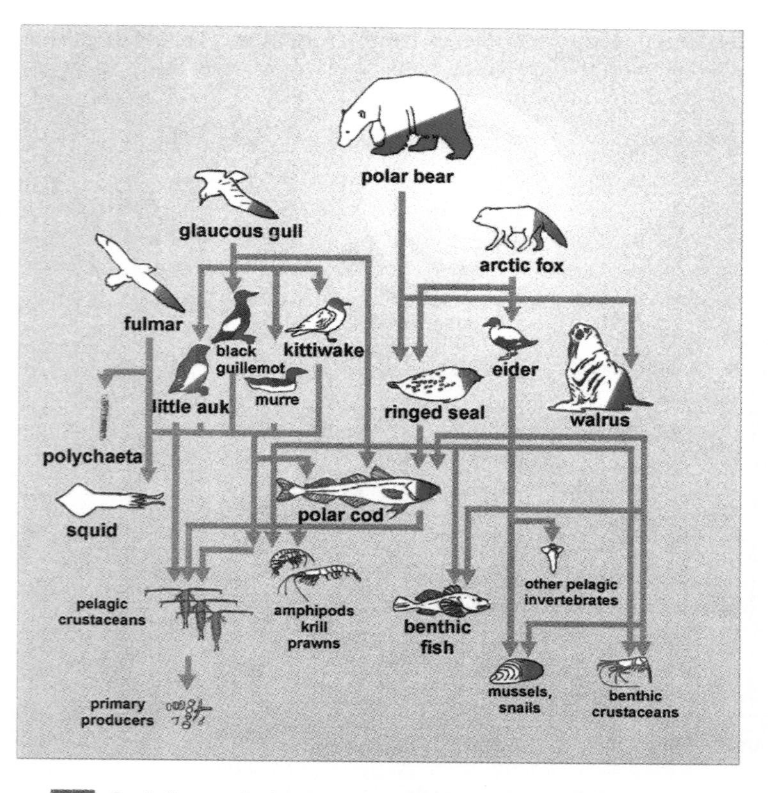

Figure 3-6. The Barents Sea ecosystem (Arctic waters) (the pattern principles from The State of the European Arctic 1997).

Biological Productivity on Different Trophic Levels

The productivity of phytoplankton, which is the initial link of all trophic chains, is the main index of biological productivity and resource potential of most large marine ecosystems. Plankton algae of the Barents Sea are represented mainly by Arctic and boreal complexes of Diatomeae, Flagellata and Peridinea. The species diversity is the highest in the Atlantic water mass and decreases to the north and east. Also toward the north and east, the share of Diatomeae increases and that of Peridinea falls. The Flagellata algae are present in mass quantities in spring and are the main component of the plankton bloom.

Estimates of annual primary production by different authors are not completely in agreement, but general features of its spatial distribution are well studied. In the Atlantic water the annual values are about 170 g C/m^2. In the coastal areas of the Kola Peninsula they are almost half, on average, and differ significantly depending on local hydrological conditions. The most probable annual values for the Arctic waters are between 60-80 g C/m^2. These data permit us to estimate the total primary production within the Barents Sea limits at about 2 billion tons.

The balance of organic matter in the upper trophic layers is estimated by the "coefficient of ecological effectiveness" or the ratio between the increment of biomass and consumed food. Characteristic values of this coefficient in marine and freshwater ecosystems are on the order of 10 percent. The summarized estimates of different authors for the Barents Sea ecosystem for the period from 1930 to the 1970s result in the following values of annual production and biomass (million tons) (Matishov *et al.*1994):

- annual production of phytoplankton: 2000 - 3000;
- annual production of zooplankton: 120 - 300;
- biomass of pelagic fishes (capelin, polar cod, herring, cod fry): 12 - 30;
- biomass of cod fishes: 3.0.

These values may vary widely depending on the abiotic factors, principally water temperature. By some estimates, during the "Arctic warming" of the 1930s, the annual production of phytoplankton increased to 8 billion tons and the biomass of planktivorous and piscivorous fishes increased correspondingly to 180 and 27 million tons. However, the contemporary state of the ecosystem is determined not so much by the natural climatic fluctuations as by the fishery which became the leading driving factor after the Second World War.

FISH FAUNA AND FISHERIES

Biology of Principal Marketable Species

Ichthyological research has been carried out in the Barents Sea since the end of the Nineteenth century. The species composition of the fish fauna is well known now, although there are many unsolved problems of species' life cycles, structures of communities and population dynamics. The full list of fish species numbers about 150. About two thirds of them are permanent inhabitants of the Barents Sea. The rest are boreal species migrating to the western and central parts of the sea in favourable hydrological conditions (Matishov and Nikitin 1997).

Not more than 30 species are of economic importance and only some of them are the objects of a regular fishery. More than 95 percent of the total catch in the Barents Sea depends on the following pelagic and bottom species (Ecology 2001): cod, haddock, capelin, herring, polar cod, sea perch, catfish, plaice, and halibut. Among anadromous fish species the Atlantic salmon is distinguished by its commercial value. There are also some invertebrates often mentioned: polar shrimp, Icelandic scallop, Kamchatka crab. The latter was introduced to the bays of the Kola Peninsula coast in the 1960s. The population developed successfully and, by recent estimates, is about 5 million specimens.

Atlantic cod *Gadus morhua morhua L.* is, for the present, the main object of the traditional fishery. This species is distinguished by a long life-span (to 20-22 years), multi-age structure of populations, quick growth and high feeding adaptability. Its migrations cover almost all of the Barents Sea. Spawning takes place along the coast of northern Norway and the northwest coast of the Kola Peninsula. Eggs, larvae and pelagic early juveniles are spread in surface water layers by currents of the Gulf Stream system over long distances in the southern part of the Barents Sea. Juvenile cod accumulate in several local areas where the fishery is prohibited. Juveniles feed mainly on masses of plankton crustaceans. Adult specimens (from 3-4 years of age) are able to migrate, depending on the location of food objects, to distant parts of the sea. Their food includes some fish species, euphausiids, and benthic animals.

Haddock, *Melanogrammus aeglefinus (L.)* and pollock, *Pollachius virens (L.)* are also connected with the Atlantic water mass. Their reproduction, growth and migration patterns are similar to those of cod. Haddock feeds mainly on bottom invertebrates (Echinodermata, molluscs, Polychaeta).

Capelin *Mallotus villosus (Muller)* is widespread practically everywhere within the Barents Sea. Its reproduction sites are situated along the southern coast of the sea from Norway to the Novaya Zemlya. Seasonal migrations of adult shoals reach the coastal waters of the Arctic archipelagos. Capelin feed on zooplankton and are, in their turn, the food for predatory fishes, sea birds and mammals.

Polar cod *Boreogadus saida (Lepechin)* is connected mainly with Arctic water masses. The greatest shoals are observed in northern and eastern parts of the sea. Spawning takes place in winter, in ice-covered and adjoining ice edge areas. Migrations of adults cover the central, eastern (near the Novaya Zemlya) and south-eastern parts of the sea. Polar cod occupies the same place in food chains as capelin.

Herring *Clupea harengus harengus L.* inhabits the southern part of the sea and reaches the Novaya Zemlya. Its larvae and juveniles are transported in the surface water layer by coastal currents to the bays of the Kola Peninsula and descend to the bottom during the period of autumn and winter convection. Older juvenile specimens (2 years of age) migrate actively in the central and eastern parts of the sea. Later, in adult ages of 2-4 years, the herring migrate to the western part of the Barents and to the Norwegian Sea. This species feeds mainly on macrozoobenthos (*Calanus*, euphausiids).

Sea perch are represented in the Barents Sea in mass quantities only by the species *Sebastes marinus (L.)* inhabiting the southern part of the sea. Main spawning areas are situated along the Norwegian coast. Juveniles occur usually in the coastal zone of the Kola Peninsula. Migrations of adults in spring and summer spread all over the southern part of the sea to 45 °E (Geese Bank); in autumn and in winter they return to coastal areas of the Kola Peninsula and north Norway. Sea perch are carnivorous fish with a wide spectrum of prey.

Three species of **catfish** are known in the Barents Sea: common catfish *Anarhichas lupus*, smaller catfish *Anarhichas minor* and blue catfish *Anarhichas latifrons*. The first of these is the most widespread because of its cold-hardiness; it inhabits relatively shallow areas along the southern coast, Bear Island and Spitsbergen. Its migrations are of the order of several hundred km from coastal to offshore zones. Main prey are fish and bottom invertebrates with hard shells such as molluscs, and Echinoderms. Catfish are planktivorous in early stages of development. Spotty catfish inhabit offshore areas with depths 50-250 m. The main spawning areas of this species are situated in the Norwegian Sea, and pelagic early developmental stages are transported to the Barents Sea where adult specimens migrate long distances. Spotty catfish feed on bottom invertebrates with hard shells and on fish. Blue catfish require the most warmth of the three, and its population is less numerous. Its spawning takes place at depths of 400-500 m in the Norwegian Sea. Blue catfish feed on fishes, jelly-fishes, comb-bearers and bottom invertebrates with soft shells.

Plaice, *Pleuronectes platessa L.* and American plaice *Hippoglossoides platessoides limandoides (Bloch)* are widespread in the southern part of the Barents Sea. The spawning and wintering areas of plaice are situated on the coastal shelf. Migration paths of adult specimens depend on hydrological conditions and cover the feeding areas of the Murmansk and Kanin Banks. Plaice feed mainly on bivalve molluscs and Polychaeta. American plaice is found almost everywhere in the Barents Sea. Main spawning areas are situated in the south-western shallows of the Barents Sea. The food consists of fish, shrimps and bottom organisms.

Black halibut, *Reinhardtius hippoglossoides (Walbaum)* inhabits south-western and central parts of the Barents Sea. Its spawning takes place in the Norwegian Sea, and eggs and larvae are brought by currents to the western part of the Barents Sea where they mature to adults and make a long migration. Halibut is a predatory fish feeding mainly on cappelin and polar cod.

Atlantic salmon, *Salmo salar L.* is an anadromous species spawning in the numerous rivers of the Barents Sea southern coast (Kola, Teriberka, Voronya, Iokan'ga and others on the Kola Peninsula; Pechora, Chesha, Indiga and others in the Nenets Autonomous

District) and migrating as adults in the Gulf Stream currents of the Barents Sea, Norwegian Sea and the North Atlantic. The age of first spawning migration varies from 3 to 12 years; the sea stage of the life cycle is from 1 to 3 years. The marine part of the population feeds intensively and reaches marketable size.

Kamchatka crab, *Paralithodes camtschaticus* is an example of a newcomer species successfully introduced by the way of artificial rearing. The first attempt at an introduction was made in 1932, but success was achieved only after the second attempt, undertaken by Murmansk Marine Biological Institute (MMBI), in the coastal waters of the Kola Peninsula in 1961. Eggs and adult animals were transported from the Far East Seas and part were released immediately to sea while part were maintained in a sea-water aquarium. The first specimen living in natural conditions was caught in 1974. At present the population has expanded along the coast in an eastern direction up to the Kanin Peninsula and Kolguev Island and in a western direction to the Lofoten Islands. By recent estimates, the size of the crab population in the Barents Sea in 2000 was about 12.5 million, and the upper limit of the population acceptable for reasons of ecosystem stability is set at 15 million (Kuzmin and Gudimova 2002).

The spawning of crabs takes place in the shallow coastal areas. Adult specimens can migrate long distances practically in all ranges of depths existing in the Barents Sea. Benthic animals provide most of their food: bivalve and Gastropoda molluscs, Echinodermata, amphipods, Polychaeta, and also fish eggs and algae. At the same time, crab larvae are included in the trophic chains as the food of predatory fish.

Development of Fisheries and Fishery Research.

The earliest knowledge of the Barents Sea fishery dates back to the fifteenth century. For a long time this activity was not reflected in the scientific literature. Only in the end of the eighteenth century and the beginning of the nineteenth century the first descriptions of the fishery and seal hunting industries were made by naturalists I. Lepekhin and N. Ozeretskovsky. The first goal-oriented expedition to the White and Barents Seas fishery areas was accomplished in 1859-1861 under the guidance of N. Danilevsky (Alekseev *et al.* 2002).

Until the end of the nineteenth century, the Russian fishery on the Barents Sea was limited to the coastal zone. The appearance of Norwegian, English and other foreign fishery vessels in offshore areas stimulated the development of the native fishery fleet and full-scale oceanographic and hydrobiological research in the Northern Basin. In 1899 the Murman Scientific and Fishery Expedition was organized under the guidance of prominent Russian scientist and civic leader Nikolay Knipovich. Research was carried out onboard the vessel "Andrey Pervozvanny" designed specially for this purpose (it was the first fisheries research vessel in world practice). The work of the Murman Expedition permitted the collection of vast amounts of data on the oceanographic conditions of the Barents Sea, bottom relief and grounds, fish, planktonic and benthic communities. Prospective fishery areas were discovered and investigated. The results of this work were widely used both by native and foreign fishermen. The fundamental research in coastal waters was carried out by Murmansk Biological Station which was founded in 1883 on the Solovetsky Islands of the White Sea and later, from 1899, continued its activity in the Kola Bay of the Barents Sea.

After the decline caused by the First World War and the revolution in Russia, fishery research was renewed in the beginning of the 1920s. The Northern Scientific and Fishery Expedition was organized in 1921. At the same time, the Floating Marine Research Institute, intended for complex oceanographic, biological and geological investigations of northern seas, was founded. This institute is famous as the scientific school from which the leaders of native oceanography and marine biology came: Nikolay Zubov, Veninamin Bogorov and others. *"Persey"* and *"Nikolay Knipovich,"* research vessels of the Institute, from the 1920s to the 1930s, conducted oceanographic surveys of all parts of the Barents Sea and made decisive input to the environmental database for this period.

In the second half of the twentieth century the intensive development of an industrial fishery was accompanied by the many-sided fundamental research, regular ship and aerial surveys, experiments on introduction of and artificial rearing of marketable species. The leading positions in these works belong to the Murmansk Marine Biological Institute (MMBI) and Polar Research Institute of Fishery and Oceanography (PINRO). Initially the investigations in this period were focussed on the estimates and forecasting of biological resources and also on the search of the most rational ways of fishery. Later, environmental problems began to gain importance. Detailed studies of chemical and radioactive contamination were started in the 1980s and have continued up to now. At present, the research activity includes geopolitical, environmental, economic information and other aspects of sustainable fisheries. Many projects are carried out in cooperation with foreign institutions, particularly Norway, and with support from ICES and other international organizations.

Present State of the Barents Sea Fishery

The Barents Sea ecosystem is one of the most productive and intensively exploited in the world's oceans. Within the Russian economic zone, it is comparable with the Far Eastern Seas taken as a whole. However, the total catches and their composition varied significantly during last decades because of inadequate fishery regulation (Figure 3-7). The catch of fish and invertebrates in the Barents region (Barents Sea

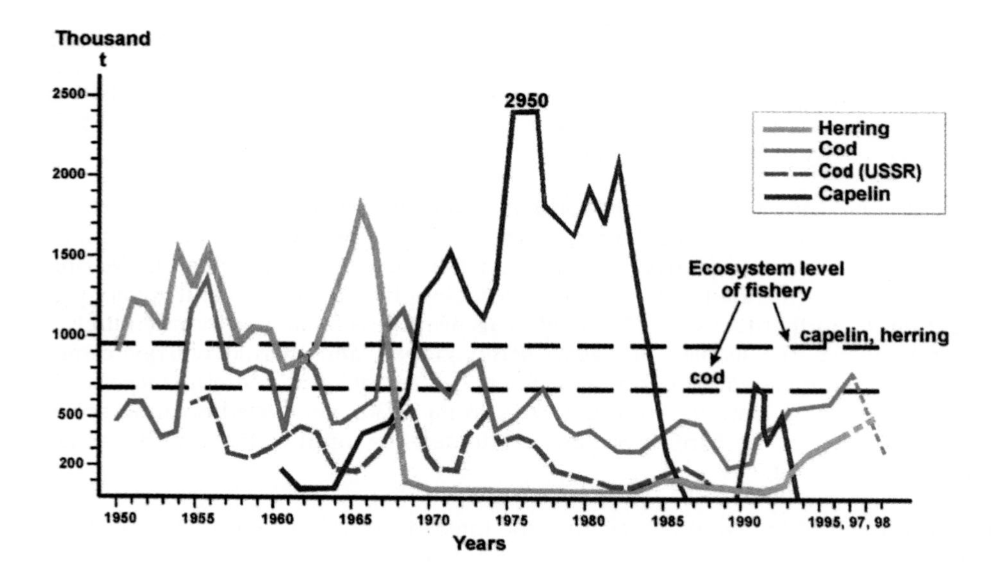

Figure 3-7. Fluctuations of the total catch of some commercial fish species (the Barents and the Norwegian seas) from ICES data.

with adjacent parts of the Norwegian and Greenland Seas with united populations of marketable fishes) reached its maximum in 1970-1980s. Record values were fixed in two successive years: 1976 – 4.3, 1977 – 4.6 million tons. Then the gradual decrease began and the lowest annual catches occurred in 1989 – 1.0 and in 1990 – 0.8 mln tons. In subsequent years the situation has improved, partly because of more effective protective measures and partly because of the general abatement of industrial activity in Russia. Annual catches from 1991 to 2000 have stabilized at the level of 2.0 –2.5 million tons which is close to the average level of 2.55 million tons for the period 1955-1999 covered by ICES statistics (http:\\www.ices.dk).

The composition of catches also varied significantly in the long run because of fluctuations in the natural environment and in the fishery withdrawal (Figure 3-8). The most impressive changes were connected with the overfishing of cod and herring in 1960-1970s. To the end of the 1970s, capelin became the main fishery object, making up to 75 percent of the total catch. Capelin was used both as a food product and as the raw material for fish flour which was widely used in stock-breeding and poultry farming. The mass catch of capelin led not only to the rapid depletion of this species' stocks but also to the deformation of the whole ecosystem pyramid at its middle and upper levels

(Figure 3-9). The reduction of sea bird populations of the bird colonies on the coasts and islands not touched by other anthropogenic impacts was especially noticeable.

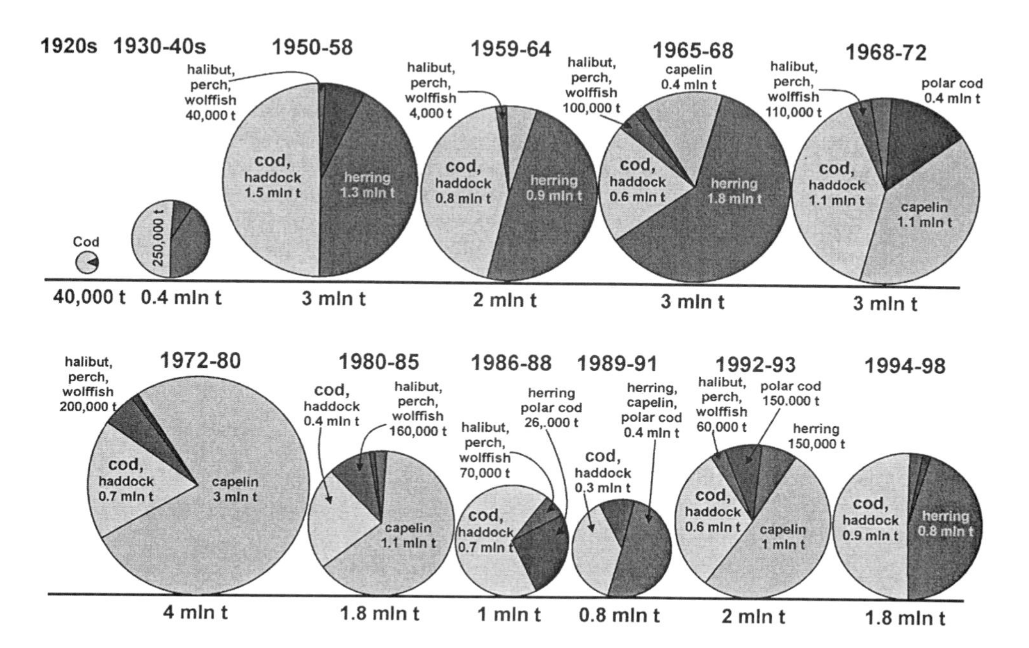

Figure 3-8. Dynamics of catch of different fish species in the Barents Sea (from PINRO and ICES and other materials) (Mln t = million tons).

The capelin fishery was totally closed between 1987 and 1990 and, after the brief renewal in 1990, again from 1994 to 1998. The present state of its population and stocks will be characterized below, along with other commercial species.

Available data on the Barents Sea fishery are not fully comparable because of differences in national statistical methodologies in Russia and Norway, as well as contradictions of approaches to the delimitation of economic zones and divergences between geographical boundaries of the seas and dividing into regions by ICES. However, general trends and proportions can be derived sufficiently well.

Cod
Cod was the most economically important of species. Its maximum stocks reached, in the 1940s, more than 4 million tons. The intensive fishery in the subsequent period and especially the overfishing of capelin, its main prey, caused the depressed state of the cod population from the end of the 1970s to the present. It was expressed in the general decrease of stocks to values below 1 million tons and in the deformation of age composition. Catches were comprised of fish not yet sexually mature.

In the beginning of the 1990s the cod stocks increased somewhat and in recent years the stocks have maintained a level of 1.1 to 1.2 million tons. The annual catches in the Barents Sea are about 250 thousand tons. The total allowable catch (TAC) in the Russian economic zone is 180 thousand tons, the actual catch in 1999 was about 100, and in 2000, was 23 thousand tons (Murmansk District Committee 2001).

Haddock
The stocks of haddock during all periods of industrial fishing were several times less then those of cod. The depression of the population occurred in the same period and for the same reasons. Average values of the fish stocks in 1950-1976 were about 500 thousand tons, in 1977-1992 – below 200 thousand tons. In 1993-1997 the stocks were restored almost to before-crisis values but, during the last several years, they fell again. The catch in the Russian economic zone in 2000 was 8.5 thousand tons or less than the half of the total allowable catch (TAC).

Sea Perch
Sea perch were caught in significant quantities (to 40 thousand tons) until the middle of the 1970s when the catches decreased sharply. The present level varies from 2 to 8 thousand tons. The industrial fishery is carried out only in Norway. Russian catches (several hundred tons annually) are received as a by-catch of the cod fishery. The stock of perches has recovered slowly after the intensive fishery of the 1970s because of the slow growth and long life cycles of this species.

Catfish
The most exploited bottom fish species at present are catfish. The total stocks of the three marketable species have decreased from the 1970s to the beginning of the 1990s from 130-140 to 50 thousand tons. The catches in the beginning of the 1990s were between 1.5 and 2.0 thousand tons; in 1997-1999 they increased to 10 to 18 thousand tons which is comparable with the TAC in the Russian economic zone (13 thousand tons). The present state of the population is still considered as depressed, so the present intensity of fishing` threatens its sustainability.

Halibut
The specialized halibut fishery in the Barents Sea has been prohibited since 1992. In the past, its maximum level reached 90 thousand tons (1970) and the stocks of that year were estimated at 300 thousand tons. In 1992 halibut stocks were at a minimum level of 45

thousand tons, but in subsequent years have increased by almost twice. Recent year-classes are relatively strong, so further restoration of the stocks is expected and the possibility of lifting the fishery ban is discussed at present.

Plaice

Plaice of the Barents Sea are the objects both of a specialized fishery and trawling by-catches. The maximum catch of all plaice species in the Barents Sea was registered in 1973 (21 thousand tons). The present state of the population is considered satisfactory. The total stock is, by recent estimation, about 40 thousand tons. The TAC for 2000 was 3.0, and the actual catch was 2.5 thousand tons.

Capelin

The fishery of mass pelagic species, as it was shown above, underwent significant changes during last three decades. The natural limits of capelin stocks before the period of intensive fishing varied from 5 to 8 million tons. During 1980-1990s the Barents Sea population fell almost to zero. Up to the present, the size of the population has increased; in 1999-2000 the total capelin biomass was close to 3 million tons. The summary Russian TAC in the Barents region is 174 and in the Russian economic zone, 38 thousand tons. The actual catch in 2000 was about 20 thousand tons.

Herring

The herring fishery within the Barents Sea is not regular because the main fishery areas are situated in the Norwegian Sea where marketable populations are concentrated. The highest level of stocks in both seas was observed in the first half of the 1950s (to 18.5 million tons; maximum annual catches from 1950-1960 were 1.5-1.6 mln tons). Overfishing, which coincided with the negative temperature trend of 1960s and especially with the deep negative anomaly of 1966, led to a decrease of stocks to values less than 0.1 million tons in the beginning of the 1970s. The herring fishery was practically stopped by all participating countries until the middle of the 1980s. The protective measures and favorable environmental conditions permitted a restoration of stocks to 11-12 million tons in after 1995, and the catches have returned almost to pre-crisis values (1.1-1.4 million tons). All these indices apply mainly to the Norwegian Sea. The Barents Sea share is several orders less, with maximum catches reaching 0.4 million tons in 1967-1968, 0.05 million – in 1985 and about 5 thousand tons in 1994.

Polar cod

Polar cod, on the contrary, are concentrated almost wholly in the Barents Sea. The polar cod fishery is of less importance than that of the two previously mentioned species. During the last three decades, the stocks of polar cod varied from about 1 million tons almost to zero. The present state of the population is satisfactory. Since the beginning of the 1990s, stocks have been between 0.5 and 1.0 million tons. In 2000 the TAC of polar cod was 35 thousand tons, almost equal to the actual catch.

In general, the fishery dynamics for the last decades display a decrease in the stock and catch level of almost all commercial fish species. Undoubtedly, the press of industrial fishing is of principal importance in comparison with other factors (climatic variations, marine pollution). The extremely negative influence of overfishing is recognized by researchers, managers and decision-makers of different countries (Matishov 1993). The trustworthiness of estimates of stocks and allowable catches is determined by the level of the current scientific knowledge. At present the ecological causes and consequences of natural fluctuations of the numbers of different interacting species are not well understood. In spite of the progress in the

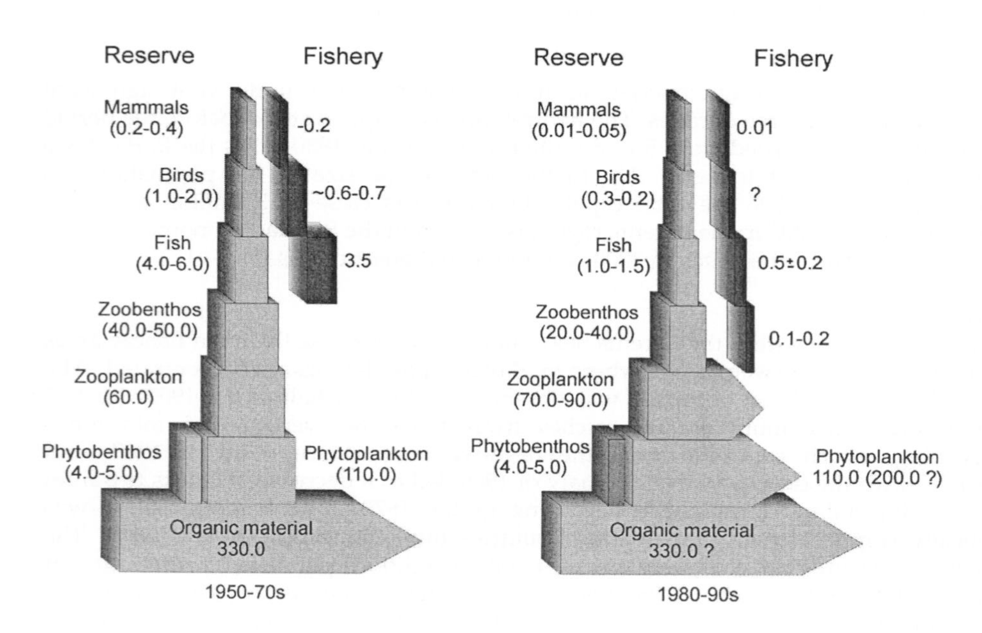

Figure 3-9. Pattern of the Barents Sea ecosystem pyramid changes in the 1950s to the 1990s (annual production and the fishery stress, million tons)

population models, researchers face great uncertainties in assessing the permissible catch. It is to be emphasized that, from the scientific viewpoint, the character of predictions in fisheries oceanography will inevitably remain preferentially stochastic. The general decrease in the stocks of major commercial fishes all over the Atlantic, widening amplitudes of climatic fluctuations, and enlargement of the role of fish food products in the diet of the population of most countries create a specific "shear" between the inertia of the market and intensification of risk factors. So far, there has been no adequate adaptation of economics, jurisdiction or fishery practices to the new conditions. A steadily decreasing volume of marine field studies aggravates the situation.

There are very serious problems of fish poaching (including the Barents Sea), throwing small cod and haddock overboard at the fishing grounds, as well as a problem of "by-catch" (absence of complex multi-species fisheries). During recent years, all these factors which escape the field of legislation become more and more "expensive" due to the concentrating of industrialized fishing activities within the Barents Sea (because of the crisis of the Russian oceanic fisheries) and, simultaneously, due to the increase of artisinal fishing during Russia's transition to a market economy. For example, the annual losses of cod and haddock due to overboard discards were, by estimates of PINRO specialists, 15 thousand tons for each species. If one takes into account the loss of larger fish, the total amount of "fish wastes" would be much higher. Since such losses are not easy to quantify, the uncertainty in risk increases. An additional contribution to the bulk of unrecorded catch is introduced with the practice of 'unregistered fishing.' Although, due to the activity of Norway and Russia, the protective measures are steadily stiffening, there is no reason to ignore the existence of such fishing entirely. The press in both countries reports that fish products are delivered to small ports which are not controlled by fishery inspection.

It must be emphasized that the current market situation, with all its drawbacks, has formed and exists in the condition of a rather favourable state of populations. According to the predictions by the experts (Borovkov *et al.* 1997), the evolution of the fisheries in the Barents Sea after 1997 would be determined by the declining stocks of demersal fishes. Development in a modern manner would only worsen the general crisis of the ecosystem of the sea, a crisis which even in the rather favourable conditions of the 1990s has persisted due to the trophic disequilibrium provoked by a collapse of the population of capelin, a main food for cod. The following measures for the ecological optimisation of the fisheries are necessary: protection of recruitment sources of the populations, trustworthy fisheries statistics, development of sustainable multi-species fisheries and aquaculture of non-traditional species, more efficient technologies of processing. Fisheries management at the local, regional and global scales has to be more adaptable, especially in conditions under which the probability of unfavourable climatic epochs and devastation of the fish stocks due to overcatch and chemical pollution becomes higher because of a combined impact of the global and regional factors.

MARINE POLLUTION AND ECOSYSTEM HEALTH

In the Barents Sea there are two complex sources of pollution (Matishov 1993):

- water-mass and atmospheric advection from external sources which is the predominant source;

- industrial activities within the basin (transport, oil and gas extraction, continental runoff, etc.).

Both factors operate in one direction and manifest no tendency toward decreasing their intensity. Thus the general ecological risk with respect to the quality of environment and biota is increasing.

Transfer of Organic and Heavy-Metal Pollutants into the Barents Sea.

The major sources of ecosystem perturbation in the Barents Sea are pollutants of various natures: oil, petroleum and petroleum products, heavy and intermediate metals, chlorinated hydrocarbons, and radionuclides. The pollutants are introduced into the Barents Sea by the riverine runoff, atmospheric transfer, dumping, ships' discharges, advection by currents, migration through soil, etc.

Atmospheric fall-out

Atmospheric fall-out is one of the major sources of chlorinated hydrocarbons, products of pyrolysis, heavy-metal compounds (including the most carcinogenic mercury and arsenic) in the Barents Sea.

The introduction of pollutants to the atmospheric circulation of the Barents Sea occurs largely from the industrial centres located at the temperate latitudes, and (what is especially important) from the regional sources. Of the latter, the largest are metallurgy plants (concerns "Severonikel" and "Pechenganikel"). These enterprises discharge annually into the atmosphere 1632 tons of copper, 2413 tons of nickel, 102 tons of cobalt, 77 tons of vanadium (Gromov 1997).

Rovinsky *et al.* (1995) have estimated that the annual flow through 1 km^2 of the surface area of the Barents Sea is 10-48 kg for sulphur, and 0.001-0.018 kg for lead. Such estimates for other pollutants are either extremely scanty or entirely absent. However, the most general representation of the volume of atmospheric fall-out may be gained by taking into account the fact that in the open Barents Sea the atmospheric concentrations of some pollutants do not exceed their background concentrations for the Kola Peninsula. So, given the available flow estimates for some microelements through the unit area it may be concluded that through 1 km^2 of the surface of the Barents Sea, not more than 0.36 kg of copper, nickel, and manganese, as well as 36 kg of iron, 10.8 kg of magnesium, and 0.004 kg of cadmium are transferred into the seawater.

Preliminary annual estimates for the organic pollutant flow from the atmosphere through the sea surface may be done based on their concentration levels in the atmospheric precipitation shown in the AMAP report (AMAP 1997) for the Barents Sea. The values obtained are between 0.2 to 0.5 kg km^{-2} for hexachlorcyclohexan, 0.1 to 0.3 kg km^{-2} for DDT, and 1.2 to 3.4 kg km^{-2} for polychlorinated biphenyls. These values for the former are comparable with the analogous estimates for the Canadian Arctic, but the levels of the concentrations of DDT and polychlorinated biphenyls are much higher. Conclusions are provisional to further research addressing the most recent changes to the data set started by MMBI in the 1970s.

Riverine runoff

The input of pollutants with the riverine runoff into the seas of the Arctic Ocean is 2-4 times higher than the atmospheric input. Of 0.5 billion cubic metres of industrial and municipal sewage water discharged into the Barents Sea, two thirds are introduced into watersheds (and then into the sea) without proper processing (Murmansk Dist. Comm. Natural Res. 2001). More complete information on the bulk of discharged pollutants introduced into the Barents Sea with the riverine runoff is virtually absent. While taking into account the data published in the annual reports of the Ministry of Ecology (from 2000 – Ministry of Natural Resources) and in the AMAP report it is possible to say only that the concentrations of hexochlorcyclohexan in the water of the Pechora and Kola rivers range from 4 to 5 ng l^{-1}, and these values are 2 times higher than those for the rivers of the Canadian Arctic. Even higher is the ratio for solid material. The analogous pattern is observed in the case of DDT and polychlorinated biphenyls. The important role of the riverine runoff in the transfer of a number of heavy metals into the Barents Sea is obvious, and that is especially true for the Kola Peninsula where the metallurgy plants are largest in the world. However, the information on the bulk of their input is absent, and therefore the problem needs additional research.

Sea currents

The waters of the Gulf Stream system are considered now as a significant source of pollutants for the Barents Sea (Savinova 1991; Matishov 1993). However, the question about the bulk of pollutants advected into the Barents Sea in the course of water and ice exchange has not been well studied. Present estimates must be considered as preliminary. On the basis of the data of the Murmansk Hydrometeorological Service, the annual input of petroleum and petroleum products to the sea may be estimated as approximately 2.4 million tons.

Industrial activities

According to the data of the experts from Murmansk Regional Committee of Ecology (Murmansk Dist. Comm. Natural Res. 2001), the waters of the Barents Sea are clear compared with the marginal seas of Europe. However, elevated concentrations of oil products (higher than the official Maximum Allowable Concentrations for the fisheries basins - 0.05 mg l^{-1}) are routinely recorded in the convergence zones and fishery grounds. In the coastal areas, in which the spawning and nursery grounds as well as the areas of intense fishing are located, the pollution levels are as a rule higher than in the open sea (Iljin *et al.* 1996). Petroleum products either sink out of the pelagic zone down to the bottom sediments or are accumulated in marine organisms through both absorption (this process is especially active in a case of algae) and feeding.

The situation will possibly worsen with the intensification of hydrocarbon extraction on the Barents Sea shelf and related transport and loading marine operations. Already now the concentrations of petroleum hydrocarbons in the important commercial algae, laminarians and fucoids, in the area of Kolguev Island and the Kanin Peninsula are 1.5 to 2 times higher than those on the Murman coast.

Transfer of Artificial Radionuclides to the Barents Sea

The major sources of artificial radionuclides introduced into the Barents Sea are atmospheric fall-out (originating from nuclear tests and the Chernobyl catastrophe), riverine runoff, advection from the west Europe plants (mainly Sellafield, Great Britain) by the Gulf Stream system of currents, discharge of fluid radioactive wastes from the objects located on the Kola Peninsula as well as accidental events accompanied by input of artificial radionuclides into the marine environment (Matishov, D.G. and Matishov, G.G. 2001). The first group of sources is characterized by powerful but short-term time impacts. The advection by river and sea waters, as well as the leaks from nuclear enterprises, atomic vessels and underwater burials are less intensive but more stable. ^{137}Cs and ^{90}Sr are the most widespread among artificial radionuclides with long half-decay periods. The total input to the Barents Sea during the atomic epoch from 1945 is (in Bk) for ^{137}Cs about 2×10^{15} from the atmosphere, 0.02×10^{15} by rivers and more then 7×10^{15} by currents; corresponding values for ^{90}Sr are 1.5×10^{15}, 0.2×10^{15} and about 2×10^{15} (Matishov *et al.* 1999). The total activity of discharged artificial radionuclides for 1961-1990 from the Kola Peninsula is estimated as 0.48×10^{15} Bk.

Since the time of the construction of the first atomic icebreaker "Lenin" and the first atomic submarine in the Barents Sea there have been two accidents which were accompanied by spills of artificial radionuclides into the marine environment:

- in the container of worked-out nuclear fuel in the Andreeva Guba of the Zapadnaya Litsa Bay (1982) where the total activity of spilt radionuclides was estimated as 37×10^{12} Bk;
- the wreck of the atomic submarine in the Ara Guba (1989) where the total activity of spilt radionuclides introduced into the marine environment was estimated as 74×10^{12} Bk.

Contamination of Marine Biota

The information on the levels of bioaccumulation of chemical and radioactive pollutants in marine organisms is of special importance. The general conclusion of the experts is that the pollution levels of the biota are rather low on the scale of absolute values and in comparison with other marginal seas. Chlorinated organic pesticides (DDT and its derivatives, α-, β-, γ-isomers of hexachlorcyclohexan, etc.) and polychlorinated biphenyls (PCB) are found in the Arctic environment and biota. These toxicants, according to the International Convention, are included in the list of most dangerous chemical pollutants. In the Barents Sea, the highest concentrations of total DDT are recorded in the Atlantic water of the Murman Coastal Current.

A special study was carried out concerning the concentrations of chlorinated hydrocarbons in tissues and organs of some commercial fishes of the Barents Sea. According to these data, the pollution of Barents Sea cod is lower than that of the cod inhabiting the Norwegian Sea. The highest concentrations of pesticides from the family of DDT and PCBs are recorded in cod liver (Savinova *et al.* 1995).

The impacts of heavy and intermediate metals on marine organisms have not been actually studied. Such studies are of special importance in relation to the beginning of industrial exploitation of petroleum and gas fields in the Barents Sea. Together with drilling wastes, toxic compounds of metals such as barium, chrome and iron will be introduced into the sea; their chronic impact will entail anomalous accumulation of these elements in marine organisms. For the Barents Sea there are fragmentary data on the content of some metals in the zooplankton of one area of the Barents Sea, and also sparse data on the concentration of metals in cod muscles.

Accumulation of radionuclides in the Barents Sea manifests certain geographic patterns which are to be studied. The algae are able to preferentially accumulate ^{40}K, while the bottom invertebrates – ^{137}Cs. The concentration of ^{137}Cs in the Barents Sea fishes, according to the available data, varies in relation to species and, possibly, to a feeding ground, food items, and environmental conditions in a given time of year. However, such studies are not numerous. According to some data, the concentration of radionuclides in the benthic organisms of the Barents Sea is rather low compared to the areas of western Europe (Matishov, D.G. and Matishov, G.G., 2001).

Therefore, at present there are rather numerous (but fragmentary, episodically collected and often geographically and seasonally non-characteristic) data on the content of various pollutants in different compartments of the Barents Sea ecosystems. The available data seem to indicate low levels of accumulation of oil hydrocarbons, carcinogenic compounds, heavy metals and pesticides in tissues and organs of the most important commercial fishes and marine invertebrates. Possibly, this is to some extent related to the high "assimilating capacity" of this basin with almost oceanic salinity which is open towards the north and the west. The current level of pollution is lower than the norms established for food products in Russia. However, the available materials cannot pretend to be statistically complete even for establishing the background pollution levels for the Barents Sea. Moreover, on the basis of these data it is very hard to define, more or less reliably, a major temporal pattern of pollution, while to do so is especially important for planning and for understanding different compensatory measures.

SOCIOECONOMIC STATE OF THE RUSSIAN COASTAL AREAS

The Barents Sea region, broadly speaking, includes catchment areas of the Barents and White Seas (the latter practically can be considered a large bay connected with the main basin by intensive water exchange and receiving runoff and pollutants from vast areas of the Russian North) (Figure 3-10). The economy and occupational state of the coastal populations are connected with the seas directly or indirectly (for example, such industrial branches as timber and apatite concentrate production are Northern Sea Route). However, the present state and perspectives of the Barents Sea ecosystem are determined primarily by three spheres of marine activity: fishing, oil and gas extraction and sea transport. The significant part of the population, including also service personnel and members of families, is occupied in the marine sector of the economy.

Population

The available statistical data permit an estimate of the total population of the catchment area at about 5 million inhabitants, with about 1.5 million of these in the coastal zone. The most prominent feature of the Barents and White Seas coasts populations is their uneven spatial distribution. More than 90 percent of the people are concentrated in two big agglomerations: Murmansk - Severomorsk on the Kola Bay of the Barents Sea (about 600 thousand) and Arkhangelsk – Severodvinsk – Novodvinsk in the Northern Dvina mouth of the White Sea (about 700 thousand). The urban population prevails absolutely. The few rural settlements on the coasts specialize in coastal fisheries and sea animal hunting. Indigenous peoples (Saami, Nenets) are occupied mainly by reindeer husbandry and freshwater fishing. Almost all Russian Saami (about 1.6 thousand) inhabit Murmansk Region). The Nenets population inhabits several territories in the Russian Arctic. Within the Barents Sea basin they are concentrated in the Nenets Autonomous District (about 6.5 thousand). exported by sea; non-ferrous metallurgy depends on the ore transportation by the Population size during the last decade decreased both because of Russia's low contemporary birth rate and because of depopulation. The latter has affected mainly the settlements where industrial activity had ceased or military bases were closed. There are several projects under

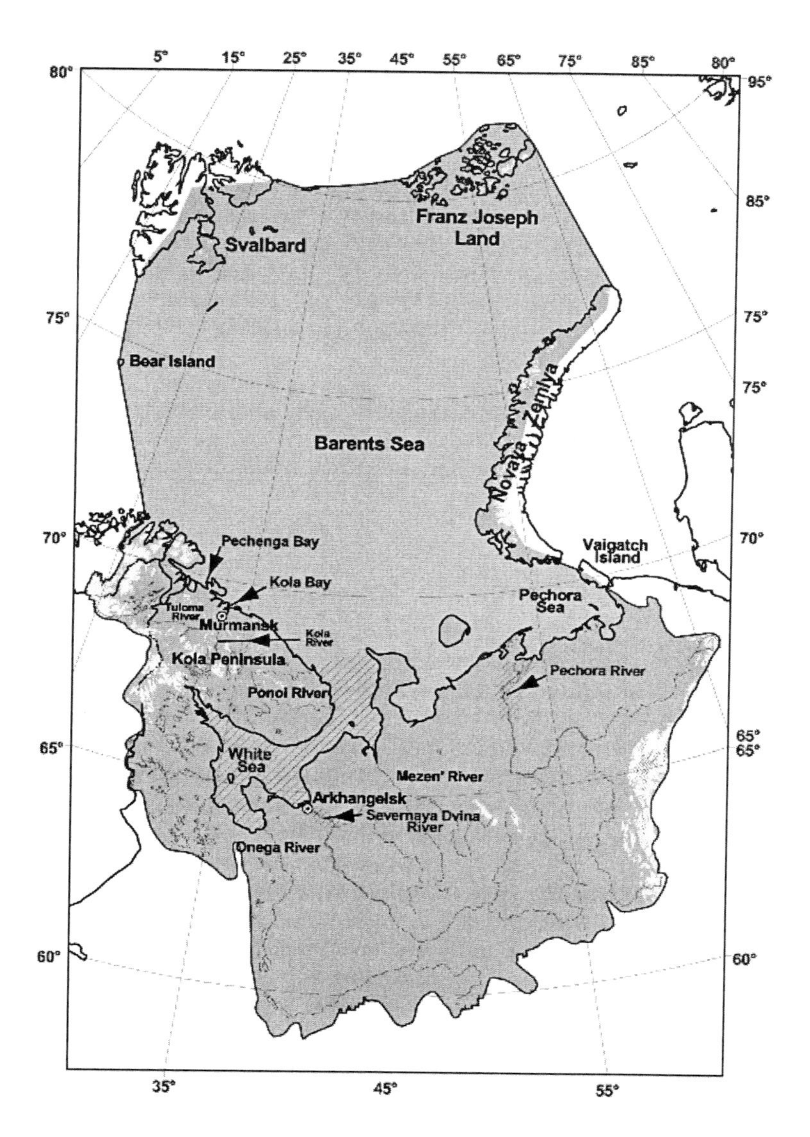

Figure 3-10. Catchment basin of the Barents and White Seas (the most impacted water areas are hatched).

discussion to organize the coastal fishery and fish processing at the sites of such abandoned settlements. The population of industrial centers of the region also decreased because of a reduction in occupation and departure of pensioners to the southern regions of Russia. This process is encouraged by authorities, and special allowances are paid to pensioners leaving the North. In particular, the population of the Murmansk Region decreased from its maximum value 1.16 million in 1991 to exactly 1 million in 1999 and continues to decrease. However, the reduction of the aged population is not very significant in comparison with other demographic trends. So the aging of the population is observed, though the Murmansk Region population remains younger than the population of other northern regions. The portion of persons aged 60 years or more increased from 6.6 percent in 1989 to 11.2 percent in 1999.

In other aspects, demographic processes in the Barents Sea area are similar to those in other regions of Russia. The disproportion of life duration between men and women (63 and 73 years correspondingly in 1997) is of the same character as that everywhere in Russia. The birth rate fell sharply from 1.3 percent in 1989 to 0.75-0.85 percent in 1994-2000. The mortality rate was the highest in 1994-1995 (to 1.2 percent) and has stabilized now at about 1.0 percent. This value is less in the Arkhangelsk Region and Karelia but higher than in northern parts of Norway and Finland.

Marine industry

The review of the marine fishery was given above. Of two other branches, sea transport and oil and gas extraction, the first influences marine ecosystems mainly indirectly. For example, the ice-breaker navigation leads to destruction of ice cover and therefore damages the habitats of seals in the coastal zone. The danger of pollution from ships is relatively small because at present oil transportation is insignificant and general cargo transportation in the Barents Sea, mainly from the Murmansk port, is on the order of 10 million tons annually.

On the other hand, the oil and gas industry is potentially very dangerous to the marine ecosystem at all stages of its development both in offshore areas and in the coastal zone. From 1970 to the 1990s, large structures are found there, and 10 fields are proposed (2 oil fields, and 8 gas and gas-condensate fields). Amongst them, a special place is occupied by the Prirazlomnoye oil field (in the shallows of the Pechora Sea) and the supergiant Shtokmanovskoye gas-condensate field (about 3 trillion cubic metres of gas). The exploitation of sea oil and gas fields will be accompanied by the creation of a new infrastructure (pipe-lines, fleets, ports). At the same time, plans for coastal (island) terminals for loading the tankers with subsequent transport of the Timano-Pechora oil to the ports of West Europe are being elaborated. Finally, the Barents Sea, as a part of the Northern Sea Route, will be used for the tankers' transit of products from the large cluster of the Yamal fields. Therefore, in the Barents Sea new ecological and industrial situations have begun to form which are to be assessed on the long-term scale.

Taking into account the great significance of the basin for the Russian fisheries as well as complicated natural conditions of the Arctic, the risk of an emerging ecological catastrophe is considered as very high by many experts and the public.

Of the potential unfavourable consequences of hydrocarbon extraction for marine ecosystems, two must be accentuated:

- accidental spills of oil hydrocarbons;
- chronic toxic input to the marine environment and biota.

The first factor is dramatically evident. In the conditions of the Barents Sea, a real risk of such pollution indeed exists, first, in the Pechora Sea (the Prirazlomnoye oil field). Moreover, a spill from a pipeline and the ensuing drift of oil here is quite likely to occur in areas of ice, and so, the consequences of spills are especially difficult to remove. Second, the risk of an accidental spill near the coasts becomes real due to the presence of large-displacement tankers. The probabilities for discharged oil patches to reach ecologically vulnerable areas on the Murman coasts (seabird and seal colonies) in case of accidents of both kinds have been calculated in MMBI using the model OILSTAT of the Norwegian company OCEANOR (Matishov *et al.* 1997).

Fortunately, the majority of the hydrocarbon fields on the shelf of the Barents Sea contain a gas condensate with a very high gas content. As practice has shown, its ecological danger is much less as compared to oil. As a result of the completion of the environmental impact assessment (EIA) for the Shtokmanovskoye gas-condensate field, we have made similar conclusions (Scientific and methodological 1997). In particular, our calculations have shown that discharge (spill) of gas condensate would not reach the coasts of either the Novaya Zemlya or the Kola Peninsula. This is due to a number of features (remoteness from coasts, volatility of raw material, strong natural variability of wind conditions and, as a consequence, the complete impossibility of a long, one-way drift).

On the whole, the problem of accidental spills of oil and gas condensate is rather more logistical than ecological. The main efforts here must be directed to creating an efficient system of safe navigation and, additionally, an intra-basin system of anti-spill measures. The insufficient state of the latter, with respect to needed scales and technologies, is a very serious problem that may dampen implementation of the first-order projects for exploration of the West Arctic shelf.

A more serious danger is represented by potential chronic pollution of the marine environment and biota. At present, in this field of (ecotoxicology) there are many more questions than answers. On the one hand, the experience of oil and gas extraction on the shelves of the Gulf of Mexico and the North Sea did not reveal serious consequences to the marine biota. Authoritative groups of experts (GESAMP 1993) have recognized the events of local, reversible negative impacts of marine oil and gas extraction on the bottom communities (which are the best bio-indicators) but do not consider it necessary to dramatize the situation. On the other hand, the lack of statistically reliable materials of long-term character, obtained not in lab conditions but, *in situ*, leaves the vastest field for ambiguous hypotheses and uncertainty.

Apparently, analysing the ecological consequences of construction and exploitation of marine oil and gas fields, it is necessary to take into consideration:

- the background state and peculiarities of the biology of marine organisms in the sites of drilling and extraction complexes;
- physical-chemical properties of the material, including its solubility in sea water at various temperature, pathways and destruction rates, capability of reacting with other compounds;
- the compartments of the ecosystem, the most vulnerable for a pollutant; its capability of accumulating or transforming in tissues of marine organisms and transit along the trophic chains;
- the resistance of marine organisms to pollutants at different stages of their life cycles and during different seasons;
- the sensitivity of different functional systems of organisms to some pollutants, determining minimum (threshold) values that provoke an effect;
- possible consequences of either short-term or long-term impacts of pollutants and their complexes on various species of the marine flora, fauna, and the whole ecosystem.

STATE OF MANAGEMENT

The management of biological resources and nature protection of the Barents Sea is carried out on three levels: interstate (Russia and Norway), national and local.

The delimitation of the continental shelf between Russia and Norway is an international problem of top priority. The disagreements are connected with different interpretations of the UN Convention on Marine Law of 1982. From the Russian point of view, the border of its possessions on the shelf must be part of the general delimitation of the Arctic Ocean into sectors (from the land border northward along the 32° longitude with the small deviation in the Spitsbergen zone). Norway proposes the delimitation along the perpendicular line to the coast. The area with disputed legal status ("loophole") is about 20 000 km^2. It is distinguished by high productivity. The fry shoals, replenishing the fish stocks, are observed here in August and September. Russia and Norway had come to an agreement on suspension of fishing in this area. However, Iceland and some other countries are continuing an illegal cod fishery here. By estimates of PINRO experts, its catches in the "loophole" during the period 1991-1994 only were about 55000 tons.

In Russia the management of marine resources is performed by federal authorities and so called subjects of the federation (regions in a wide sense, including districts, republics, autonomous districts and two federal cities). By the Constitution of Russia the following issues fall within the joint governance of federal authorities and subjects of the federation:

- issues of possession, use and disposal of land, depths, water and other natural resources;
- delimitation of the state property;
- nature use, environmental protection and provision of environmental safety.

The principle of joint management is concretized in federal laws "On the Continental Shelf of RF", "On depths" and, concerning the living water resources – in the law "On the Animal World" (all adopted in 1995). By the law "On the Continental Shelf" all shelf resources are in the exclusive management of federal authorities. Federal programs and plans of extraction of mineral resources are elaborated with participation by subjects of the federation in case they foresee the use of a coastal infrastructure of corresponding regions.

Biological resources, according to legislation, are divided into federal and local. The latter include the resources of internal basins (except salmon). The management and disposal of sea bio-resources is in the complete authority of federal authorities (at present – of the State Committee on Fishery). Regions can only submit proposals and participate in the elaboration of federal programs.

Fishing on the shelf and in territorial seas is permitted based on licences issued by the Committee on Fishery with participation of the State Committee on Natural Resources, the Defence Ministry, the Federal Border Service and the Federal Customs Service. Regional authorities organize the collection of applications for fishing and submit the proposals on the distribution of quotas.

The imperfection of the fishery management system impedes the development of the coastal fishery. In the Murmansk Region this problem is intensified also because of restrictions on visits to the border territories which includes almost all of the northern coast of the Kola Peninsula.

Economic principles of marine bio-resources use are confirmed and supplied in the law "On Internal Sea Waters, Territorial Sea and adjoining zone of the Russian Federation" (1998). The principle of pay for the living resources is established here, including their auction sale by the rules determined by the Russian government. Certain exclusions are foreseen for indigenous people of the North, for whom a "special regime of nature use" can be established by agreement with regional authorities and local self-government bodies.

The state of marine resources management as a whole can be characterised by the following conclusions:

1) the legislative process so far developed, includes some important laws on several issues under discussion ("On the Fishery and Protection of Water Objects," "On the Russian Federation Arctic Zone" and others);
2) the centralisation of nature use is kept and strengthened in some aspects;
3) the mechanism of interaction between central and regional authorities on marine resources use is elaborated insufficiently; the agreements on the division of competence between the center and regions, that have been concluded earlier, actually became invalid;
4) there are gaps in the legal regime of 12-mile zone of territorial waters which is of special importance for the organisation of coastal fishery on the local level.

REFERENCES

Alekseev, A.P., S.I. Nikonorov, V.P. Ponomarenko. 2002. 150 years of scientific and fisheries researches in Russia [Алексеев А.П., Никоноров С.И., Пономаренко В.П. 150 лет научно-промысловых исследований в России] Russian Scientific Conference "Historical Experience of Scientific and Fisheries Researches in Russia". Moscow: VNIRO Publ. 7-16 (In Russian).

Arctic Pollution Issues: A State of the Arctic Environment Report. AMAP, Oslo, 1997, 186p

Atlas of the oceans. 1980. Terms, definitions, reference tables. [Атлас океанов. Термины, понятия, справочные таблицы]. Moscow. 156p (In Russian).

Belkin, I., S. Levitus, J. Antonov, S.A. Malmberg. 1997. On the North Atlantic "Great Salinity Anomalies". NOAA Technical Rep. N 5, 28 p

Borovkov, V.A., M.S. Shevelev, V.N. Shleinik. 1997. Marine ecosystems of the Barents Sea at the present stage [Боровков В.А., Шевелев М.С., Шлейник В.Н. Морские экосистемы Баренцева моря на современном этапе]. Abstracts of presentations of the I Congress of Russian Ichthyologists. Astrakhan, p. 133 (In Russian).

Ecology of the Barents Sea commercial fish species. 2001. [Экология промысловых видов рыб Баренцева моря] Apatity: KSC Publ.. 461p

GESAMP Impact of oil and related chemicals on the marine environment. 1993. GESAMP Reports and Studies. London, IMO, No. 50, 123p

Gromov, S.A. 1997. Heavy metal pollution in the Russian Arctic/ The AMAP International Symposium on Environmental Pollution in the Arctic. Extended abstracts, vol.1. Tromso, Norway. June 1-5, 1997. 142-145

Hydrometeorology and hydrochemistry of the USSR seas. 1990. V.1 Barents Sea. Part 1. Hydrometeorological conditions. [Гидрометеорология и гидрохимия морей СССР. Т.1 Баренцево море. Ч.1. Гидрометеорологические условия]. Leningrad: Hydromet. Publ. 280p (In Russian).

Iljin G.V., T.L. Schekaturina, V.S. Petrov. 1996. Comparative characteristics of hydrocarbon composition of Southern Barents Sea bottom sediments [Ильин Г.В., Щекатурина Т.Л., Петров В.С. Сравнительная характеристика углеводородного состава донных отложений южной части Баренцева моря] // Oceanologiya, vol.36, No.5, 787-792. (In Russian).

Kamshilov, M.M., E.A. Zelikman, M.I. Roukhiyajnen. 1958. Plankton of the Murman coastal zone // Patterns of accumulation and migration of marketable fishes in the Murman coastal zone and their connection with biological, hydrological and biochemical processes [Камшилов М.М., Зеликман Э.А., Роухияйнен М.И. Планктон прибрежья Мурмана // Закономерности скоплений и миграций промысловых рыб в прибрежной зоне Мурмана и их связь с биологическими, гидрологическими и гидробиологическими процессами] Moscow-Leningrad: Academy Publ. 59-102.

Kuzmin, S.A.and E.N. Gudimova. 2002. Introduction of the Kamchatka (red king) crab to the Barents Sea Peculiarities of biology, perspectives of fishery [Кузьмин С.А., Гудимова Е.Н. Вселение камчатского краба в Баренцево море. Особенности биологии, перспективы промысла]. Apatity: KSC Publ. 236p (In Russian).

Matishov, G.G. 1993. Anthropogenous destruction of the ecosystems in the Barents and the Norwegian Seas. Apatity, KSC Publ.. 116p

Matishov, G.G., V.V. Denisov, S.L. Dzhenyuk, S.F. Timofeev. 1999. Marine ecosystems of the Barents Sea influenced by the natural and anthropogenic factors (current conditions and problems). In: Global changes and the Barents Sea region. Proceedings of the First Int. BASIS Res. Conference. Spb, Russia, Feb. 22-25 1997. Muenster, Germany 105-134

Matishov G.G., V.V. Denisov, A.N. Zuev, O. Johansen, and C.Toro. 1999. Application of mathematical models in the Barents Sea Watch system for the forecasting of emergency ecological situations on the oil and gas deposits of the Barents and Kara shelves // 3rd International Conference «Mastering of the Shelf of the Russian Arctic Seas». St-Petersburg, 23-26 September. 432-433

Matishov G.G. and B.A. Nikitin, eds. 1997. Scientific and methodological approaches to the estimation of the gas condensate extraction impact on the Arctic Seas ecosystems (on the example of the Stockman gas-condensate deposit) [Научно-методические подходы к оценке воздействия газонефтедобычи на экосистемы морей Арктики (на примере Штокмановского проекта)]. Apatity: KSC Publ. 393p

Matishov, G.G., S.F. Timofeev, S.S. Drobysheva, V.M. Ryzhov. 1994. Evolution of ecosystems and biogeography of European Arctic Seas. [Матишов Г.Г., Тимофеев С.Ф., Дробышева С.С., Рыжов В.М. Эволюция экосистем и биогеография морей Европейской Арктики]. Saint-Petersburg: Nauka Publ. 222p (In Russian).

Matishov, G.G.and L.G. Pavlova. 1990. General ecology and paleogeography of the polar oceans. [Матишов Г.Г., Павлова Л.Г. Общая экология и палеогеография полярных океанов]. Leningrad: Nauka Publ. 224p (In Russian).

Matishov, D.G. and G.G. Matishov. 2001. Radiational ecological oceanology [Матишов Д.Г., Матишов Г.Г. Радиационная экологическая океанология] Apatity, KSC Publ. 418p. (In Russian).

Nikulin, G.N., Roldugin, V.K., Beloglazov, M.I., Karpechko, and A.Yu. 2000. Observations on the main ozone content on the Kola Peninsula in 1998 and 1999 [Никулин Г.Н.,

Ролдугин В.К., Белоглазов М.И., Карпечко А.Ю. Наблюдения за общим содержанием озона на Кольском полуострове в 1998 и 1999 гг.] Meteorologiya i gidrologiya. N 11. 96-99. (In Russian).

Rovinsky F. Ya., G.M. Chernogaeva, S.G. Paramonov. 1995. The role of river runoff and atmospheric transport in pollution of Russian Northern Seas [Рогинский Ф.Я., Черногаева Г.М., Парамонов С.Г. Роль речного стока и атмосферного переноса в загрязнении севрных морей России] // Meteorologiya i gidrologiya, No.9, 22-29. (In Russian).

Savinova T.N. 1991. Chemical pollution of the northern seas. Ottawa: Canadian Translation of Fisheries and Aquatic Sciences. No.5536. 174p

Savinova, T.N., G.W. Gabrielsen, S. Falk-Petersen. 1995. Chemical pollution in the Arctic and Sub-Arctic marine ecosystems: an overview of current knowledge. NINA Fagrapport 1. Trondheim: NINA-NIKU. 68p

State of the environment of the Murmansk District in 2000 [Состояние окружающей среды Мурманской области в 2000 г.]. 2001. Murmansk: Murmansk District Committee on Natural Resources. 186p

Zenkevich, L.A. Biology of the USSR seas. [Зенкевич Л.А. Биология морей СССР]. 1963. Moscow-Leningrad, Academy Publ. 739p (In Russian).

Large Marine Ecosystems of the World
G. Hempel and K. Sherman (Editors)

4

The Scotian Shelf

K.C.T. Zwanenburg

INTRODUCTION

This essay presents a brief history of the Scotian Shelf large marine ecosystem. It describes eastern and western portions of the shelf and gives an overview of its long-term exploitation. The paper discusses the fisheries management objectives for the ecosystem and the degree to which these have been achieved. It reviews recent changes made to the management system to address identified shortcomings. Finally it presents an overview of ongoing research that is moving us toward the ability to manage human activities, including fishing, within a large marine ecosystem context. The essay is written from a somewhat ichthyocentric point of view, both because of the biases of the author and the fact that the management of fisheries has the longest history and is the most developed of the management systems in place.

Physical and Biological Setting

The Scotian Shelf LME covers about 185,000 km^2 of continental shelf (< 400 m) off the southeast coast of Canada (Figure 4-1). The shelf is characterized by complex bottom topography with numerous offshore shallow (< 100 m) banks and deep (> 200 m) basins. It is separated from the southern Newfoundland Shelf by the Laurentian Channel and borders the Gulf of Maine to the southwest. Surface circulation is dominated by a southwestward flow with low salinity waters from the Gulf of St. Lawrence discharging onto the Scotian Shelf on the south side of the Cabot Strait. Part of this flow rounds Cape Breton to form the southwestward moving Nova Scotia Current on the inner half of the shelf. The remainder flows along the Laurentian Channel, turns at the shelf break, and eventually enters the Gulf of Maine through the Northeast Channel. The Nova Scotia Current enters closer inshore. The amplitude of the annual cycle in sea surface temperature (16°C, Petrie *et al.* 1996) in the region from the Laurentian Channel to the Middle Atlantic Bight is the largest anywhere in the Atlantic Ocean (Weare 1977) and one of the largest in the world except for semi-enclosed seas.

The vertical structure of the water column on the Scotian Shelf undergoes large seasonal variation. In winter, strong winds and cold air temperatures result in rapid heat loss and water becomes vertically mixed to depths of 50-150 m. In

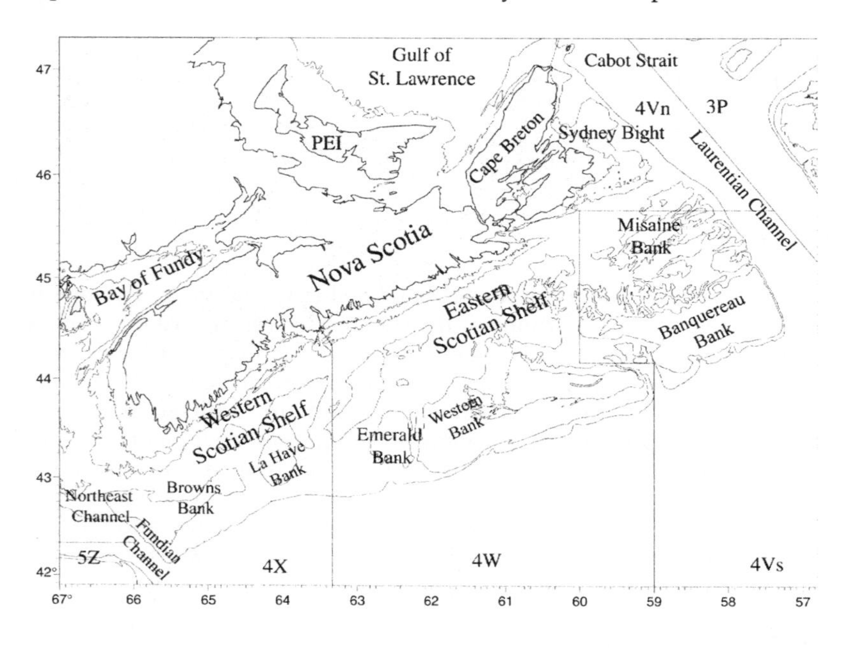

Figure 4-1. Map of the Scotian Shelf. The vertical line at approximately 63° 20'is the boundary between the eastern and western shelf regions. The names of the features referred to in the physical description of the systems are also given.

spring and summer, solar heating combined with reduced salinity (due to ice melt and advection), make the surface waters less dense and lead to rapid stratification, trapping the cold, winter-mixed water below. In the central and western shelf relatively warm water of offshore origin (slope water, a mixture of Gulf Stream and Labrador slope waters) moves in along the bottom beneath the winter-cooled waters because the offshore water is more saline and hence more dense. The winter-chilled cold intermediate layer (CIL), with temperatures varying from 2-5°C, is sandwiched between warmer surface and bottom layers. A significant difference between the western and eastern shelf is that on most of the eastern Scotian Shelf bottom topography prevents the warm offshore waters from penetrating inshore and the CIL (< 5°C) extends to the bottom depths > 200-m). Therefore, bottom temperatures vary from 2-4°C in the northeast to 8-10°C in the

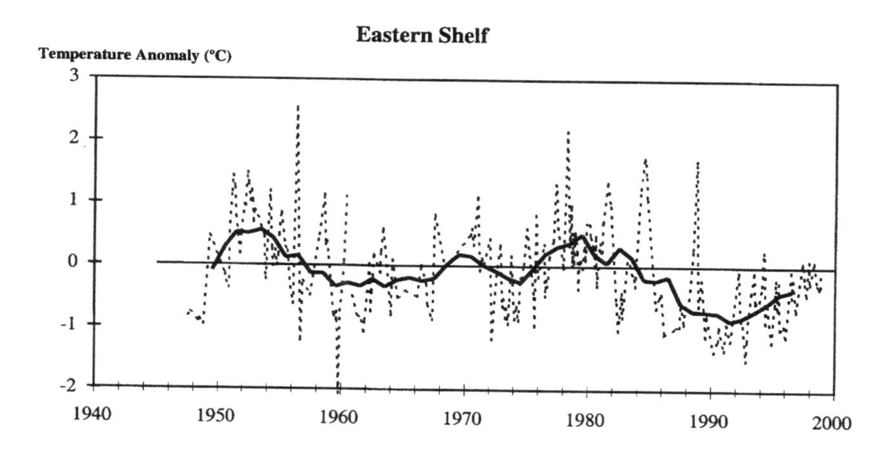

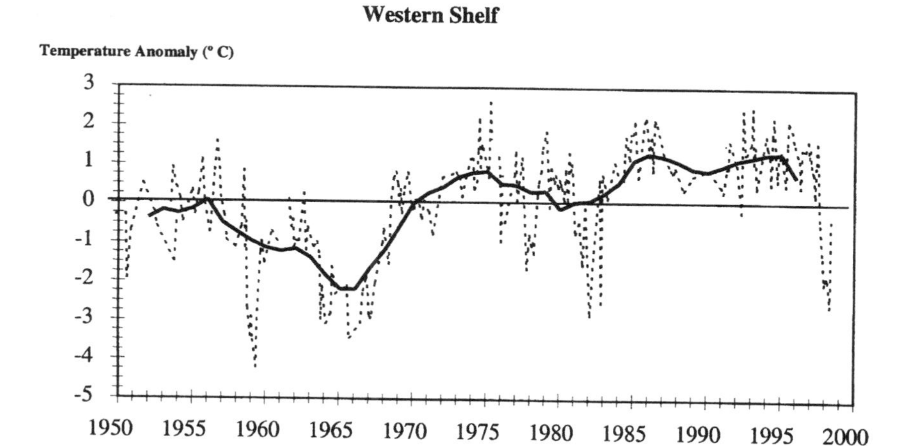

Figure 4-2. Long-term near bottom temperature anomalies for the eastern and western Scotian Shelf. The dark line represents the 5-yr running annual mean anomaly and the dashed line represents the annual values. The eastern shelf series is from Misaine Bank and the western shelf series is from Emerald Basin.

deep basins in the central and west (Zwanenburg *et al.* 2002). There are few seasonal variations at these depths.

Annual changes in near bottom temperatures on the Scotian Shelf are among the most variable in the North Atlantic. Near bottom waters of the western shelf generally remained warmer than average from the 1970s to 1997 (Figure 4-2). The highest sustained temperature anomalies in the approximate 50-year record were observed in the mid-1990s. In 1998, cold Labrador Slope water again appeared with a subsequent lowering of temperatures in Emerald Basin by over 3°C (Drinkwater *et al.* 2000). From the late-1960s to the mid-1970s, bottom temperatures in the northeast oscillated near or above average. They rose above normal around 1980 but by the mid-1980s, temperatures fell sharply. Below approximately 50 m, temperatures have generally remained colder than normal and the coolest in the approximately 50-year record occurred in the early 1990s. In recent years, the waters have been warming and are now approaching normal.

The eastern and western parts of the Scotian shelf large marine ecosystem are considered closely linked but separate because of differences in physical environment and in fish and invertebrate communities. The different temperature regimes also result in differences in growth rates for animals of the same species that have populations on both parts of the shelf. The eastern shelf fishes have traditionally been dominated by cod (*Gadus morhua*), redfish (*Sebastes spp*), haddock (*Melanogrammus aeglefinus*) and American plaice (*Hippoglossoides platessoides*), while the western shelf fishes are dominated by dogfish (*Squalus acanthias*), haddock, redfish and Pollock (*Pollachius virens*). Dogfish biomass on western shelf has increased to about 60 percent of total demersal fish biomass over the past 20 years. For the eastern shelf we observe significant increases in the abundance of cold-water species concurrent with the negative temperature anomaly of the last 15 years. We observed significant increases in the numbers of capelin (*Mallotus villosus*), turbot (*Reinhardtius hippoglossoides*), northern shrimp (*Pandalus borealis*), snow crab (*Chiononectes opilio*), and sand lance (*Ammodytes dubius*). Although the increase in these species is partly attributable to the negative temperature anomaly, the low biomass of cod, a predator of all but turbot, must also be contributing.

Zwanenburg *et al.* (2002) show that over the past 3 to 4 decades, there have been changes in the abundance and distribution of many marine species on the Scotian shelf. In both areas, biomass of demersal fishes and the average size of demersal fishes (Figure 4-3) have decreased significantly (Zwanenburg 2000), particularly in the past 15-20 years. Some demersal fish populations have declined precipitously (particularly true of the commercially exploited species) while at the same time the abundance of other species, both invertebrates and small pelagic fishes and marine mammals has increased. These changes suggest that there may have been changes in the trophic structure of these ecosystems.

While the changes in fish abundance and species composition can be relatively well described, information on other trophic levels is less available. There are no synoptic time series of primary, secondary, or benthic production for the Scotian Shelf. Some data for specific years or for limited areas of the shelf are available but these are not sufficient to describe long-term dynamics. Information on marine mammals is equally sparse with the exception of pinnipeds, in particular grey seals, whose population has been increasing at a rate of some 12 percent per annum since the early 1960s (Zwanenburg and Bowen 1990). Large migratory pelagic fishes (tunas, sharks, and swordfish) are seasonal transitory residents in these systems whose dynamics are described at ocean basin scales rather than the Scotian Shelf scale. Seabirds form an important component of both systems, however there is little information on changes in distribution or abundance.

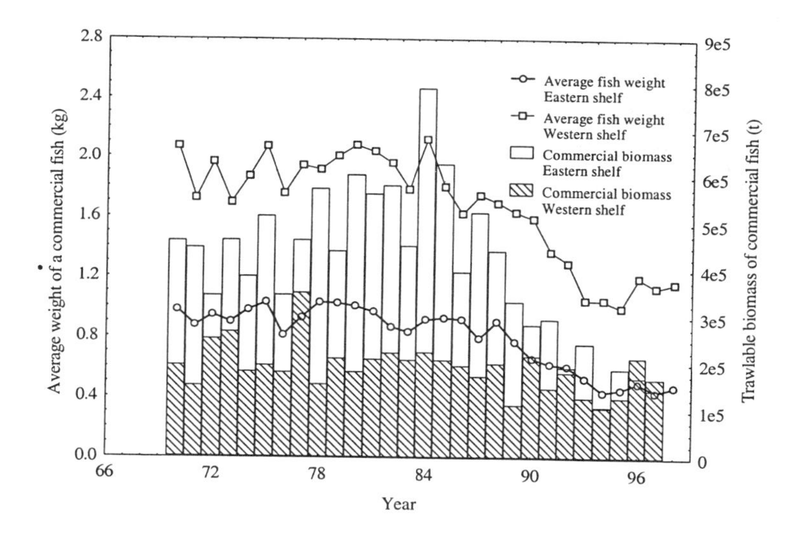

Figure 4-3. Average individual weight and estimated trawlable biomass of commercially exploited demersal fish on the eastern and western Scotian Shelf.

Fishing and other human activities such as oil and gas exploration and extraction are conducted within this dynamic physical and biological framework. Domestic fishing effort in both areas increased rapidly from 1977 until the early 1990s when low spawning stock biomass and low recruitment prompted a closure of the eastern shelf cod fishery, and severe restrictions on the western shelf. The fisheries on the eastern Scotian Shelf remain essentially closed to date and show no evidence of rebuilding due to continued low recruitment. Why did these fish populations decline so rapidly in the past 15 years, after supporting fisheries for centuries and what were the relative contributions of human activities, notably fishing, and changes in environmental conditions to these declines?

Exploitation

First Nations people have harvested the Scotian shelf for several thousand years; although, commercial fishing did not start until the mid-1500's. In 1602, Samuel de Champlain reports meeting a Basque fisherman called Savalet making his 42nd voyage to the Scotian Shelf (Innis 1954). At one voyage per year, he would have started fishing there in 1560 making him one of the first Europeans to fish there. By 1709 Nova Scotia was exporting about 10,000 t of cod and 4,000 t of mackerel and herring (Figure 4-4). By 1806 this had increased to about 40,000 t, and to over 100,000 t by the 1880s. In 1973, total landings of fish from the Scotian Shelf exceeded 750,000 t. In the early 1990's many of the east coast cod fisheries, including that of the eastern Scotian shelf, were closed.

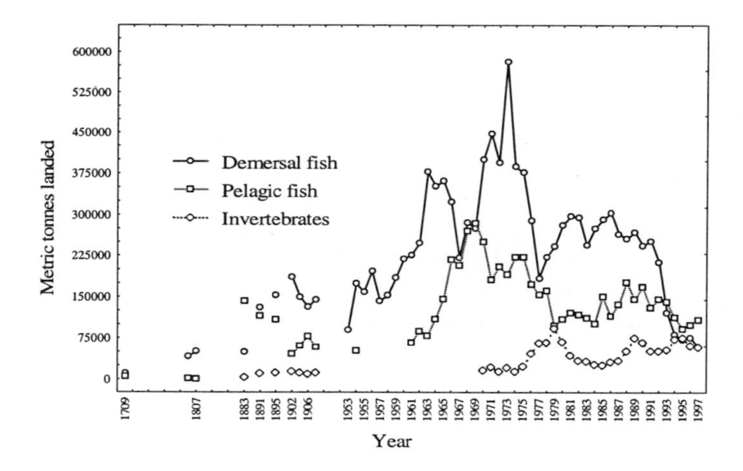

Figure 4-4. Landings of fish and invertebrates from the Scotian Shelf.

On the eastern shelf, total landings of demersal fish (mainly gadoids) declined from a maximum of 450,000 metric tons (t) in 1973 to less than 15,000 t in 1997 (Figure 4-5). A moratorium on fishing, especially for cod, was imposed in 1993 and remains in effect. Landings of pelagic fish (mainly Atlantic herring, *Clupea harengus* and Atlantic mackerel, *Scomber scombrus*) declined from 120,000 t in 1970 to 30,000 t in 1997, with a recent increase in landings. Landings of invertebrates have increased from less than 4,000 t in 1985 to 30,000 t in 1997, mainly due to increased landings of snowcrab and northern shrimp. For the western shelf pelagic species, mainly (Atlantic herring and Atlantic mackerel) dominate the landings which have fluctuated between about 75,000 and 155,000 t from 1970 to 1997. Landings in 1997 reached about 80,000 t.

Demersal fish landings declined from 110,000 t in 1992 to 40,000 in 1996. The fisheries for the commercially exploited gadoids of primary interest, Atlantic cod, haddock, and pollock remain active on the western shelf although total allowable catches (TACs) are low relative to the documented history of these fisheries. Invertebrate landings showed a variable increase from about 10,000 t in 1970 to over 50,000 t in 1994 and a subsequent decline to about 30,000 t in 1997.

With the decline in the more traditional commercial species since the middle of the 1980s there has been a tendency, in both areas, towards increased landings of formerly less utilized species [monkfish (*Lophius americanus*), cusk (*Brosme brosme*), white hake (*Urophycis tenuis*), and several species of skate (*Raja spp.*)].

Management and Protection

From 1950 to 1977 Scotian shelf fisheries were regulated by the International Commission for the Northwest Atlantic Fisheries (ICNAF). A lack of regulatory control and the limited development of the Canadian domestic fleet led to the establishment of Canada's 200-mile exclusive economic zone in 1977 bringing Scotian shelf fisheries under Canadian jurisdiction. Post-1977 fisheries management had two broad objectives; the development of the Canadian fishing industry, and best use of fish resources (Parsons 1993, Halliday and Pinhorn 1996). The immediate intentions were to rebuild depleted stocks, and subsequently to prevent growth and recruitment over-fishing. The economic objectives of developing the Canadian fishing industry after 1982 were to: 1) maintain an economically viable fishery with a normal business failure rate without government subsidies, 2) to maximize employment in the industry (with the proviso that everyone should make a reasonable income), and 3) to maximize Canadian participation without competitive interference from foreign activities (Kirby 1982).

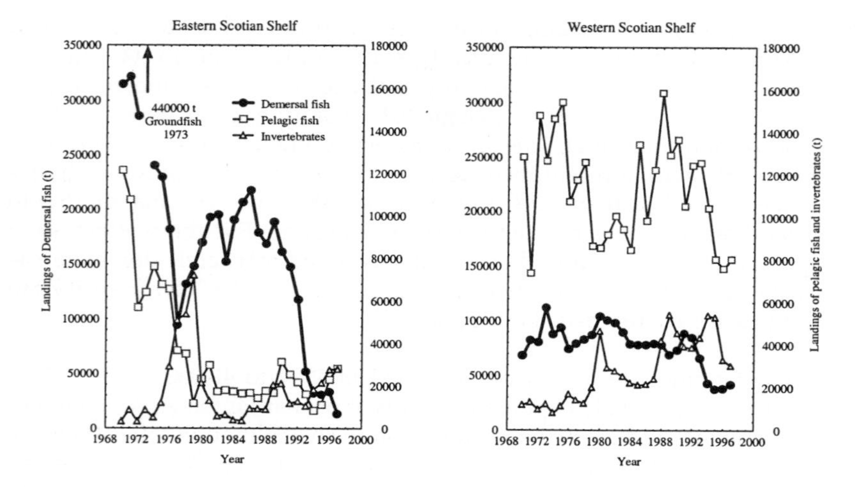

Figure 4-5. Landings of fish and invertebrates from the eastern and western Scotian Shelf for the period 1970 – 1997. Figure 4-5. Landings of fish and invertebrates from the eastern and western Scotian Shelf for the period 1970 – 1997.

Canada's implicit conservation strategy was to ensure that utilization of fish resources was sustainable. It was considered that if exploitation of individual marine fish populations remained at or below a target fishing mortality rate (F) of $F_{0.1}$ that the fisheries would be sustainable in perpetuity. An $F_{0.1}$ corresponds to that F (calculated from a yield-per-recruit curve) at which one unit of increase in fishing effort (proportional to F) will give an increase in yield $1/10^{th}$ that of a unit of effort on the unexploited stock. The primary objective of scientific advice to managers was to estimate F relative to $F_{0.1}$ for each fished population, and then to recommend removing more or less fish biomass. Managers implemented this advice by changing total allowable catches (TACs) for each population. Although some considerations were given to minimum levels of spawning stock biomass (SSB) they were not operationalized because of essentially non-predictive relationships between estimates of SSB and recruitment. The long-term over-exploitation of fish resources contributing to a number of fishery closures in 1993, attests to the failure in achieving the sustainability objective. It is not possible to determine whether or not the $F_{0.1}$ strategy would have been successful because F was rarely if ever below $F_{0.1}$ (Angel *et al.* 1994, Halliday and Pinhorn 1996).

Angel *et al.* (1994) concluded that ambiguity in management objectives (favouring development and exploitation over conservation), an ineffective management process, poor estimates of fish population status, and over-capacity in the harvesting sector, all contributed to overexploitation. Conflicts between conservation and economic objectives resulted in trade-offs favouring the latter including reversion to "by-catch" fisheries rather than closure when quotas for any one species were reached. The assumption that economic constraints would restrain fishing effort prior to recruitment overfishing was not borne out. Management was ineffective at protecting vulnerable species in multispecies fisheries because of incentives for mis-reporting, or high-grading. Government subsidization of the fisheries ensured that they remained viable at high effort levels to the point of stock collapse. Stock sizes were often overestimated leading to over-exploitation. Finally there was a lack of consideration of minimum spawning stock biomass and failure to protect individual spawning components. To these shortcomings we should also add a less than adequate understanding of how exploited populations respond to exploitation in a variable environment. Our understanding of how changing conditions affect productivity of fish populations remains rudimentary.

Of the 42 recommendations made by the 1993 Task Force on Incomes and Adjustments in the Atlantic Fishery (DFO 1993), 13 relate to reducing capacity in the fish harvesting and processing sectors. The response was a plan for the government to buy-out 50 percent of harvesting capacity. For the Scotian Shelf to date, a total of 350 licenses (14 percent) have been bought out, leaving a significant over-capacity in place.

Burke *et al.* (1996) make recommendations to improve fisheries management in eight key areas, strategic planning, catch and effort monitoring, fishing entitlements, enforcement, service delivery, and resource analysis. Specific recommendations include: 1) delegation of decision making and planning to users with a clear division of responsibilities between users and government, 2) augmenting output controls such as TACs with input controls like effort limitations, 3) increasing sanctions for violators 4) perpetuation of existing industry resource shares to provide incentives for industry to meet conservation objectives, and 5) definition of minimum SSB for each stock coupled with spawning area closures to ensure maintenance of genetic diversity within stocks.

Sinclair *et al.* (1999a) review the progress toward implementation in each of these areas. The establishment of community quotas (for some gear sectors) has transferred significant management responsibilities to fishermen. These boards are responsible for allocations of quota within communities and for the development of annual harvest plans, which require approval by DFO prior to the fishery. Regional advisory committees which were responsible for these issues now develop Integrated Fisheries Management Plans dealing with overarching issues such as like authorized fishing gears, closed areas, and minimum reporting requirements. Sinclair *et al.* (1999a) also highlight the adoption of the United Nations Agreement on the Conservation and Management of Straddling Fish Stocks and Highly Migratory Fish Stocks, and the Food and Agriculture Organization Code of Conduct for Responsible Fishing. These require the explicit incorporation of the precautionary approach and of limit reference points within fisheries management. Although neither has yet been operationalized, there is work underway to define them (see below). On a broader scale there is also a stated desire to incorporate ecosystem level objectives in fishery management.

Another significant change in the management and scientific investigation of the Scotian Shelf is that the degree and quality of communication between resource users and resource scientists has improved. The Fishermen and Scientists Research Society (FSRS) was established in 1994 to promote effective communication between fishermen, scientists, and the general public and to conduct collaborative research relevant to the long-term sustainability of marine fisheries. The Society now has about 200 fishermen and scientist members, manages a comprehensive annual survey of fishes on the Eastern Scotian Shelf (the sentinel survey), and is involved in a wide range of research projects in collaboration with DFO, non-governmental organizations and universities in the region. The FSRS has served as a model for other collaborative initiatives between the fishing industry and resource scientists. The most far-reaching of these are a series of surveys for long-term monitoring of commercially exploited fish. They collect detailed geo-referenced information on fishing practices, catch rates of targeted and by-catch species, and a host of biological information at broader spatial and temporal scales than standard DFO surveys. They provide information essential to understanding both the ecological impacts of fishing, and the functioning of exploited ecosystems.

Effective communication and co-operation is also evident between the oil and gas industry and regulatory agencies. The Sable Island Offshore Energy Project (SOEP) is currently the largest gas extraction project on the Scotian Shelf with a projected life span of 25 years. SOEP has established the Sable Offshore Energy Environmental Monitoring Advisory Group (SEEMAG) with members representing the industry, regulatory agencies, and user groups. This group designed and regularly reviews results of the effects monitoring program in public fora. One of the most significant problems being deliberated is the long-term impact of voiding "produced water" which contains an array of potentially toxic substances, during the gas extraction process.

Canada's Oceans Act (1997) commits it to development of an integrated strategy for ocean management. The development of this strategy is ongoing and is addressing, among others, issues like conservation of biodiversity, marine environmental quality, and integrated planning and management. DFO has been designated the lead agency in this process in co-operation with a host of other federal and provincial departments, and aboriginal governments. The definition and implementation of an integrated management plan for the eastern Scotian Shelf, as well as the development of marine protected areas on the Scotian Shelf, are current areas of research for the newly formed Oceans Act Office.

Research

Objectives

Awareness of the need to incorporate ecosystem objectives into fisheries management has been growing in Canadian fisheries management over the past decade. Some ecosystem considerations presently incorporated in fisheries management plans include: 1) restrictions on rockweed harvests to ensure sufficient habitat for juvenile stages of a number of marine organisms, 2) denial of an application to fish krill because of its role as prey for many harvested species, 3) consideration of the role of grey seal (*Halichoerus grypus*) predation on cod population dynamics, (Zwanenburg and Bowen 1990), 4) measures to reduce by-catches of harbour porpoise, 5) studies on the impacts of trawling and clam dredging on benthic habitat, and 6) gear modifications to reduce by-catch (Sinclair *et al.* 1999b).

The definition of ecosystem objectives for the management of human activities (including fishing) on the Scotian Shelf and elsewhere is now a requirement under Canada's Oceans Act. To define reasonable objectives it is essential that we improve our understanding of how the ecosystem of the Scotian Shelf has changed over time and how both eastern and western areas function. Descriptions of physical and biological changes provide the context within which fisheries and other human activities have taken place and to which they may be contributing. Disentangling the fishery and environmental effects leading to ecosystem change has, so far, been inconclusive (Zwanenburg 2000). The Comparative Dynamics of Exploited Ecosystems Project (CDEENA) is a multidisciplinary project being undertaken to better understand the changes that have occurred in the Scotian shelf ecosystem. Several modelling approaches (mass balance, trophodynamic) are being used to understand how the system is structured, its dynamics, how it has changed in space and time, the relative effects of fishing and environmental changes, and the implications of changes on the fisheries.

Global objectives of ecosystem level management have been articulated (Agenda 21, of the United Nations Convention on Economic Development [UNCED]) and contain broad references to the maintenance of biodiversity and habitat productivity. Although the specific ecosystem considerations to be incorporated remain undefined beyond these generalities, it is likely that they will require modifications to or expansion of our present monitoring activities. Systematic trawl surveys conducted in support of fisheries management (1950s to the present, see Halliday and Koeller 1980) are the richest and most consistent source of monitoring information available. The proposed Atlantic Zone Monitoring Program will augment the scope of these long-standing surveys to include increased monitoring of physical and biological oceanographic information. Long-term synoptic data on primary and secondary production are not available. Information on the distribution and abundance of benthic invertebrates and small pelagic fishes, and the distribution and concentrations of pollutants are sparse or non-existent, and must be considered in these expanded monitoring programs. Our understanding of the linkages between primary and secondary production and the production of fishes (both demersal and pelagic) and benthic invertebrates is unsatisfactory. Although work is in progress to address this shortcoming (CDEENA), it is likely that this will be a long-term (decadal) process, especially given the paucity of even simple diet information linking predators to prey. Information on the more complex linkages among primary/ secondary production, benthic production, and fish production, required to make predictions about the ecosystem level effects of exploitation or changes in environmental conditions, is even less available.

The Precautionary approach, and Limit reference points

The definition of the specific ecosystem considerations to be incorporated in management of human activities on the Scotian shelf, and improvement in our understanding of ecosystem functioning will be long-term undertakings. This is a reality that must not preclude immediate actions to ensure that human activities do not become the prime forcing variable of the dynamics of the Scotian Shelf ecosystems. In response, a number of research initiatives are attempting to define and operationalize the Precautionary Approach as proposed by the Food and Agriculture Organizations Code of Conduct for Responsible Fishing (FAO 1995 a, b). The precautionary approach is 1) objectives are set, 2) strategies (plans to achieve objectives) are implemented, 3) unacceptable outcomes are defined (limit reference points), 4) uncertainty is taken into account, 5) system performance is monitored (effective and measurable indicators) and 6) that agreed on corrective actions are taken if limits are approached.

For the Scotian Shelf this work has focused on the development of multiple indicators for fish stock or ecosystem status, rather than on the definition of limit reference points. The development of multiple indicators based on the traffic light approach of Caddy (1999) considers the state of an array of variables or indicators relevant to the status of individual fish stocks, or at a higher level organization, the status of the ecosystem as a whole. To evaluate the status of the system, the state of each variable is evaluated relative to a limit reference point (if these have been defined) or to its historical dynamic range. The state of each variable can then be judged as either good, bad, or intermediate (green, red or yellow) and the integration of the states of all variables gives an indication of the status of the system. The value of this approach is that it allows us to incorporate a broad array of indicators into the determination of ecosystem status. It also allows for both qualitative and quantitative measures to be evaluated in a single framework within which judgements on reliability, accuracy, or importance of each indicator can be explicitly defined. The traffic light approach could allow for truly integrated evaluation in that it could incorporate stock indicators, ecosystem indicators, economic and social indicators, and indicators of regulatory compliance. Within such a framework ecosystem considerations or objectives can be defined as limit or target values for integrals of exploitation rate, integrated community size structure (Gislason and Rice 1998) , trophic level balance or others. Although the traffic light approach has significant appeal and potential, its development is far from complete.

At present, work is proceeding on the definition of status indicators, limit reference points for these, and methods of integration which reflect the relative reliability, accuracy, and importance of each indicator. Although the array of indicators may be broad, they should share some characteristics. They must be measurable, interpretable relative to the status of the system, and sensitive to changing states of the system. For individual fished stocks the potential status indicators being considered include, estimated biomass (Sequential Population Analysis [SPA] and survey based), catchability, F, total mortality rate, recruitment, size/age selection patterns, size /age composition of the population and removals, fish condition, growth rate, size at maturity, geographic distribution (density and area occupied) of the population (Figure 4-6). The indicators proposed for evaluation of ecosystem status are less well defined but include primary productivity, secondary productivity, balance among demersal/pelagic/benthic productivity, integrated community size spectrum, K-dominance, diversity, and evenness. Environmental conditions would have impacts on both the productivity of individual species and on the ecosystem as a whole. For example, water temperature may be thought of as the thermostatic control which regulates the rates of a host of processes and interactions. The flexibility of the traffic light approach will also allow for the definition and evaluation of indicators of the ecosystem effects of human activities such as incidental mortalities of commercially exploited fishes, by-catch and incidental mortalities of non-commercial species, and the incidental mortality of protected species. Economic and social outcome indicators could include the number of licenses, commercial catch rates, and gross revenues from the fisheries, while compliance indicators would relate to the accuracy of landings statistics and area of capture, discarding, dumping, and high-grading rates, and rates of gear violations which have a direct impact on other status indicators (e.g. use of liners or small hooks).

The traffic light approach has a number of features lacking in the present system of stock or ecosystem evaluation. It requires that the characteristics of all indicators such as measurability, interpretability, and limit reference points (or dynamic range) be clearly defined before they are considered. This forces all

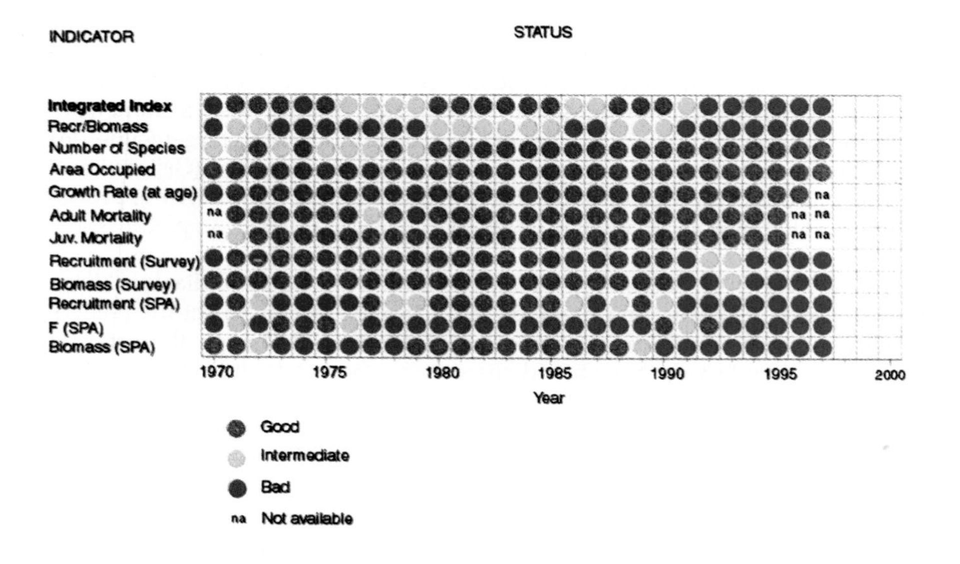

Figure 4-6. The traffic light approach applied to eastern Scotian Shelf cod stock status (illustration only). The status of each indicator is judged against either a limit reference point or its historical range (Good, Intermediate, or Bad). The integrated index, in this case, is the weighted sum of all indicators where Good (green)= 1, Intermediate (yellow) = 2, and Bad (red) = 3. Both the relative weighting of each indicator, and the method of integration can be altered in this realization of the process.

participants in the status determination process to formalize and evaluate their *information.* At present this is not the case in that opinions expressed at public meetings may at times carry the same weight as formal analyses of data. The traffic light approach will also allow us to investigate the effects of variable weighting schemes for the array of indicators, on the overall integrated view of system status. The method provides a framework in which the divergent opinions of user or regulatory groups can be formalized by giving variable weights to system status indicators and examining the effects of these weighting schemes on system status relative to objectives.

CONCLUSIONS

The Scotian Shelf is a physically and biologically dynamic large marine ecosystem. Large inter-annual fluctuations in temperature and in the abundance and distribution of fishes, invertebrates, and marine mammals have been documented. Exploitation, mainly in the form of commercial fisheries, operates against this dynamic background and has contributed to these dynamics especially by reducing the abundance of commercially exploited species. The relative contribution of fishing and environmental changes to the overall dynamics remains an area of active research. Management of human activities has been mainly in the form of fisheries management and has focussed on maintaining constant exploitation rates for individual fish populations and ensuring the economic viability of the fishing industry. More recently the inclusion of some ecosystem level objective into fisheries management has been made explicit, but the definition of overarching ecosystem objectives for management such as maintenance of biodiversity and habitat productivity is ongoing. The precautionary approach to management has been adopted, and the traffic light approach to system evaluation is being developed.

ACKNOWLEDGEMENTS

Ralph Halliday and Alida Bundy provided reviews and insightful comments on an earlier draft of this paper. Thanks to Bob Mohn and Gerry Black for their permission to use the traffic light figure.

REFERENCES

Angel, J.R., D.L. Burke, R.N. O'Boyle, F.G. Peacock, M. Sinclair, and K.C.T. Zwanenburg. 1994. Report of the Workshop on Scotia-Fundy Groundfish Management from 1977 to 1993. Can. Tech. Rep. Fish. Aquat. Sci.: 1979 vi+175p

Burke, D.L., R.N. O'Boyle, P. Partington, and M. Sinclair. 1996. Report of the second workshop on Scotia-Fundy groundfish management. Can. Tech. Rep. Fish. Aquat. Sci. 2100: vii+247p

Caddy, J.F. 1999 Deciding on precautionary management measures for a stock, based on a suite of Limit Reference Points (LRP's) as a basis for a multi-LRP harvest law. NAFO Sci. Coun. Studies, 32:55-68

Canada's Oceans Act. 1997. http://www.mar.dfo-mpo.gc.ca/science/OceansAct

DFO (Department of Fisheries and Oceans Canada), 1993. Charting a new course: towards the fishery of the future. Task force on Incomes and Adjustments in the Atlantic Fishery. Miscellaneous Publication, Catalogue No. Fs 23-243/1993, 199p

Drinkwater, K.F., E. Colbourne and D. Gilbert. 2000. Overview of environmental conditions in the Northwest Atlantic in 1998. NAFO Sci. Coun. Studies 33:39-87

FAO. 1995a. Precautionary approach to fisheries. Part 1. Guidelines on the precautionary approach to capture fisheries and species introductions. FAO Fisheries Technical Paper No. 350, Part 1. Rome, FAO. 1995, 52p

FAO. 1995b. Precautionary approach to fisheries Part 2: Scientific papers. FAO Fish. Tech. Paper No. 350, Part 2. Rome, FAO. 1996, 210p

Gislason, H., and J. Rice. 1998. Modelling the response of size and diversity spectra of fish assemblages to changes in exploitation. ICES J. of Mar. Sci. 55: 362-370

Halliday, R.G., and P.A. Koeller. 1980. A history of Canadian groundfish trawling surveys and data usage in ICNAF Divisions 4TVWX. In: Doubleday W.G. and D. Rivard, eds. Bottom Trawl Surveys. Can. Spec. Pub. Fish. Aquat. Sci. 58: 27-41

Halliday, R.G., and A.T. Pinhorn. 1996. North Atlantic fishery management systems: A comparison of management methods and resource trends. J. Northwest Atlant. Fish. Sci. 20:21–34

Innis, H.A. 1954. The Cod Fishery; the history of an international economy. University of Toronto Press 522p

Kirby, J.L. 1982. Navigating troubled waters: a new policy for the Atlantic fisheries. Supply and Services Ottawa. 379 p

Parsons, L.S. 1993. Management of marine fisheries in Canada. Can. Bull. Fish. Aquat. Sci. 225, 784p

Petrie, B., K. Drinkwater, D. Gregory, R. Pettipas and A. Sandström. 1996. Temperature and salinity atlas for the Scotian Shelf and the Gulf of Maine. Can. Tech. Rep. Hydrogr. Ocean Sci. 171, 398 p

Sinclair, M., R. O'Boyle, L. Burke, and S. D'Entremont. 1999a. Incorporating Ecosystem Objectives within fisheries management plans in the Maritimes region of Atlantic Canada. ICES CM 1999/ Z:03. 20p

Sinclair. M., R.N. O'Boyle, D.L. Burke, and F.G. Peacock. 1999b Groundfish Management in Transition within the Scotia-Fundy Area of Canada. ICES

J. Mar.Sci. 56: 1014-1023

Weare, B.C. 1977. Empirical orthogonal analysis of Atlantic Ocean surface temperatures. Quart. J. R. Met. Soc. 103: 467-478

Zwanenburg, K. C. T. 2000. The effects of fishing on demersal fish communities of the Scotian Shelf, ICES J. Mar. Sci. 57:503-509

Zwanenburg, K.C.T., and W.D. Bowen. 1990. Population trends of the grey seal (*Halichoerus grypus*) in Eastern Canada. In: W.D. Bowen, ed. Population biology of sealworm (*Pseudoterranova decipiens*) in relation to its intermediate and seal hosts. Can. Bull. Fish. Aquat. Sci. 222

Zwanenburg, K.C.T., D. Bowen, A. Bundy, K. Drinkwater, K. Frank, R.N. O'Boyle, D. Sameoto, and M. Sinclair. 2002. Decadal changes in the Scotian Shelf large marine ecosystem. In: Sherman, K. and H.-R. Skjoldal, eds. Large Marine Ecosystems of the North Atlantic. Elsevier Science. New York and Amsterdam. Chapter 4, 105-150. 449p

Large Marine Ecosystems of the World
G. Hempel and K. Sherman (Editors)

5

Assessment and sustainability of the U.S. Northeast Shelf Ecosystem

Kenneth Sherman, Jay O'Reilly and Joseph Kane

INTRODUCTION

The United States' Northeast Shelf ecosystem encompasses 260,000 km^2, extending over four sub-areas: the Gulf of Maine, Georges Bank, Southern New England and the Mid-Atlantic Bight to Cape Hatteras and from the estuaries of the coast seaward to the edge of the continental shelf (Figure 5-1). It is among the most productive of the 64 Large Marine Ecosystems (LMEs) located around the coastal margins of the world's ocean basins. The Northeast Shelf has been studied continuously since the 1860s. Results of early studies of the shelf, its physics, chemistry, geology, and fisheries can be found in monographs of the U.S. Fish Commission (now the National Marine Fisheries Service), U.S. Geological Survey, Woods Hole Oceanographic Institution, and the Bingham Oceanographic Collection at Yale University. An excellent synthesis of the early work from the turn of the century through 1970 is given by Emery and Uchupi (1972). The early research on shelf productivity and biology ranks among the best in the world, and can be found in the papers of Bigelow (1926, 1927), Fish (1936), Redfield *et al.* (1941), Redfield (1937), Bigelow *et al.* (1940), and Riley (1941). More recent summaries of the shelf ecosystem are given in Cohen (1976), O'Reilly and Busch (1984), Backus (1987), O'Reilly and Zetlin (1998), Sherman *et al.* (1988, 1996, 2000) and (NOAA 1998).

For three and a half centuries the ecosystem has supported large and important fisheries, extending from the export trade in salted cod of the early colonists to the overfishing of the total fin-fish biomass in the late 1960s and early 1970s and on to turning the corner from stock depletion toward fish stock recovery beginning in the mid-1990s. Excessive fishing mortality imposed on the resources by European factory fleets precipitated the passage of legislation by the United States that established a U.S. exclusive Fishery Management Zone (FMZ) in 1976 and the exclusion of foreign fisheries by 1980. Fishery yields of the 1990s landed

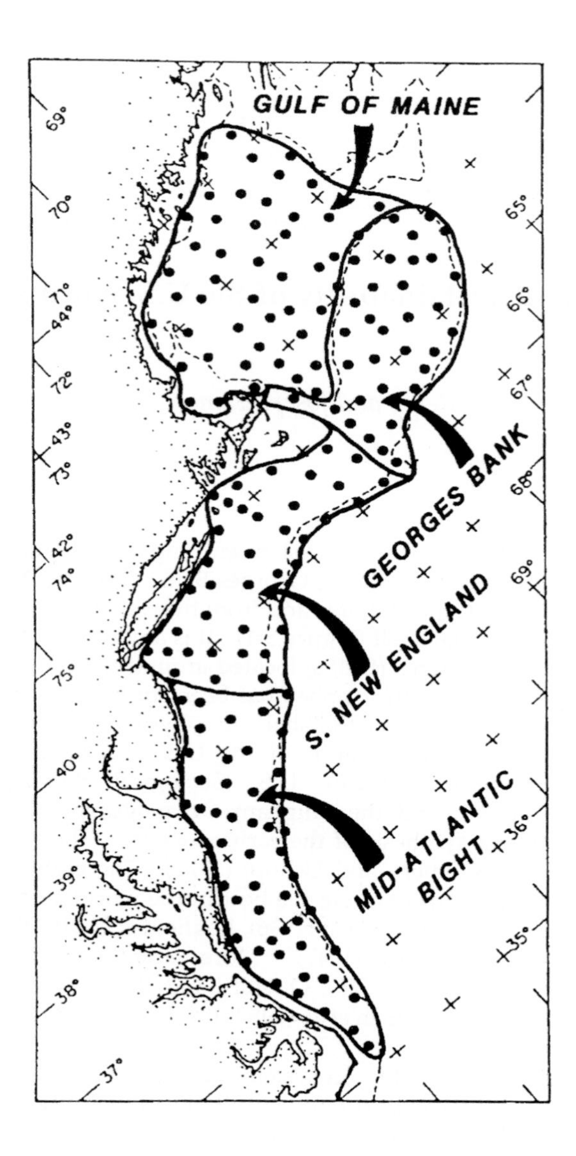

Figure 5-1. The United States Northeast Shelf ecosystem and subareas including the Gulf of Maine, Georges Banks, Southern New England and the Mid-Atlantic Bight showing sampling locations monitored seasonally for spatial and temporal changes in plankton, fish, and physical oceanography of the system since 1977.

by US fishermen supported a harvesting and processing industry valued at $1,000,000,000 annually to the economy of the coastal states from Maine to North Carolina. A recent analysis of a broad scope of fisheries, tourism, shipping, and other economic activity associated with the entire ecosystem using an input-output approach estimates the value of goods and services of the ecosystem at $376 billion with an employment impact on 4 million persons (Hoagland *et al.* 1999). The fisheries component of the ecosystem has, however, been subjected to intensive fishing effort, particularly from the 1960s through the 1990s. During this three-decade period, several demersal species (e.g. cod, haddock, flounder) were significantly overfished, resulting in declines in spawning biomass well below their long-term potential yield (LTPY) levels (Murowski *et al.* 1999).

HYDROGRAPHY AND CIRCULATION

The first major description of the hydrography of two of the four sub-areas of the NE Shelf ecosystem—Gulf of Maine and Georges Bank—was by Bigelow (1927). For the other two sub-areas of the shelf ecosystem—Southern New England and the Mid-Atlantic Bight—the early descriptions of hydrography were by Bigelow (1933) and Bigelow and Sears (1935). They reported a general seasonal pattern of late summer and early autumn thermal stratification, winter water-column mixing, and spring onset of thermocline formation. This general hydrographic pattern has not changed from the 1920s through the late 1990s as described by more contemporary investigators (Brooks 1996, Lynch *et al.* 1997, Loder *et al.* 2001, Pershing *et al.* 2001). The mean circulation in the shelf region was first described in a comprehensive manner by Bumpus and Lauzier (1965) and Bumpus (1973) on the basis of data from the return of drift bottles and sea-bed drifters. The flow is generally toward the west-southwest (Figure 5-2) although it may reverse at times, particularly in summer (Bumpus 1969). Direct current measurements also show a mean westerly flow at 5 to 10 cm s^{-1} that approximately parallels the isobaths (Beardsley *et al.* 1976). The same measurements, however, show that the variation is considerably larger than the mean so that over short periods of time the flow may be in any direction.

The mass balance for the shelf region includes the mixing of about 157 km^3 of fresh runoff with slope water to form 2.4 to 3.4×10^3 km^3 of shelf water annually (Wright 1976, Bush 1981). In addition, water enters the region from south of Nantucket as part of the general westward flow on the shelf. As a result, a substantial volume of the shelf water must be removed from the shelf each year. A continuous loss of water occurs near Cape Hatteras. Also, some exchange probably occurs continuously at the shelf/slope front. In addition, detached parcels of shelf water have been observed in the slope water region. Wright (1976) has suggested that these parcels may occur frequently. A mechanism that may remove large volumes

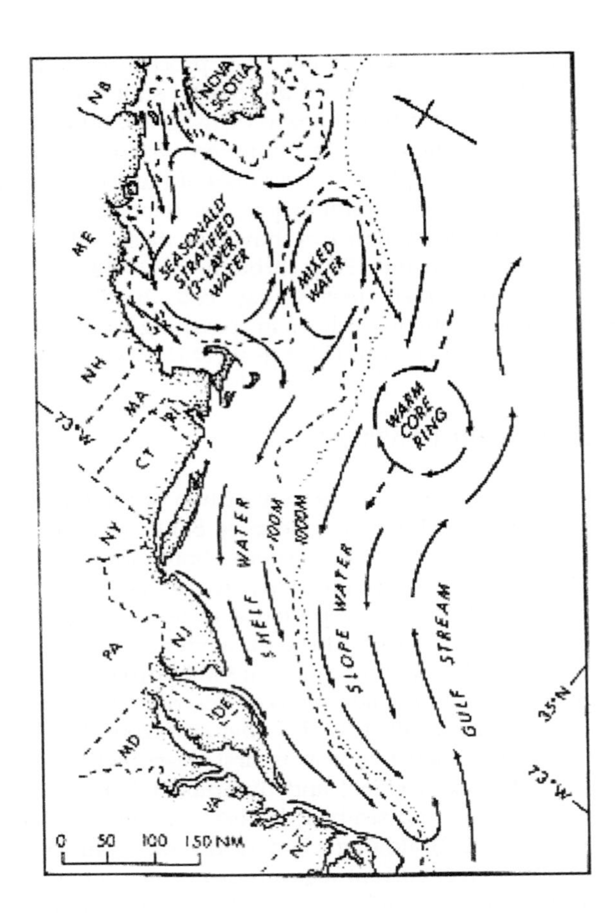

Figure 5-2. Mean nontidal surface circulation in the Gulf of Maine, Georges Bank, southern New England, and the Middle Atlantic Bight. The northern sector of the shelf ecosystem is characterized by a cyclonic gyre and a seasonally stratified three-layered water-mass system over the deep basins of the Gulf of Maine and mixed water with an anticyclonic gyre over the shoal bottom of Georges Bank. Further south, the waters move southwesterly along the broad shelf of southern New England to the narrower, gently sloping shelf plain of the Middle Atlantic Bight. (Reprinted from M.C. Ingham, R.S. Armstrong, J.L. Chamberlin, S.K. Cook, D.G. Mountain, R.J. Schlitz, J.P. Thomas, J.J. Bisagni, J.F. Paul, and C.E. Walsh. 1982. Summary of the Physical Oceanographic Processes and Features Pertinent to Pollution Distribution in the Coastal Waters of the Northeastern United States, Virginia to Maine. NEMP Oceanography Summary Rept. NEMP-IV-82-C-004. Washington, DC: U.S. Department of Commerce.)

of water from the shelf is the passage of warm core rings. These clockwise-rotating features originate from meanders in the Gulf Stream that pinch off and then move westward through the slope region. On the east side of the rings, a tongue of cooler shelf water is often seen extending off the shelf and around the ring (Sherman *et al.* 1996). Incursions of cool, low salinity water originating from the Newfoundland-Labrador and Scotian Shelf ecosystems represent a source of interannual variability in water-column structure (Pershing *et al.* 2001).

PRIMARY PRODUCTIVITY

Phytoplankton primary productivity and chlorophyll *a* of the Northeast Shelf ecosystem have been monitored periodically by the NOAA-NMFS Northeast Fisheries Center on shelf-wide cruises from Cape Hatteras to Nova Scotia since March 1977. The following summary is based on measurements made on those cruises and reports by O'Reilly and Busch (1984), O'Reilly *et al.* (1987), O'Reilly and Zetlin (1998), and Gregg *et al.* (2002).

Estimates of annual total phytoplankton primary production from Cape Hatteras to Nova Scotia are presented in Figure 5-3 by region, using data from 1,089 productivity stations. The regions selected are based on the recurring seasonal patterns of chlorophyll distribution along the continental shelf. Annual production on the shelf ranges from 10,834 to 21,043 kJ m^{-2} yr^{-1} (260-505 gCm^{-2} yr^{-1}) with the annual average of 350 gCm^2yr^{-1}. The areas of highest estimated production on the shelf occur on the central, shallow portion of Georges Bank [18,960 kJ m^{-2} yr^{-1} (445 $gCm^{-2}yr^{-1}$)] and along the coast between New Jersey and North Carolina [21,043 kJ $m^{-2}yr^{-1}$ (505 $gCm^{-2}yr^{-1}$)] which correspond to the areas with consistently high chlorophyll *a* concentrations (O'Reilly and Zetlin 1998).

Riley's (1941) work on Georges Bank produced estimates of net production lower than the values presented here. His results, however, were based on data collected using the dissolved-oxygen method with surface water, and therefore would be expected to yield lower estimates than the ^{14}C method. The areas of the shelf with the lowest estimated annual production include the outer shelf area between Cape Hatteras, the southern edge of Georges Bank and near-shore Gulf of Maine, and the mid-shelf area between Delaware Bay and Chesapeake Bay. (Figure 5-3).

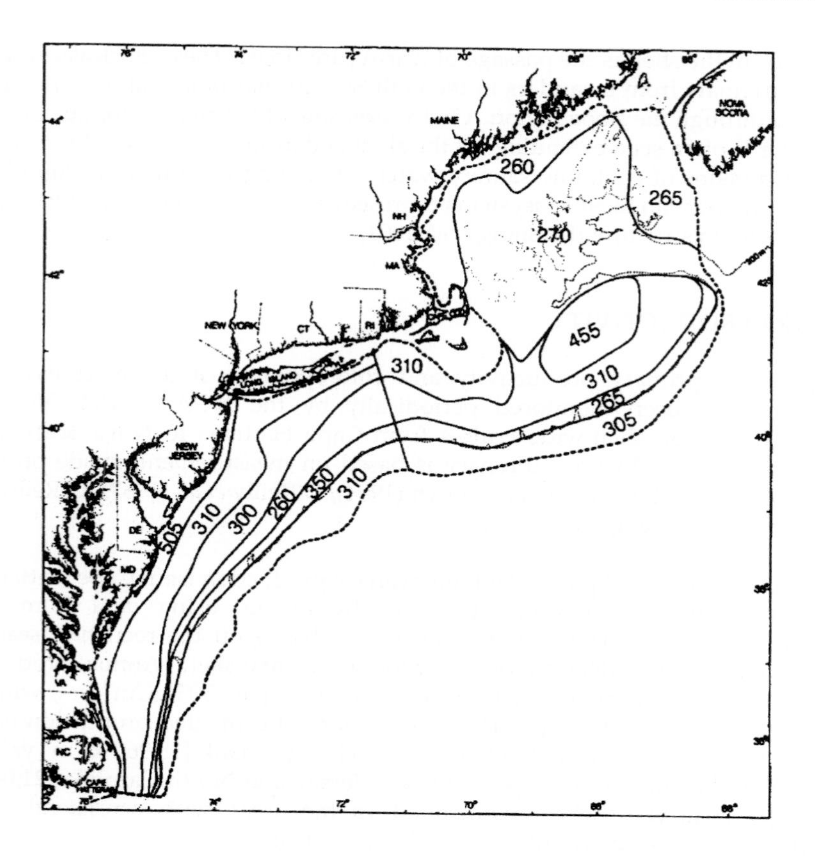

Figure 5- 3. Estimated annual primary production in the Northeast Shelf ecosystem

NUTRIENTS

Georges Bank

The annual cycle of nitrate concentrations on Georges Bank has a peak in February, with concentrations declining in spring due to utilization by phytoplankton, then falling to a low in summer (Pastuszak *et al.* 1981). As autumn begins, concentrations increase and continue to do so until a winter maximum is again reached. Water on both sides of Georges Bank at and below 150 m contains high nutrient levels due to the influx of slope water into the Gulf of Maine (Ramp and

Vermersch 1978). It has been suggested by Sigaev (1978) from analysis of a ten-year data base, and by Schlitz and Wright (1980), that upwelling of the deeper water occurs. Schlitz and Cohen (1984) have attributed about 40 percent of the high nutrient concentrations on the northern side of the Bank to continuous upwelling along the northern side, transport of upwelled water east of Nantucket and Cape Cod, and to deep winter mixing. Most of the remainder may be due to regeneration of nutrients in the water column, since recycling from sediments has been estimated by Thomas *et al.* (1978) as only 6.9 percent of the total regeneration in March and 2.5 percent in July.

During February, nitrate is high at the surface on southeastern Georges Bank (9 μg-at 1^{-1}) and in the vicinity of Nantucket Shoals (10 μg-at 1^{-1}). In spring, nitrate is very low on top of the Bank, but increases to the north and south (Cohen and Wright 1978). During April and May, generally all nutrient concentrations are lower than during February. Nitrate ranges from 0.5 to 4 μg-at 1^{-1} in the near-surface water and from 2 to 4 μg-at 1^{-1} at 50 m. Phosphate and silicate are low (0.1-0.3 μg-at 1^{-1}) to a depth of 100 m (Pastuszak *et al.* 1981). In September and October, concentrations are higher than in summer, but are still low (0-1 μg-at 1^{-1}) in the shallow portion of the Bank. High concentrations occur in the eastern portion (3-10 μg-at 1^{-1} nitrate and silicate) and at 30 to 50 m depth on the northern and southern flanks (1-11 μg-at 1^{-1}). From November to January nutrient concentrations begin to increase, with high concentrations in deeper waters to the north and the south. Maximum concentrations are reached during January and February.

Middle Atlantic Area

In the Middle Atlantic area in March, Walsh et al. (1978) measured more than 5 μg-at 1^{-1} nitrate in well-mixed waters off New York. By summer when stratified conditions exist in the water column, nitrate diminishes to undetectable concentrations in the euphotic zone. Walsh *et al.* (1978) suggested that nutrients are limiting to the phytoplankton in the Middle Atlantic area during summer, because the nitrogen must be derived from recycled sources including benthos, zooplankton, and bacterioplankton. Forty percent of the nitrogen demand for production of phytoplankton is thought to come from regeneration mechanisms, even though there is evidence of cross-shelf advective input of nutrients during spring and summer when upwelling winds are present [see Walsh (1981) for a more complete discussion of pelagic-demersal coupling, and longshore cross-shelf patterns from the Middle Atlantic Bight to Georges Bank].

Eppley (1980) and O'Reilly and Busch (1984), as well as Walsh (1981), imply that regeneration plays an important role in providing nitrogen in shelf ecosystems. During March, there are strong inshore-offshore gradients in surface waters, then in April the water column becomes well mixed, with ammonium nitrogen isopleths generally vertical. Concentrations decrease from 4.8 μmol 1^{-1} in the Raritan-Hudson estuary to 0.4 μmol l^{-1} at mid-shelf, with low levels (0.2 μmol l^{-1}) in deep water. As succession of the phytoplankton from spring to summer communities occurs, concentrations of ammonium nitrogen continue to decrease. In May, surface concentrations are low and reflect vertical stratification, with highest concentrations at 40 to 80 m in water about 100 to 140 km from the coast. In June, concentrations are similar to those in May, and highest near the seabed at the inshore apex area (4-9 μmol l^{-1}); however, by October there are again high levels in the surface waters near the estuary (Waldhauer *et al.* 1981).

ZOOPLANKTON

The zooplankton of the Northeast Shelf Ecosystem has been examined for seasonal trends in abundance and species composition since the late 1970s (Sherman *et al.* 1983,1988; Kane 1993, Jossi and Goulet 1993), and during four earlier periods. The first period is distinguished by the classic measurements of zooplankton volumes and species made by Bigelow during the second decade of the century (1912 through 1920). The second period includes the volume measurements made by Bigelow and Sears from the late 1920s through the 1940s. The third stage covers the late 1930s to the 1960s, with the biomass and species demographic studies of Fish (1936a,b), Clarke and Zinn (1937), Clarke (1940), Redfield (1941), Clarke *et al.* (1943), Riley and Bumpus (1946), Deevey (1952, 1956, 1960a,b) and Grice and Hart (1962). The fourth period includes the more contemporary measurements of biomass variability (Sherman *et al.* 2002, Judkins *et al.* 1980), and the study of Durbin and Durbin (1996) who examined the regulatory processes of the Northeast Shelf zooplankton (e.g. temperature, food, and predation).

During the decade 1977 to 1987, samples were collected by the NOAA-NMFS, Northeast Fisheries Science Center's Marine Resources, Monitoring, Assessment and Prediction (MARMAP) Program, using 0.333 mm mesh bongo nets towed obliquely through the water column to a maximum depth of 200 m. Gear was towed at 1.5 – 2.0 knots for at least 5 minutes at 200 sampling stations in a grid network at 35 km spacing within the entire ecosystem, six or more times per year (Sibunka and Silverman 1984,1989). Since 1987, sampling with bongo nets has been continual at 6x/year at approximately 120 locations over the spatial extent of the ecosystem through 2002. In addition to point-samplings with nets, the Northeast Fisheries Science Center (NEFSC) staff has completed monthly Continuous Plankton Recorder (CPR) transects across the Gulf of Maine from 1961 through 2002. This transect series has been augmented by a second transect from New York to Bermuda, across the Mid-Atlantic Bight of the Northeast Shelf Ecosystem, from 1976 to 2002.

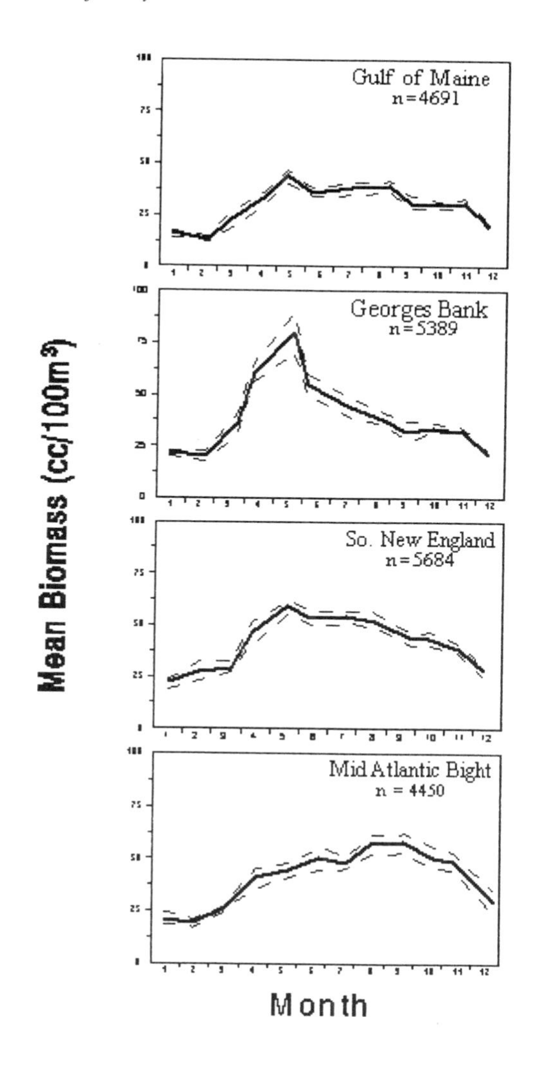

Figure 5-4. The annual cycle of zooplankton biomass in the four subareas of the Northeast Shelf ecosystem. The solid line is the time series monthly mean sample displacement volume and the dashed lines represent the 95% confidence interval.

ZOOPLANKTON DIVERSITY AND BIOMASS

The zooplankton biodiversity during the MARMAP surveys of the shelf included 394 taxa, with 50 dominant in at least one location in one or more seasons, including copepods, chaetognaths, barnacle larvae, cladocerans, appendicularia, doliolids, brachyuran larvae, echinoderm larvae, and thaliaceans (Sherman *et al.* 1988). The pattern of zooplankton biomass is based on the examination of 20,214 sample displacement volumes. The mean annual values of volumetric displacement are characterized by trends depicting annual cycles that differ among the four sub-areas (Figure 5-4).

In the Gulf of Maine, biomass peaks during spring (44 cc/100 m^3) and remains high through the summer (36-39 cc/100 m^3). The biomass declines in autumn (September) to a winter low (January-February). On Georges Bank, the spring increase in biomass peaks in May at a level that is nearly twice the spring peak in the Gulf of Maine, followed by a decline that continues through autumn to a winter minimum (< 20.2 cc/100 m^3). The waters of Southern New England maintain a relatively high biomass from May through August (55-60 cc/100 m^3). The annual decline in biomass extends from late August through autumn to a winter minimum. Further south in the Mid-Atlantic Bight, the annual peak is not reached until late August and September (60cc/100 m^3) followed by a decline from November until the annual minimum in February (19 cc/100 m^3).

Temporal observations of zooplankton volumes on the shelf, reported by Kane (1993) for the Georges Bank area of the Shelf Ecosystem and Jossi and Goulet (1993) for the Gulf of Maine and the Mid-Atlantic Bight, reveal inter-annual variability in zooplankton abundance. Examination of Continuous Plankton Recorder (CPR) data for the transect across the Mid-Atlantic Bight from 1976 through 1990 revealed a declining trend for 1977 through 1986 ($p=0.026$). However, the inter-annual variability observed from net tows for the entire time-series was not statistically significant as a declining trend ($p=0.30$). The abundance of zooplankton on the northern CPR transect across the Gulf of Maine generally increased in abundance from 1961 through 1990 ($p=0.054$), interrupted by a downward trend during the MARMAP decade ($p=0.06$). Based on MARMAP net sampling for Georges Bank, Kane reported volumes higher than the 10-year median biomass for 1977, 1978, and 1979, and lower for 1982, 1983, and 1984. According to Kane, the higher volumes may have contributed to the increases in sand lance, *Ammodytes* spp., during the late 1970s. In contrast, Payne *et al.* (1990) reported an inverse relationship between sand lance and the copepod, *Calanus finmarchicus* during the population explosion of sand lance, suggesting density-dependent grazing control of the standing stock of zooplankton. This observation, however, was limited to a relatively restricted area of the Northeast Shelf Ecosystem known as the Stellwagen Bank area (Payne *et al.* 1990). On Georges Bank where sand lance, herring, and mackerel are present in relatively high numbers in spring, zooplankton biomass in spring (measured as displacement volume per 100 m^3) was high during the population explosion of sand lance, 1971-1981, with median annual volumes during six of these years equal to or exceeding the 30-year (1972 through 2001), long-term median value of 43 cc/100 m^3 during six of these years (Figure 5-5). Furthermore, contemporary levels of chlorophyll *a* in northeast surface waters scanned by the SeaWiFS satellite ocean color sensor (1998-2002) are comparable to historical estimates based on CZCS data (1979-1985) reprocessed with modern algorithms (Greg *et al.* 2002), and with historical *in situ* measurements (O'Reilly and Zetlin 1998).

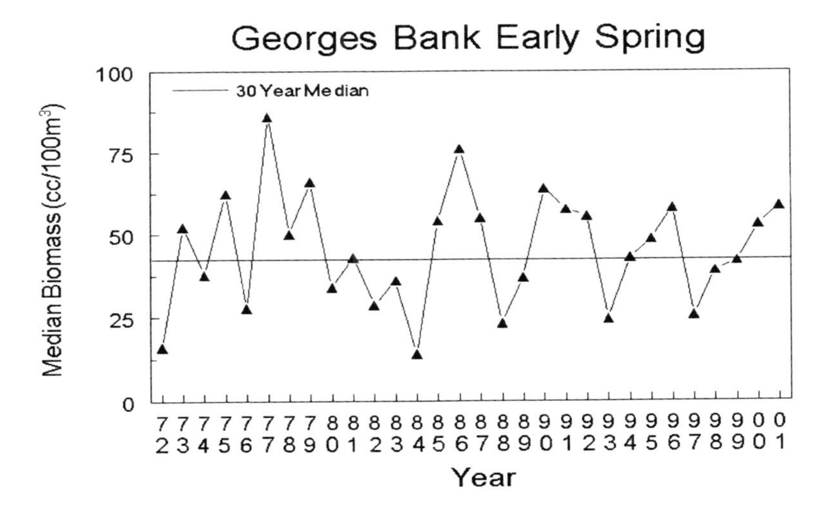

Figure 5-5. 30-year median zooplankton biomass on Georges Bank

This is illustrated in Figure 5-6. Both periods exhibit some annual variability in the mean surface chlorophyll, however, it is noteworthy that the lowest median chlorophyll estimate and the lowest zooplankton biomass volume were observed in 1983 (Figure 5-7). Chlorophyll and zooplankton values are among the highest in the time series from the late 1990s to 2001. Unfortunately phytoplankton measurements were not made by Northeast Fisheries Service Center from 1985 through 1998.

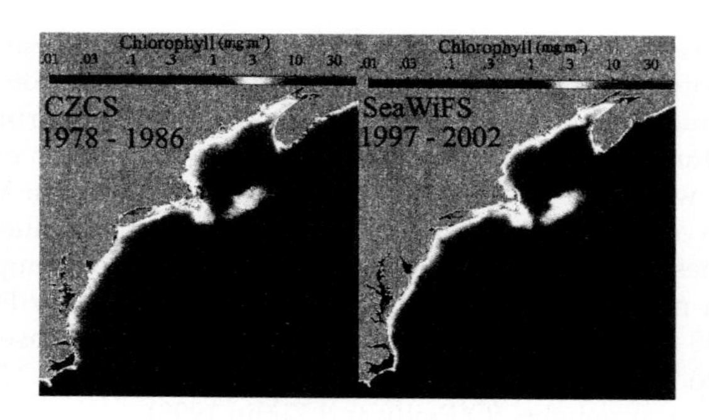

Figure 5-6. Comparison of surface chlorophyll values based on the Coastal Zone Color scanner sensing data for 1978-1986 and SeaWiFS sensing data for 1997 to 2002.

The MARMAP zooplankton time-series was examined for shifts in species composition using the annual time-series of log abundance of five zooplankton species—*Calanus finmarchicus, Pseudocalanus* spp., *Centropages typicus, Metridia lucens*, and *Centropages hamatus*—for four regions: Georges Bank, Gulf of Maine, Mid-Atlantic Bight, and Southern New England, for the period 1977-1987. Annual time-series were constructed by averaging all samples within a year and also by taking the maximum sample within a year. In terms of temporal behavior, there were no real differences between these two approaches, so the average values were analyzed. The methods described by Solow (1994) were used to extract an overall trend from the five species for each of the four regions. There was no significant trend for the Mid-Atlantic Bight (p=0.785) or Southern New England (p = 0.128).

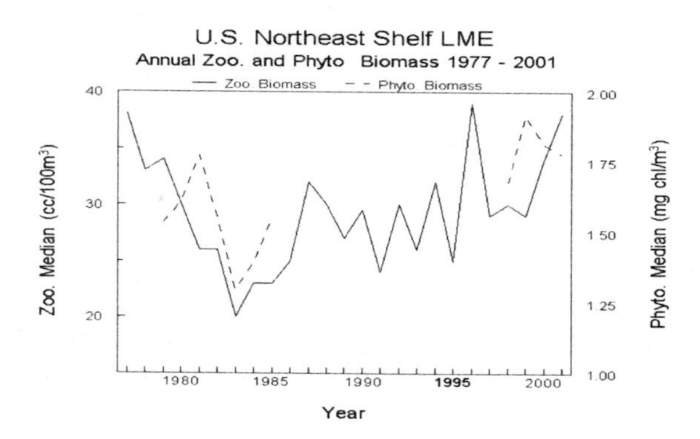

Figure 5-7. Annual zooplankton and phytoplankton biomass of the Northeast Shelf ecosystem, 1977-2001.

There were significant trends for Georges Bank ($p = 0.031$) and the Gulf of Maine ($p = 0.014$). The correlations between these trends and the five species are given in Table 5-1. From an examination of biodiversity, it appears that the dominant copepod community may be undergoing a shift toward an increasing abundance of *Centropages hamatus* on Georges Bank and in the Gulf of Maine. Furthermore, the zooplankton biomass of the ecosystem has been steadily rising since 1983 to levels at or above the long-term median from 1996 to 2001 (Figure 5-5).

Table 5-1. Correlations between the five copepod species and the time trends found by the minimum/maximum autocorrelation factor analysis for Georges Bank and the Gulf of Maine subareas of the Northeast Shelf ecosystem

Species	Georges Bank	Gulf of Maine
Calanus finmarchicus	-0.32	-0.64
Pseudocalanus spp.	-0.82	-0.54
Centropages typicus	-0.44	-0.35
Metridia lucens	-0.69	-0.77
Centropages hamatus	0.70	0.76

Source: A.R. Solow. 1994. Detecting change in the composition of a multispecies community. Biometrics 50:556-565

FISH AND FISHERIES

From the mid-1960s through the 1990s the biomass of the principal groundfish species within the ecosystem declined significantly from overfishing of the spawning stock biomass (Murawski *et al.* 1999). The decline was measured in both a trend in reduced commercial landings and a parallel decline in catches (kg/ tow) from the fisheries independent bottom trawl surveys of the Northeast Shelf ecosystem conducted by the Northeast Fisheries Science Center of NOAA's National Marine Fisheries Service (Figure 5-8).

The declining trends in catch and spawning stock biomass of the demersal fish stocks are the subject of considerable effort to "turn the corner" from overfishing to a rebuilding campaign based on significant reductions in fishing effort. Four areas representing 5000 NM of former principal fishing grounds were closed in 1994 to fishing gears capable of catching groundfish. In addition, in 1994, the days-at-sea time allowed for commercial fishery operations was reduced to 50 percent of pre-1994 levels. New regulations were also implemented to increase minimum net mesh sizes, to initiate a moratorium on new vessel entrants, and to mandate vessel and dealer reporting of catches. New regulation limited catches of species designated as "depleted," and total allowable catch levels (TACs) were implemented. To further reduce fishing effort on the depleted groundfish species, the U.S. Congress approved a plan for "buying-out" 79 groundfishing vessels from fishing license and vessel owners. These management actions have resulted in significant reductions in exploitation rates and a positive biological response in increasing spawning stock biomass (ssb) and in recruitment levels of haddock and yellowtail flounder (Figures 5-9a and 5-9b). Since the 1994 reductions in exploitation rate, the passage of the Sustainable Fisheries Act of 1996 has placed additional requirements for rebuilding the depleted groundfish stocks of the Northeast Shelf ecosystem including cod, hake, pollock and several flounder species.

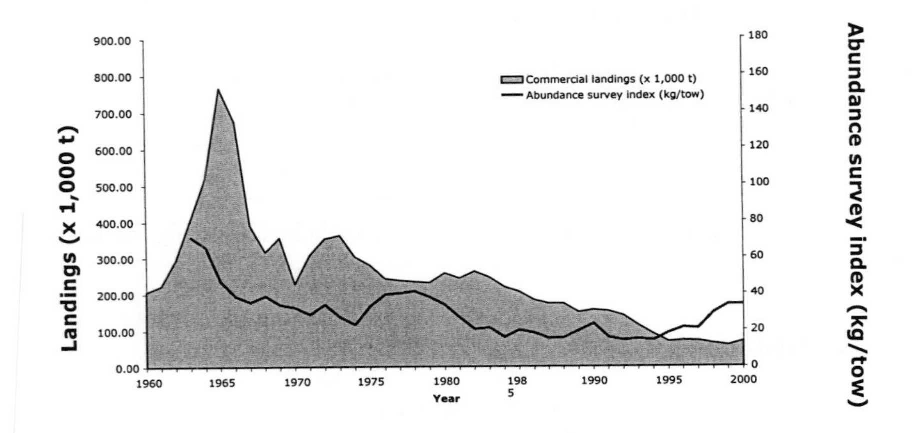

Figure 5-8: Landings in metric tons (t) and abundance index of principal groundfish and flounders, 1960-2000

In addition to the recovery of groundfish stocks, good progress has been made in the recovery of the large pelagic fish biomass of Atlantic herring and mackerel stocks inhabiting the Northeast Shelf ecosystem. Prior to 1967, both species were the targets of very heavy fishing mortality by European factory class vessels operating within the boundary of the ecosystem. Following the passage of the Magnusson Fishery Conservation Management Act of 1976, foreign vessels were excluded from fishing off the northern coast of the United States with the exception of several joint venture operations with US fishing interests. Since 1976, in the absence of any heavy market demand in the United States for either mackerel or herring as a table-food-fish, the stocks of both species have recovered from a state of depletion to robust spawning stock levels constituting an estimated combined biomass of 5.5 million metric tons.

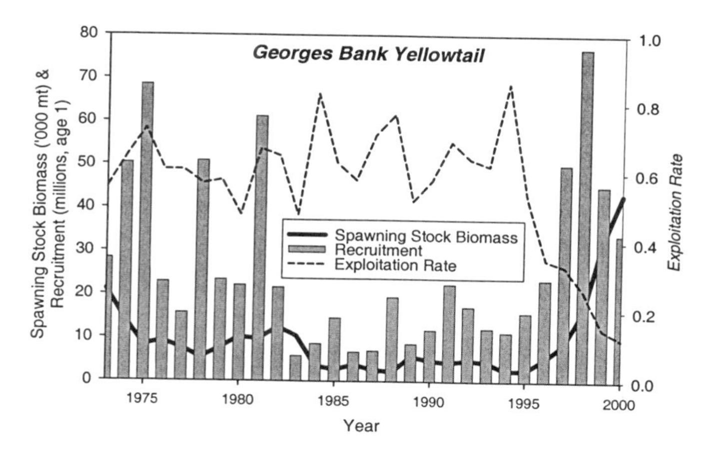

Figure 5-9a. Late 1990s George's Bank yellowtail recovery trend in spawning stock biomass (ssb) and recruitment in relation to reductions in exploitation rate

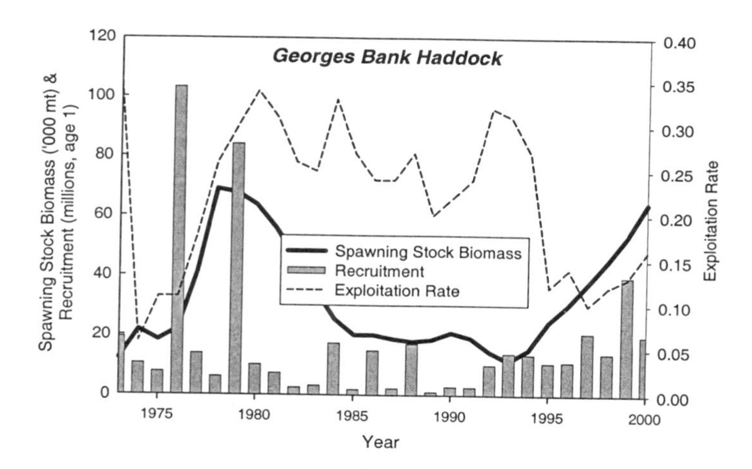

Figure 5-9b. Late 1990s George's Bank haddock recovery trend in spawning stock biomass (ssb) and recruitment in relation to reductions in exploitation rate

POLLUTION AND ECOSYSTEM HEALTH

River basin drainage into the NE Shelf ecosystem is undergoing some degree of improvement in the removal of toxics, including heavy metals (Windom 1996). Most of the wastewater plants discharging into the Northeast U.S. Shelf ecosystem are providing secondary treatment, and recent regulations have begun to address overflows from combined sewage/stormwater discharge systems that introduce inadequately treated wastewater into the ecosystem. Large-scale biological consequences from existing levels of toxics were unlikely. Levels of lead and DDT have declined. Elevated levels of toxics are limited to the vicinities of large urban outfalls, with a few relatively small areas that can be described as biological "hot-spots" (O'Connor 1996). Evidence of neoplasms in fish and shellfish has been found, and periodic advisories have been issued to limit the human consumption of striped bass, bluefish, and blue crabs because of potential health risks from chlorinated hydrocarbons that have been detected in tissues of these species.

Diagnostics are improving, and states continue to issue advisories on red tides if they present any risk to human health. Also, research is underway to improve surveys of pollution effects on individual organisms and populations (White and Robertson 1996). With regard to eutrophication, Malone and Conley (1996) indicate, based on case history studies in the Hudson River and Chesapeake Bay systems, that in the former there is no consistent pattern of any surplus production attributable to measured increased eutrophication. In the offing of the Hudson, in some years, it is possible to recognize the effects of eutrophication in the discharge plume that contributed periodically to anoxia, significant benthic mortalities and economic losses of shellfish in the 1970s and 1980s. It has been demonstrated by Jaworski and Howarth (1996) that near-coastal waters have been subjected to increased eutrophication. Phytoplankton blooms in the Chesapeake seem to vary in extent and intensity, but at the present time, no evidence of encroachment into the relatively open waters of the Mid-Atlantic Bight sub-area of the Northeast U.S. Shelf ecosystem is apparent.

Within the NE Shelf ecosystem, there is evidence of increasing frequency of biotoxin-related mortality of marine mammals, along with recent closures of offshore shellfish beds because of increased levels of paralytic shellfish poisoning (White 1996). Heavy metal concentrations in demersal fish, crustaceans, and bivalve mollusks continue to be monitored as biological indicators (Schwartz *et al.* 1996) in relation to development of strategies for mitigation of pollution stress on the ecosystem. General trends of pollutant loading for heavy metals appear to be decreasing, while the introduction of excessive nutrient loadings is increasing (Jaworski and Howarth 1996).

Coastal sites used to dump sludge from New York City waste treatment plants have shown signs of repopulation and improved water quality (Ingham 1996) following the cessation of sludge dumping on the continental shelf, indicating that mitigating actions can lead to a measurable degree of recovery of stressed benthic communities. Sites contaminated with PCBs, however, do not show rapid recovery following mitigation actions to eliminate the disposal of PCBs. Residence times of PCBs in sediments of New Bedford Harbor are quite long, and adverse effects on shellfish species continue to be detected several years after banning the disposal of PCBs in coastal waters (Capuzzo 1996).

Human intervention significantly impacts the NE Shelf ecosystem. Harmful blooms of planktonic algae have been increasing (Sherman 2000). Several algal species have been implicated in oxygen depletions leading to shellfish mortalities. Other algal species have been responsible for massive die-offs of bay scallops and other mollusks in embayments along the southern New England coast (Epstein 1996). Whether the algal blooms are the result of excessive nitrates and phosphates entering coastal waters from wastewater is a question that is being investigated by state and federal scientists.

ECOSYSTEM LEVEL RESPONSE TO PELAGIC PERTURBATIONS

Examination of the temperature information for the Northeast Shelf Ecosystem during the last two decades indicates that the temperatures from 1978 through 1982, a period of population depression for the principal pelagics, was intermediate between a slightly warmer period of the mid-1980s (1983 to 1986) and moderately cooler waters of the mid-1960s (Holzwarth and Mountain 1992). The influence of temperature on the survival of early developmental stages of sand lance, mackerel, and herring is not well understood and requires further study. The positive and negative correlations between temperatures and herring and mackerel dominance reported by Skud (1982) are not evident in more recent studies. Both species show population declines and increases attributed to excessive fishing effort in the 1970s (Anthony 1993, Murawski 1991). Excessive fishing mortality has been significantly reduced for the principal pelagic species, and that reduction is considered the principal factor in their recovery (Murawski 1991). The lowest period of larval herring abundance from 1971 through 1992 occurred from 1976 to 1984, spanning the intermediate and warm temperature periods. This is in contrast to the warming and cooling correlation and analysis of changes in herring distribution reported by Skud (1982), and lends support to the predominant influence of excessive fishing mortality on the decline of both herring and mackerel abundance. The recovery trend for haddock and yellowtail flounder is a direct response to the reduction in fishing effort on these species (Fogarty and Murawski 1999).

Based on the analysis of zooplankton volume measurements, it was found that the zooplankton decline at Georges Bank was more variable than for the other three sub-areas ($p = 0.10$). The persistent downward trend in the late 1970s and early 1980s coincided with the increase in abundance level of the principal pelagics and the increase in biomass of another pelagic zooplanktivore, the butterfish (*Peprilus triacanthus*). During the years 1977, 1978, and 1979, the highest numbers of zooplankters, based on CPR sampling, coincided with the first three years of the seven-year depressed state of the herring and mackerel biomass. However, during the decade of the 1990s the zooplankton biomass of the ecosystem was above the annual median level of the 25-year monitoring time-series (Figure 5-5). From a bottom-up examination of the ecosystem, the zooplankton component has been sufficiently robust to sustain the recovery of both herring and mackerel to a level of 5.5 mmt and support the initiation of haddock and yellowtail flounder recovery. In addition to this biomass, a robust zooplankton prey field is needed to support the recovery of the depressed gadoid and flounder stocks during their larval stages, posing another significant biomass to be considered in estimating the carrying capacity of the ecosystem. Whether the decline in zooplankton biomass of the late 1970s and early 1980s reflects a response to increasing predation by the growing biomass of the pelagic fish species, or a biofeedback response to an environmental signal remains an important unanswered question. However, the results of the recent zooplankton analysis based on 25 years of NE Shelf ecosystem volume data, when coupled with evidence of relatively high mean annual primary productivity of 350 gCm^2yr^{-1} suggests that the bottom of the carrying capacity of the food web for the fish component of the ecosystem remains robust.

The characteristics of multi-decadal variation in total zooplankton abundance appear to agree with the evolving biodiversity stability hypothesis, wherein individual species undergo variability in abundance, while allowing for greater stability in the multi-decadal levels of zooplankton biomass. In this regard, the species shifts and the inter-annual and decadal variability observed in zooplankton biomass appear to have allowed for sufficient residual sustainability in biodiversity and abundance to support the recovery of zooplanktiverous herring (Figure 5-10a) and mackerel (Figure 5-10b) stocks of the Northeast Shelf ecosystem from their very low levels in the mid-1970s to the unprecedented level of five and a half million metric tons in recent years. Assessments of the effects of the recent outburst of the herring and mackerel stocks on the biomass and biodiversity of zooplankton during the recovery of depleted haddock and yellowtail flounder stocks of the Northeast Shelf ecosystem are continuing.

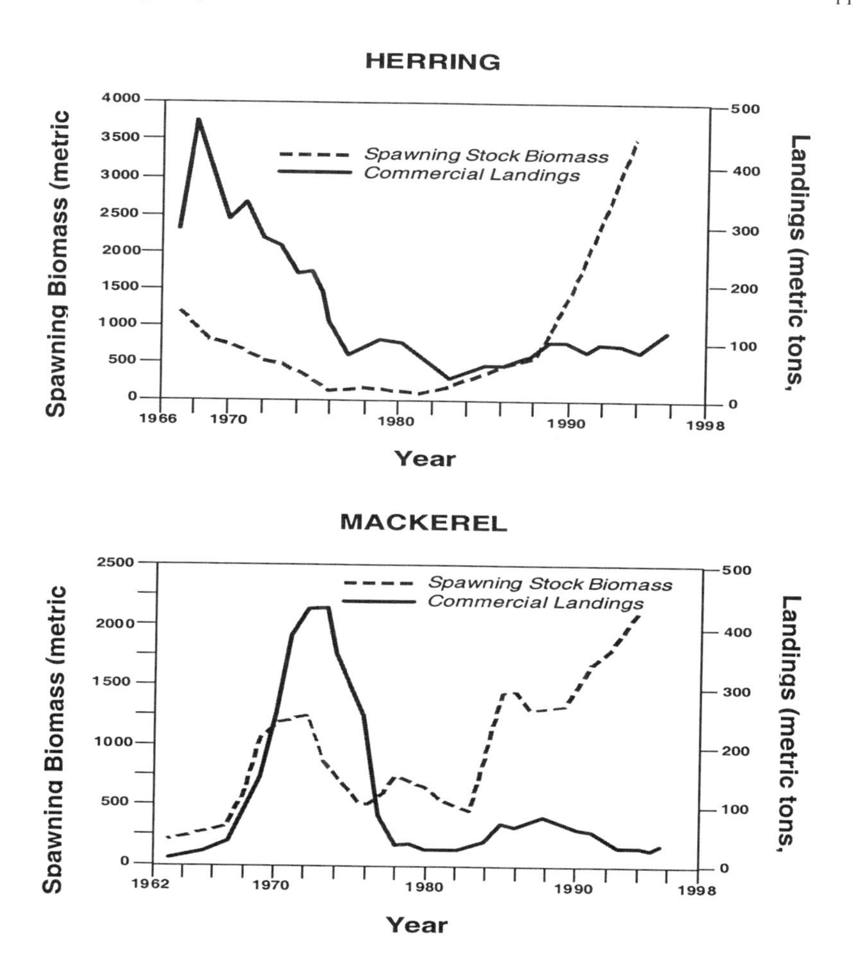

Figure 5-10. Top (a) - Atlantic herring commercial landings and spawning stock biomass, 1967 through 1996 (thousand metric tons). Bottom (b)- Atlantic mackerel landings and spawning stock biomass, 1963 through 1996 (thousand metric tons).

SUSTAINABILITY AND MANAGEMENT

A bottom up examination of key ecosystem indicators provides evidence of system resilience and robustness. The persistence in primary productivity at a level of an estimated 350 gCm^2yr^{-1} and a zooplankton biomass at or higher than the long term median value since the mid-1990s are important reference points for characterizing the base of the food web of the Northeast Shelf ecosystem as stable at a level of productivity sufficient to support the increasing levels of spawning stock biomass of both the pelagic and demersal fish components of the ecosystem. The temperature signal of the first two decades, while showing evidence of interannual variability, is not indicative of any long term persistent upward or downward trend indicative of an oceanographic regime shift of the kind described for the North Pacific (Brodeur and Ware 2002) or the Northeast Atlantic (Reid and Beaugrand 2002).

The Northeast Shelf ecosystem continues to be highly productive at the lower end of the food web. Model simulation based on biological and economic data indicates the need for further reduction in fishing effort if long-term sustainability of preferred high-demand and high-priced species is to be realized (Edwards and Murawski 1996). Options for implementing an ecosystem-based management regime is presently the topic of interest to marine managers who recognize the need for improving present management practices.

The legislative authority for proceeding toward ecosystem-based management for the Northeast Shelf ecosystem can be found inseveral Public Laws. Foremost among these are the Magnuson Fishery Conservation and Management Act (P.L. 94-265), the Marine Mammal Protection Act (P.L. 103-238), the National Environmental Policy Act (P.L. 91-190), the Federal Water Pollution Control Act (P.L. 92-500), and the National Coastal Monitoring Act (H.R. 2130 Title V 1992). Within each of these acts, the Federal Government is directed, as the nation's steward of natural resources, to collaborate closely with the coastal states in moving forward to implement more coherent and integrated ecosystem-based management for the coastal environment and fisheries within the ecosystem. This theme has been explored by the U.S. National Research Council and found to be both feasible and desirable (NRC 1990, 1999).

For the important fisheries supported by the ecosystem, the reductions of fishing effort as part of the rebuilding program for depleted groundfish stocks has proven successful. The actions taken to recover depleted fish stocks are consistent with two of the targets declared by the governments participating in the World Summit on Sustainable Development (WSSD) held in Johannesburg in 2002. They established 2010 as the target year for countries to introduce ecosystem based assessment and management practices, and agreed to move toward the recovery of the world's depleted fish stocks to maximum sustainable levels by 2015. It is clear from the actions taken by the fisheries management councils (New England and Mid Atlantic councils) responsible for the fish and fisheries of the Northeast Shelf ecosystem and the Secretary of Commerce on the advice of the National Marine Fisheries Service, that significant progress is being made consistent with the WSSD targets.

Further actions required to mitigate stress on the ecosystem should be based on: (1) quantitative science, (2) consensus of user groups, and (3) a framework for governance. Substantial progress has already been made in addressing two of these issues: (1) Federal agencies in concert with Regional Fisheries Management Councils, have acknowledged their stewardship responsibilities and initiated action to mitigate stress on ecosystem components; (2) an extensive information base on the resources of the ecosystem is available to user groups on electronic networks. Considerable scientific effort by Federal and State governments is funded and underway in the region, along with pertinent research activities in academia and the private sector. A major unfinished task is the development of a governance mechanism for ensuring that future management actions are planned and executed with the participation and concurrence of the principal constituent groups throughout the area encompassed by the ecosystem. The various jurisdictions responsible for the management of the NE Shelf ecosystem's rich fisheries, biodiversity, productivity, nutrient inputs and other capital assets will need to develop more effective means for governance of the ecosystem at all government levels, including municipal. State, regional, and national, if the large populations living in urban centers along the Northeast coast are to realize the long-term sustainable benefits to be derived from the goods and services of one of the most productive large marine ecosystems in the world.

REFERENCES

Anthony, V.C. 1993. The state of groundfish resources off the northeastern United States. Fisheries 18(3):12-17

Backus, R.H., ed. 1987. Georges Bank. MIT Press. Cambridge MA.

Beardsley, R.C., W.C. Boicourt, and D.V. Hansen. 1976. Physical oceanography of the Middle Atlantic Bight. In: M. Grant Gross, ed. Middle Atlantic continental shelf and the New York Bight. Limnol. Oceanogr. Special Symposia 2:20-34

Bigelow, H.B. 1926. Plankton of the offshore waters of the Gulf of Maine. Bull. U.S. Bur. Fish. 40 (Part II):1-509

Bigelow, H.B. 1927. Physical oceanography of the Gulf of Maine. Bull.U.S. Bur. Fish. 40(Part II):511-1027

Bigelow, H.B. 1933. Studies of the waters of the continental shelf, Cape Cod to Chesapeake Bay. I: The cycle of temperatures. Pap. Phys. Oceanogr. 2(4). 135p

Bigelow, H.B. and M. Sears. 1935. Studies of the waters of the continental shelf, Cape Cod to Chesapeake Bay. II: Salinity. Pap. Phys. Oceanogr. Meteorol. 4(1):1-94

Bigelow, H.B., L.C. Lillick and M. Sears. 1940. Phytoplankton and planktonic protozoa of the offshore waters of the Gulf of Maine. I. Numerical distribution. Trans. Am. Philos. Soc., N.S. 31 (Part III):149-191

Brodeur, R.D. and D.M. Ware. 1992. Long-term variability in zooplankton biomass in the subarctic Pacific Ocean. Fish. Oceanogr. 1:32-38.

Brooks, D.A. 1996. Physical oceanography of the shelf and slope seas from Cape Hatteras to Georges Bank: A brief overview. In: Sherman K., N.A. Jaworski and T.J. Smayda, eds. The Northeast Shelf Ecosystem: Assessment, Sustainability and Management. Blackwell Science, Inc. Cambridge, MA. 47-74

Bumpus, D.F. 1969. Reversal in the surface drift in the Middle Atlantic Bight area. Deep-Sea Res. 16 (Suppl.):17-23

Bumpus, D.F. 1973. A description of the circulation on the continental shelf on the east coast of the United States. Progr. Oceanogr. 6:111-157

Bumpus, D.F. and L. Lauzier. 1965. Surface circulation on the continental shelf off eastern North America between Newfoundland and Florida. Serial Atlas of the Marine Environment, Folio F. Am. Geogr. Soc.

Bush, KL. 1981. Middle Atlantic Bight transports determined from a salinity-heat box model using historical hydrographic and meteorological data. Thesis, University of Delaware, Newark, DE. 247p

Capuzzo, J.E.McD. 1996. Biological effects of contaminants on shellfish populations in coastal habitats: A case history of New Bedford, Massachusetts. In: K. Sherman, N.A. Jaworski, and T.J. Smayda, eds. The Northeast Shelf Ecosystem: Assessment, Sustainability, and Management. Blackwell Science, Inc. Cambridge MA. 457-466

Clarke, G.L. 1940. Comparative richness of zooplankton in coastal and offshore areas of the Atlantic. Biol. Bull. 78:226-255

Clarke, G.L. and D.J. Zinn. 1937. Seasonal production of zooplankton off Woods Hole with special reference to *Calanus finmarchicus*. Biol. Bull. Woods Hole 76:371-383

Clarke, G.L., E.L. Pierce and D.F. Bumpus. 1943. The distribution and reproduction of *Sagitta elegans* on Georges Bank in relation to hydrographical conditions. Biol. Bull. Woods Hole 85:201-226

Cohen, E.B. 1976. An overview of the plankton communities of the Gulf of Maine. Int. Comm. Northwest Atl. Fish., Selected Pap. 1:89-105

Cohen, E.B. and W.R. Wright. 1978. Changes in the plankton on Georges Bank in relation to the physical and chemical environment during 1975-76. Int. Counc. Explor. Sea. C.M.1978/L:27

Deevey, G.B. 1952. Quantity and composition of the zooplankton of Block Island Sound. 1949. Bull. Bingham Oceanogr. Collect. Yale Univ. 13:120-164

Deevey, G.B. 1956. Oceanography of Long Island Sound, 1952-1954. V. Zooplankton. Bull. Bingham Oceanogr. Collect. Yale Univ. 15:113-155

Deevey, G.B. 1960a. The zooplankton of the surface waters of the Delaware Bay region. Bull. Bingham Oceanogr. Collect. Yale Univ. 17(Article 2). 5-53

Deevey, G.B. 1960b. Relative effects of temperature and food on seasonal variations in length of marine copepods in some Eastern American and Western European waters. Bull. Bingham Oceanogr. Collect. Yale Univ. 17(Article 2:54-86)

Durbin, E.G. and A.G. Durbin. 1996. Zooplankton dynamics in the Northeast Shelf ecosystem. In: K. Sherman, N.A. Jaworski, and T.J. Smayda, eds. The Northeast Shelf Ecosystem: Assessment, Sustainability, and Management. Blackwell Science, Inc. Cambridge, MA. 129-152

Edwards, S.F. and S.A. Murawski. 1996. Potential benefits from efficient harvest of New England groundfish. In: K. Sherman, N.A. Jaworski, and T.J. Smayda, eds. The Northeast Shelf Ecosystem: Assessment, Sustainability, and Management. Blackwell Science, Inc. Cambridge, MA. 511-526

Emery, K.O. and E. Uchupi. 1972. Western North Atlantic Ocean topography, rocks, structure, water life and sediments. Am. Assoc. Pet. Geol. Mem. 17:1-532

Eppley, R.W. 1980. Estimated phytoplankton growth rates in the central oligotrophic oceans. In: P. Falkowski, ed. Primary Productivity in the Sea. Plenum Press, New York. 231-242

Epstein, P.R. 1996. Emergent stressors and public health implications in large marine ecosystems: an overview. In: K. Sherman, N.A. Jaworski, and T.J. Smayda, eds. The Northeast Shelf Ecosystem: Assessment, Sustainability, and Management. Blackwell Science, Inc., Cambridge, MA. 417-438

Fish, C.J. 1936a. The biology of *Calanus finmarchicus* in the Gulf of Maine and Bay of Fundy./ Biol. Bull. 70:118-141

Fish, C.J. 1936b. The biology of *Pseudocalanus minutus* in the Gulf of Maine and Bay of Fundy. Biol. Bull. 70:193-216

Fish, C.J. 1936c. The biology of *Oithona similis* in the Gulf of Maine and Bay of Fundy. Biol. Bull. 70:168-187

Fogarty, M.J. and S. Murawski. 1999. Large scale disturbance and the structure of marine systems: Fishery impacts on Georges Bank. Ecological Applications 8(1):S6-S-22.

Gregg, W.W., M.E. Conkright, J.E. O'Reilly, F.S. Patt, M. Wang, J. Yoder and N. Casey-McCabe. 2002. NOAA/NASA CZCS Re-analysis Effort. Applied Optics 41(9):1615-1628

Grice, G.D. and A.D. Hart. 1962. The abundance, seasonal occurrence and distribution of the epizooplankton between New York and Bermuda. Ecol. Monogr. 32:287-308

Holzwarth, T. and D. Mountain. 1992. Surface and bottom temperature distributions from the Northeast Fisheries Center spring and fall bottom trawl survey program, 1963-1987; with addendum for 1988-1990. Northeast Fish. Sci. Center Ref. Doc. 90-03. 77p

Hoagland, P., D. Jin, E. Thunberg, and S. Steinback. 1999. Economic activity associated with the Northeast Shelf large marine ecosystem: application of an input-output approach. Marine Policy Center, Woods Hole Oceanographic Institution and Social Sciences Branch, NEFSC/NMFS, Woods Hole, MA 36p

Ingham, M.C. 1996. Effects of closure of a continental shelf dumpsite. In: K. Sherman, N.A. Jaworski, and T.J. Smayda, eds. The Northeast Shelf Ecosystem: Assessment, Sustainability, and Management. Blackwell Science, Inc. Cambridge MA. 441-456

Jaworski, N.A. and R. Howarth. 1996. Preliminary estimates of the pollutant loads and fluxes into the Northeast Shelf ecosystem: Assessment, Sustainability, and Management. Blackwell Science, Inc. Cambridge MA. 351-357.

Jossi, J.W. and J.R. Goulet Jr. 1993. Zooplankton trends: U.S. Northeast Shelf ecosystem and adjacent regions differ from northeast Atlantic and North Sea. ICES J. Mar. Sci. 50:303-313

Judkins, D.C., C.D. Wirick and W.E. Esaias. 1980. Composition, abundance and distribution of zooplankton in the New York Bight, September 1974-September 1975. Fish. Bull.U.S. 77:669-683

Kane, J. 1993. Variability of zooplankton biomass and dominant species abundance on Georges Bank, 1977-1986. Fish. Bull., U.S. 91:464-474

Loder, J.W., J.A. Shore, C.G. Hannah and B.D. Petrie. 2001. Decadal-scale hydrographic and circulation variability in the Scotia-Maine region. Deep-Sea Res. II 48, 3-35

Lynch, D.R., M.J. Holboke and C.E. Naimie. 1997. The Maine coastal current: spring climatological circulation. Cont. Shelf Res. 17(6):605-634

Malone, T.C. and D.J. Conley. 1996. Preliminary estimates of the pollutant loads and fluxes into the Northeast Shelf ecosystem. In: K. Sherman, N.A. Jaworski and T.J. Smayda, eds. The Northeast Shelf Ecosystem: Assessment, Sustainability, and Management. Blackwell Science, Inc. Cambridge, MA. 351-357

Murawski, S.A. 1991. Can we manage our multispecies fisheries? Fisheries 16(5):5-13

Murawski, S.A. *et al.* 1999. New England Groundfish. Our Living Oceans: Report on the Status of U.S. Living Marine Resources, 1999. U.S. Dep. Commer., NOAA Tech. Memo. NMFS-F/SPO-41. Washington DC.

National Research Council (NRC). 1990. Managing Troubled Waters: The Role of Marine Environmental Monitoring. National Academy Press. Washington, DC.

National Research Council (NRC). 1999. Sustaining Marine Fisheries. National Academy Press. Washington, DC.

New England Fishery Management Council. 1994. Amendments 5 and 6 to the New England Groundfish Management Plan, Peabody, MA.

NOAA. 1998. Status of Fishery Resources off the Northeastern United States for 1998. NOAA Technical Memorandum NMFS-NE-115. Woods Hole, MA: Northeast Fisheries Science Center. 149p

O'Connor, T.P. 1996. Coastal sediment contamination in the Northeast Shelf large marine ecosystem. In: K. Sherman, N.A. Jaworski, and T.J. Smayda, eds. The Northeast Shelf Ecosystem: Assessment, Sustainability, and Management. Blackwell Science, Inc. Cambridge MA. 239-257

O'Reilly, J.E. and C. Zetlin. 1998. Seasonal, horizontal, and vertical distribution of phytoplankton chlorophyll *a* in the northeast U.S. continental shelf ecosystem. U.S. Dept. Commerce, NOAA Tech. Rep. NMFS 139, Fishery Bulletin. 120p

O'Reilly, J.E. C. Evans-Zetlin and D.A. Busch. 1987. Primary Production. In: Georges Bank. R.H. Backus, ed. MIT Press, Cambridge, MA. 220-233

O'Reilly, J.E. and D.E. Busch. 1984. Phytoplankton primary production on the Northwest Atlantic shelf. Rapp. P.-V. Réun. Cons. Int. Explor. Mer 183:255-268

Overholtz, W.J., S.A. Murawski and K.L. Foster. 1991. Impact of predatory fish, marine mammals, and seabirds on the pelagic fish ecosystem of the northeastern USA. ICES Mar. Sci. Symp. 193:198-208

Pastuszak, M. W.R. Wright and L. Despres-Patanjo. 1981. One year of nutrient distribution in the Georges Bank region in relation to hydrography, 1975-1976. Int. Counc. Explor. Sea. C.M. 1981/C:7

Payne, P.M., D.N. Wiley, S.B. Young, S. Pittman, P.J. Clapham and J.W. Jossi. 1990. Recent fluctuations in the abundance of baleen whales in the southern Gulf of Maine in relation to changes in selected prey. Fish. Bull., U.S. 88:687-696

Pershing, A. J., C. H. Greene, *et al.* (2001). "Oceanographic responses to climate in the northwest Atlantic." Oceanography 14: 76-82

Pershing, A. J., C.H. Greene, C. Hannah, D.G. Mountain, D. Sameoto, E. Head, J. W. Jossi, M. C. Benfield, P. C. Reid, T. G. Durbin. 2001. Gulf of Maine/Western Scotian Shelf ecosystems respond to changes in ocean circulation associated with the North Atlantic Oscillation. *Oceanography*. 14:76-82

Ramp, S.R. and J.A. Vermersch Jr. 1978. Measurements of the deep currents in the Northeast Channel, Gulf of Maine. Int. Counc. Explor. Sea. C.M. 1978/C:40

Redfield, A.C. 1941. The effects of the circulation of water on the distribution of the calanoid community in the Gulf of Maine. Biol.Bull. 80:86-110

Redfield, A.C., H.P. Smith, B. Ketchum. 1937. The cycle of organic phosphorus in the Gulf of Maine. Biol.Bull. 73 (3):5421-443.

Reid, P.C.and Beaugrand, G. 2002. Interregional biological responses in the North Atlantic to hydrometeorological forcing. In Sherman, K.; Skjoldal, H.R. (Ed.): Large Marine Ecosystems of the North Atlantic: Changing states and sustainability. Large Marine Ecosystems Series. Elsevier: Amsterdam, The Netherlands. 27-49

Riley, G.A. 1941. Plankton studies. IV. Georges Bank. Bull. Bingham Oceanogr. Collect. Yale Univ. 7:1-73

Riley, G.A. 1947. Seasonal fluctuations of the phytoplankton population in New England coastal waters. J. Mar. Res. 6:114-125

Riley, G.A. and D.F. Bumpus. 1946. Phytoplankton-zooplankton relationships on Georges Bank. J. Mar. Res. 6:54-73

Schlitz, R.J. and E. Cohen. 1984. A nitrogen budget for the Gulf of Maine and Georges Bank. Biol. Oceanogr. 3:203-222

Schlitz, R.J. and W.R. Wright. 1980. Evidence of upwelling at the northern edge of Georges Bank. IDOE Int. Symposium on Coastal Upwelling (abstract)

Schwartz, J.P. N.M. Duston and C.A. Batdorf. 1996. Metal concentrations in winter flounder, American lobster, and bi-valve mollusks from Boston Harbor, Salem Harbor, and coastal Massachusetts: A summary of data on tissues collected from 1986 to 1991. In: K. Sherman, N.A. Jaworski and T.J. Smayda, eds. The Northeast Shelf Ecosystem: Assessment, Sustainability, and Management. Blackwell Science, Inc. Cambridge, MA. 285-312

Sherman, B. 2000. Marine ecosystem health as an expression of morbidity, mortality and disease events. Mar. Pol. Bull. 41(1-6):232-254

Sherman K., J. Kane, S. Murawski, W. Overholtz and A. Solow. 2002. The U.S. Northeast Shelf Large Marine Ecosystem: Zooplankton trends in fish biomass recovery. In: K. Sherman and H.R. Skjoldal, eds. Large Marine Ecosystems of the North Atlantic: Changing States and Sustainability. Elsevier Science.195-215

Sherman, K. and A.M. Duda. 1999. An ecosystem approach to global assessment and management of coastal waters. Mar. Ecol. Prog. Ser. 190:271-287

Sherman K., J.R. Green, A. Solow, S. Murawski, J. Kane, J. Jossi, and W. Smith. 1996a. Zooplankton prey field variability during collapse and recovery of pelagic fish in the Northeast Shelf ecosystem. In: K. Sherman, N.A. Jaworski and T.J. Smayda, eds. The Northeast Shelf Ecosystem: Assessment, Sustainability and Management. Blackwell Science, Inc. Cambridge MA. 217-236

Sherman, K., N.A. Jaworski and T.J. Smayda. 1996b. The Northeast Shelf Ecosystem: Assessment, Sustainability, and Management. Blackwell Science, Inc. Cambridge MA. 564p

Sherman, K., M. Grosslein, D. Mountain, D. Busch, J. O'Reilly and R. Theroux. 1988. The continental shelf ecosystem off the northeast coast of the United States.

In: H. Postma and J.J. Zilstra, eds. Ecosystems of the World 27: Continental Shelves. Elsevier, Amsterdam, The Netherlands. 279-337

Sherman, K., J.R. Green and L. Ejsymont. 1983. Coherence in zooplankton of a large northwest Atlantic ecosystem. Fish. Bull. U.S. 81:855-862

Sigaev, I.K. 1978. Intra-year variability of geostrophic circulation on the continental shelf off New England and Nova Scotia. Int. Comm. Northwest Atl. Fish., Selected Pap. 3:97-107

Skud, B.E. 1982. Dominance in fisheries: A relation between environment and abundance. Science 216:144-149.

Solow, A.R. 1994. Detecting change in the composition of amultispecies community. Biometrics 50:556-565

Studholme, A., J.E. O'Reilly and M. Ingham, eds. 1995. Effects of the Cessation of Sewage Sludge Dumping at the 12 Mile Site: Proceedings from the 12 mile dumpsite symposium, Long Branch, NJ. 18-19 June 1991. NOAA Technical Report TR 10056, Fishery Bulletin. 257p

Thomas, J.P., J.E. O'Reilly, C.N. Robertson and W.C. Phoel. 1978. Primary productivity and respiration over Georges Bank during March and July 1977. Int. Counc. Explor. Sea. C.M. 1978/L:37

Waldhauer, R., A. Matte, A.F.K. Draxler and J.E. O'Reilly. 1981. Seasonal ammonium-nitrogen distribution across the New York Bight shelf. In: L.L. Ciaccio and A.C. Christini, eds. Water Conference, Ramapo College of New Jersey, May 1-2, 1980. 274-286

Walsh, J.J. 1981. Shelf-sea ecosystems. In: A.R. Longhurst, ed. Analysis of Marine Ecosystems. Academic Press, London. 159-196

Walsh, J.J.,. T.E. Whitledge, F.W. Barvenik, C.O. Wirick, S.O. Howe, W.E. Esaias and J.T. Scott. 1978. Wind events and food chain dynamics within the New York Bight. Limnol. Oceanogr. 23:659-683

White, A.W. 1996. Biotoxins and the health of living marine resources of the Northeast Shelf ecosystem. In: K. Sherman, N.A. Jaworski, and T.J. Smayda, eds. The Northeast Shelf Ecosystem: Assessment, Sustainability, and Management. Blackwell Sciencve, Inc. Cambridge, MA. 405-416

White, H.H. and A. Robertson. 1996. Biological responses to toxic contaminants in the Northeast Shelf large marine ecosystem. In: K. Sherman, N.A. Jaworski and T.J. Smayda, eds. The Northeast Shelf Ecosystem: Assessment, Sustainability and Management. Blackwell Science, Inc. Cambridge, MA. 259-283

Windom, H.L. 1996. Riverine contributions to heavy metal inputs to the Northeast Shelf ecosystem. In: K. Sherman, N.A. Jaworski, and T.J. Smayda, eds. The Northeast Shelf Ecosystem: Assessment, Sustainability and Management. Blackwell Science, Inc. Cambridge, MA. 313-325

Wright, W.R. 1976. The limits of shelf water south of Cape Cod 1941-1972. J. Mar. Res. 34:1-14

Yoder, J.A., S.E. Shollaert, J.E. O'Reilly. 2002. Climatological phytoplankton chlorophyll and sea surface temperature patterns in continental shelf and slope waters off the northeast U.S. coast. Limnol.Oceanogr. 47(3):672-682.

Large Marine Ecosystems of the World
G. Hempel and K. Sherman (Editors)

6

The Yellow Sea LME and Mitigation Action

Qisheng Tang

A new era in ocean use was initiated when, in 1982, the United Nations Law of the Sea Convention established Exclusive Economic Zones (EEZ) extending up to 200 miles from the base lines of the territorial seas, and including almost 95 percent of the annual global biomass yields of usable marine living resources. Coastal states have sovereign rights to explore, manage, and conserve the marine resources of the zones. However, an EEZ is a relatively narrow zone, and water and economically important living marine resources exchange freely throughout a large marine ecosystem regardless of political boundaries. The results of activities and processes in the EEZ of one coastal state can affect resources in the EEZs of other coastal states. Where multi-jurisdictional conditions exist, as in many coastal seas, holistic conservation and management regimes are essential. A conceptual framework to enable such conservation and management, the "Large Marine Ecosystems (LMEs)" approach, provides the basis for developing new strategies for marine resources management and ecosystem sustainability (Sherman and Alexander 1986, 1989; Sherman, Alexander and Gold 1990, 1991, 1993; Sherman, Jaworski and Smayda 1996; Sherman, Okemwa and Ntiba 1998; Kumpf, Steidinger and Sherman 1999; Sherman and Tang 1999). During the United Nations Conference on the Environment and Development (UNCED) in Brazil in 1992, the declaration on the ocean recommended that nations of the globe: (1) prevent, reduce, and control degradation of the marine environment; (2) develop and increase the potential of marine living resources; and (3) promote the integrated management and sustainable development of coastal areas and the marine environment. LMEs as global units of ocean space and principal assessment and management units for coastal ocean resources have a broad application to marine management, especially in coastal seas where the LME approach is most likely to assist in improving the sustainable use of transboundary resources and ecosystem management.

The Yellow Sea is a semi-enclosed shelf sea with distinct bathymetry, hydrography, productivity, and trophically dependent communities. Shallow, but rich in nutrients and resources, the sea is most favorable for coastal and offshore fisheries, and it has well-developed multispecies and multinational fisheries. Over the past several decades, the resource populations in the sea have changed greatly, and significant changes to the structure of the fisheries have resulted from non-sustainable fishing, greatly reducing catch-per-unit-effort. Many valuable marine resources are threatened by both land and sea-based sources of pollution and by the unsustainable exploitation of natural resources. Loss of biomass, biodiversity and habitat have resulted from extensive economic development in the coastal zone. Therefore, in order to promote sustainable exploitation of the sea, implementation of effective management strategies is an important and urgent task.

The purpose of this chapter is to describe the Yellow Sea as an LME, emphasizing the changing states of living resources in the ecosystem. Detailed information on the ecological characteristics, and changes in indices of productivity and biomass yields and their causes are reported. Suggestions for mitigation actions of effective ecosystem management of the Yellow Sea LME are offered in the final section.

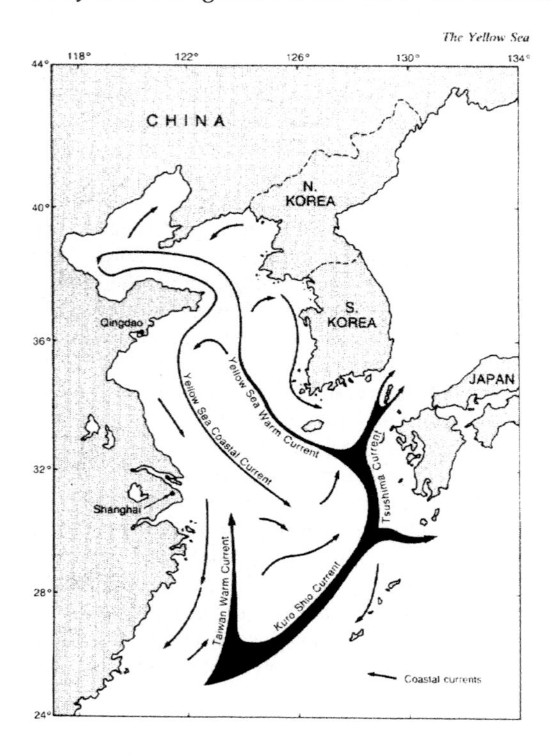

Figure 6-1. Schematic diagram of the major current system in winter in the Yellow Sea (from Gu *et al.* 1990)

ECOLOGICAL CHARACTERISTICS AND EXPLOITATION

The Yellow Sea is located between the North China continent and the Korean Peninsula, being separated from the West Pacific Ocean by the East China Sea in

Peninsula, being separated from the West Pacific Ocean by the East China Sea in the south, and linked with Bohai Sea, which is an arm of the Yellow Sea, in the north. It covers an area of about 400,000 km^2, with a mean depth of 44 m and most of the area shallower than 80m. The central part of the sea, traditionally called the Yellow Sea Basin, ranges in depth from 70m to a maximum of 140m.

The general circulation of the Yellow Sea is a basin-wide cyclonic gyre comprised of the Yellow Sea Coastal Current and the Yellow Sea Warm Current. The Yellow Sea Warm Current, a branch of the Tsushima Warm Current from the Kuroshio Region in the East China Sea, carries water of relatively high salinity (> 33 PSU) and high temperature (> 12°C) northward along 124°E and then westward, flowing into the Bohai Sea in winter. This current, together with the coastal current flowing southward, plays an important role in exchanging the waters in this semi-enclosed sea (Figure 6-1).

The water temperature in the shallow regions of the Yellow Sea varies seasonally according to the influence of the continental climate; freezing occurs in winter along the coast in the northern part, while the water temperature may rise in summer to that of the subtropical sea (27-28°C) (Figure 6-2). Below 50m, the Yellow Sea Cold Water Mass forms seasonally. This cold water mass is characterized by low temperature, with the bottom temperature lower than 7°C in its central part. It is believed to be the remnant of local water left over from the previous winter due to the effect of cold air from the north (Ho *et al.* 1959; Guan 1963). Stratification is the strongest in summer, with a vertical temperature gradient greater than 10°C/10m. All rivers have peak runoff in summer and minimum discharge in winter, and this alternation has important effects on the salinity of the coastal waters. The sea annually receives more than 1.6 billion metric tons (t) of sediments, mostly from the Yellow River and Changjiang (Yangtze) River, which have formed large deltas.

The Yellow Sea lies in the warm temperate zone, and the fauna of the Yellow Sea is recognized as belonging to a sub-East Asia Province of the North Pacific temperate zone (Cheng 1959; Liu 1959, 1963; Dong 1978; Zhao *et al.* 1990). The communities are composed of species with various ecotypes. Warm temperate species are the major component of the biomass and account for about 60 percent of the total biomass of resource populations; warm water species and boreal species account for about 15 percent and 25 percent, respectively. Fish are the main living resource and about 280 fish species are found. Of these, 46 percent are warm temperate forms, 45 percent are warm water forms, and 9 percent are cold temperate forms. Because most of the species inhabit the Yellow Sea throughout the year, the resource populations of the fauna have formed an independent community. The diversity and abundance of this community are lower than that of the community found in the South China Sea and the East China Sea. The Shannon-Wiener diversity index (H') and the Simpson ecological dominance index (C) of the resource populations were determined to be 2.3 and 0.34, respectively (Tang 1988).

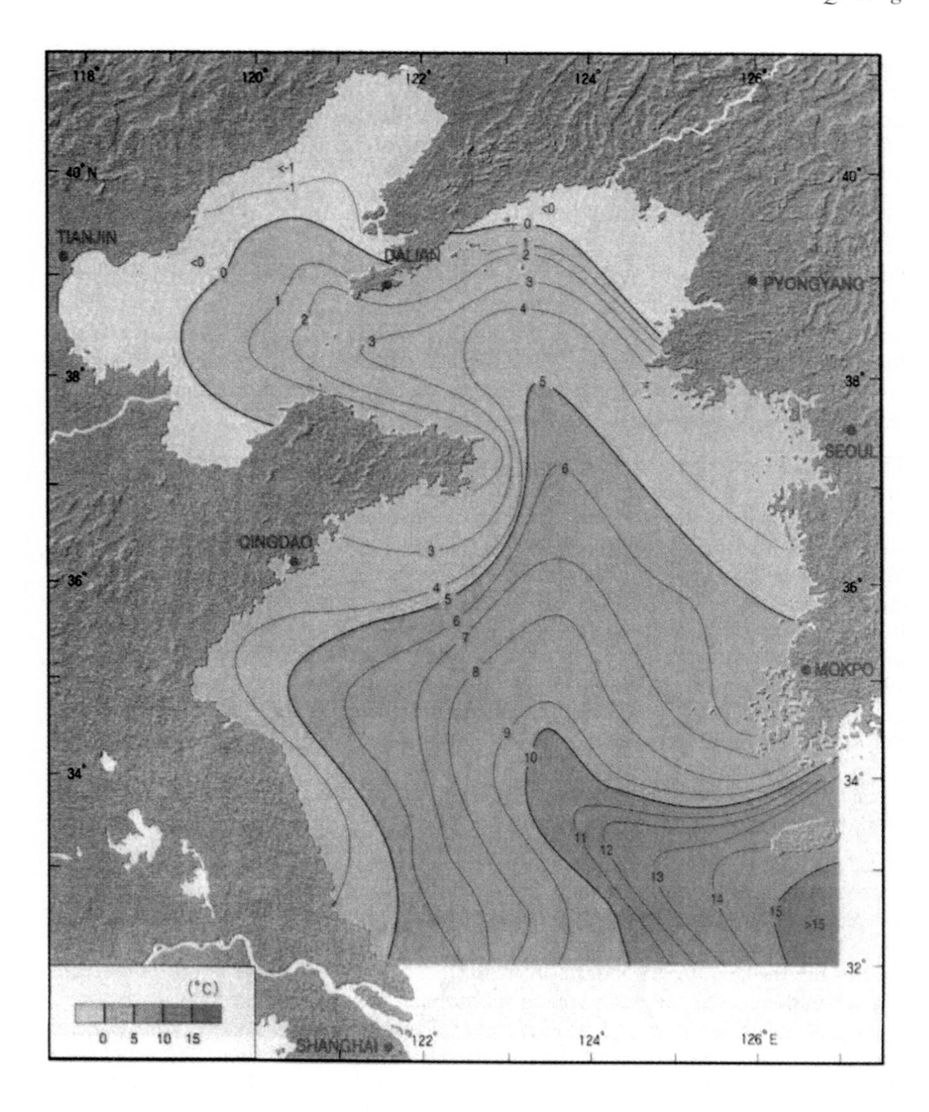

Figure 6-2A. February surface distributions of water temperature in the Yellow Sea (from Lee *et al.* 1998).

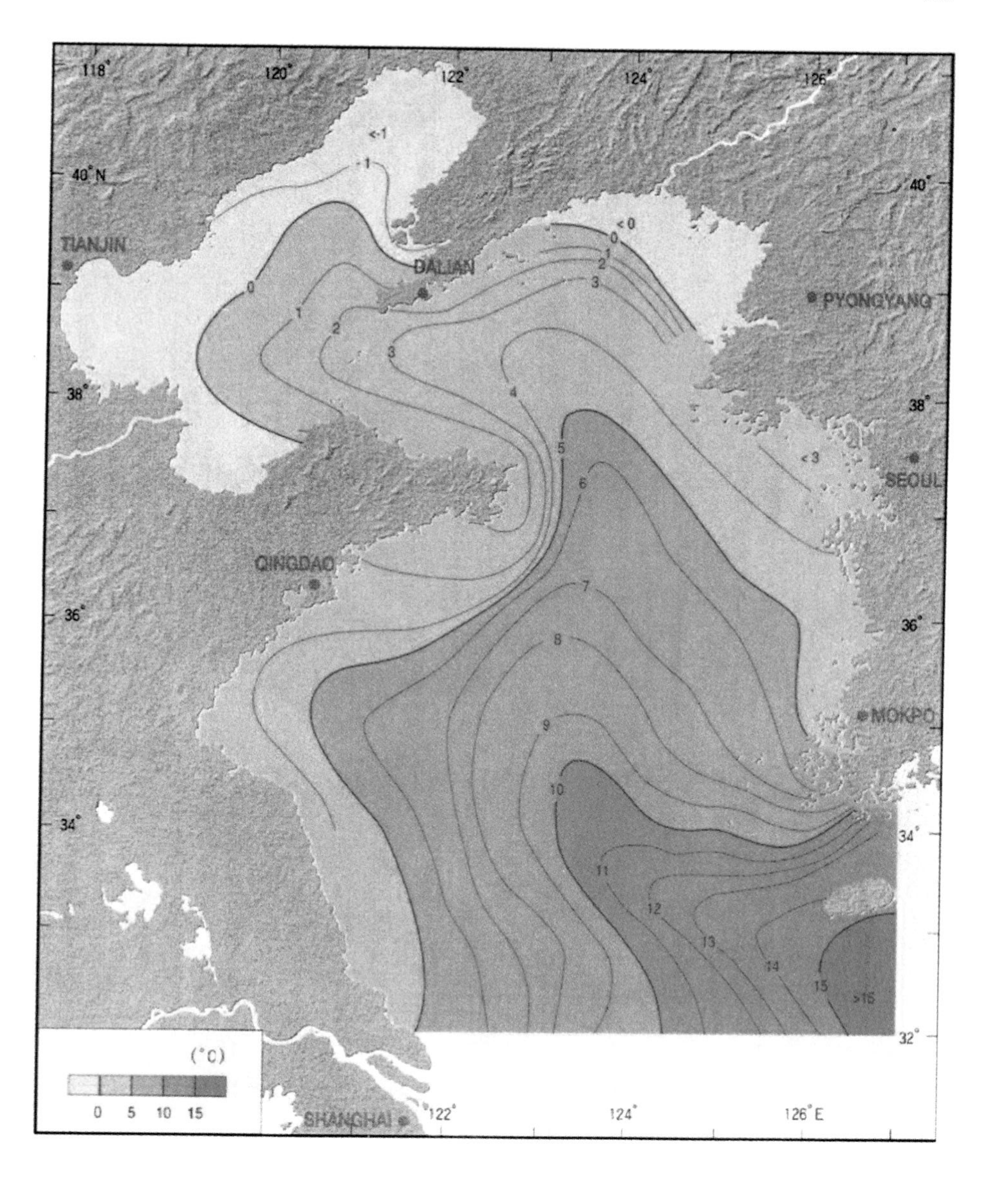

Figure 6-2B. February bottom distributions of water temperature in the Yellow Sea (from Lee *et al.* 1998).

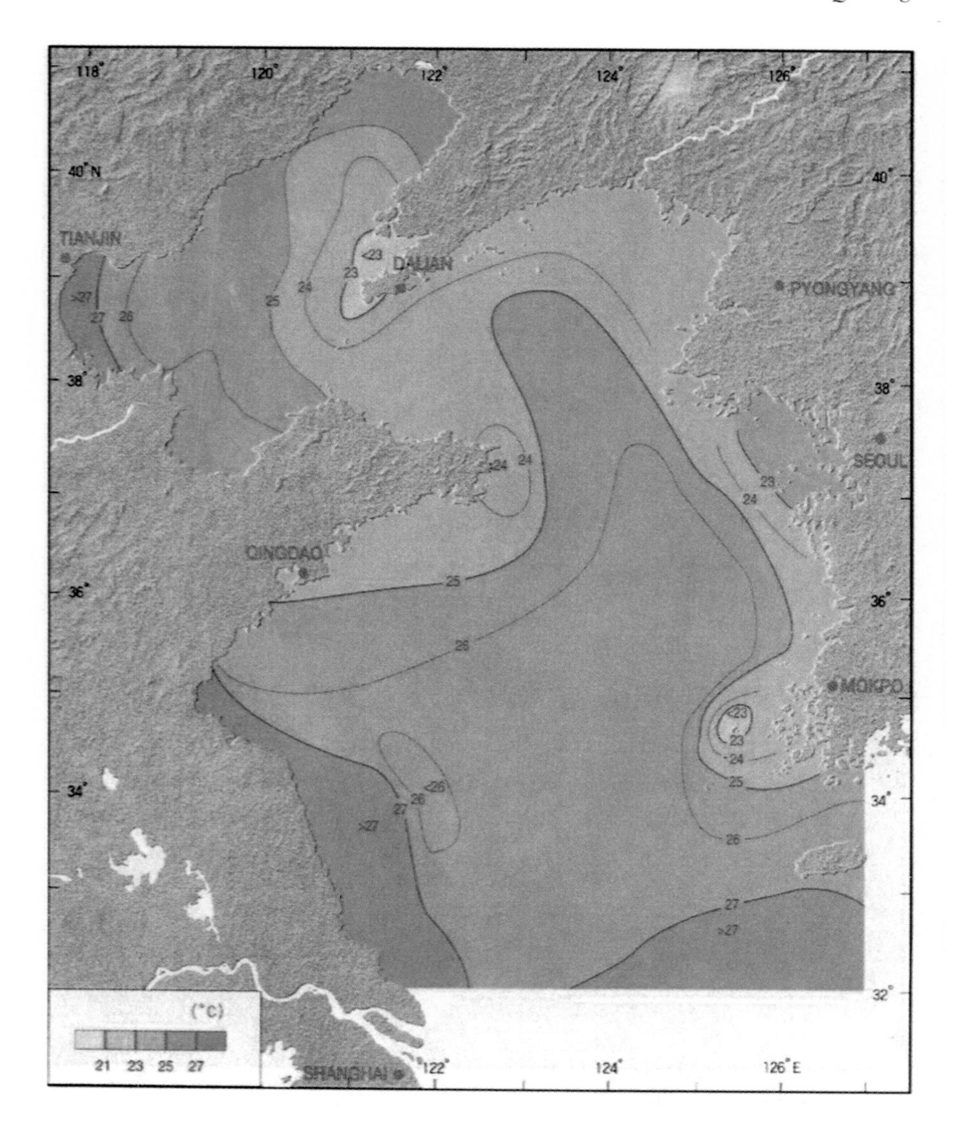

Figure 6-2C. August surface distributions of water temperature in the Yellow Sea (from Lee et al. 1998).

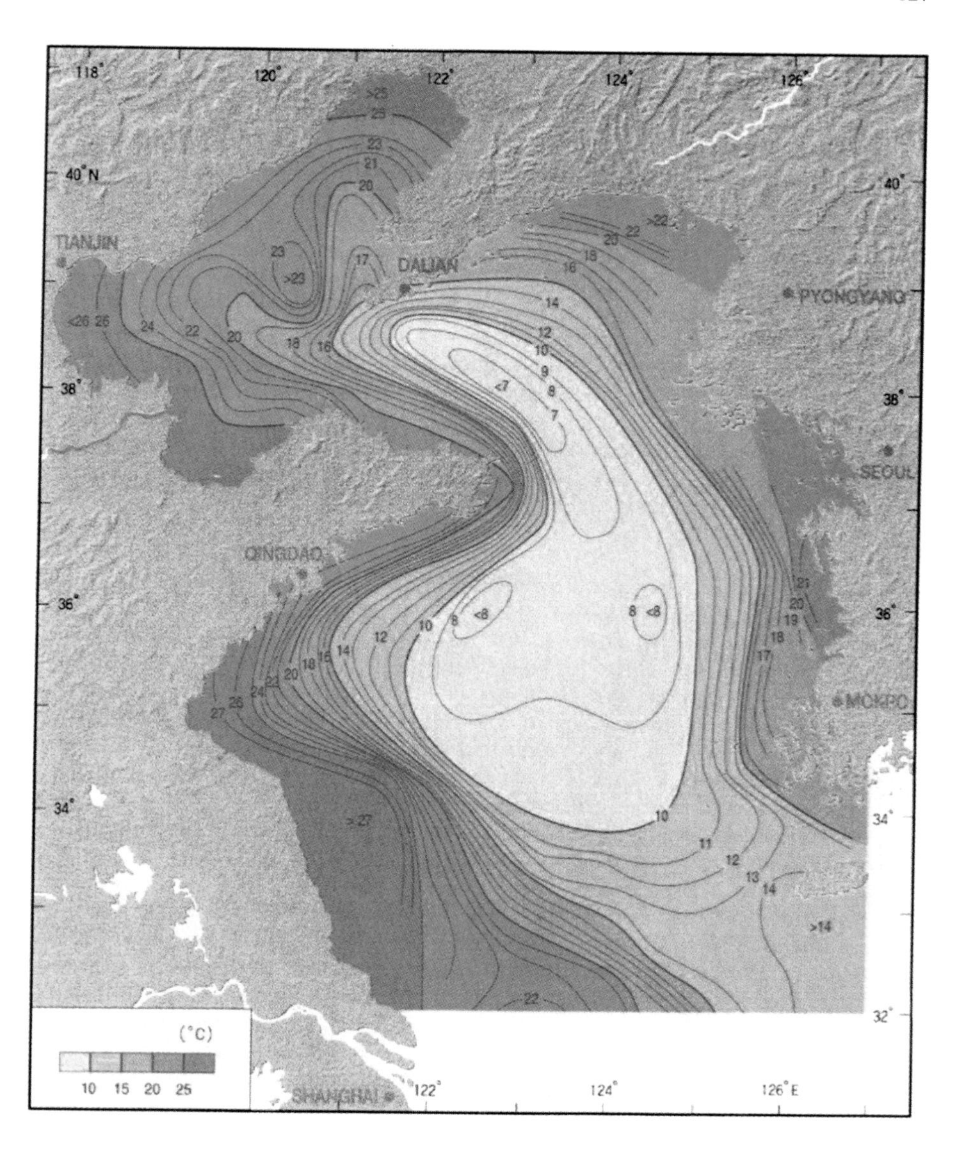

Figure 6-2D. August bottom distributions of water temperature in the Yellow Sea (from Lee *et al.* 1998).

Marked seasonal variations characterize all components of the communities, and are possibly related to the complex oceanographic conditions. Turbidity and sediment type appear to be the major parameters that affect the distribution of planktonic and benthic organisms in the coastal waters of the Yellow Sea. The habitat of resource populations in the Yellow Sea can be divided into two groups: near shore and migratory. When water temperature begins to drop significantly in late autumn, most resource populations migrate offshore toward deeper and warmer water and concentrate mainly in the Yellow Sea Basin. There are three overwintering areas: the mid-Yellow Sea, 35 to 37°N, with depths of 60 to 80m; the southern Yellow Sea, 32 to 35°N, with depths about 80m; and the northern East China Sea. The cold temperate species (e.g., cod, herring, flatfish, and eel-pout) are distributed throughout the mid-Yellow Sea, and many warm temperate species and warm water species (e.g., skates, gurnard, *Saurida elongata*, jewfish, small yellow croaker, spotted sardine, penaeid shrimp, southern rough shrimp, and cephalopods) are also found there from January to March. In the southern Yellow Sea, all species are warm temperate and warm water species (e.g., small yellow croaker, *Nibea albiflora*, white croaker, jewfish, anchovy, *Setipinna taty*, butterfish and chub mackerel).

The Yellow Sea food web is relatively complex, with at least four trophic levels. There are two trophic pathways: pelagic and demersal (Figure 6-3). Japanese anchovy and macruran shrimp (e.g., *Crangon affinis* and southern rough shrimp) are key species. About 40 species eat anchovy, including almost all of the higher carnivores of the pelagic and demersal fish, and the cephalopods. Japanese anchovy is an abundant species in the Yellow Sea, with an annual biomass estimated at 2.5 to 4.3 million metric tonnes in 1986-1996. *Crangon affinis* and southern rough shrimp, which are eaten by most demersal predators (about 26 species), are also numerous and widespread in the Yellow Sea. These species occupy an intermediate position between major trophic levels, and they interlock the food chains to form the Yellow Sea food web.

The resource populations in the Yellow Sea are multispecies in nature. Approximately 100 species are commercially harvested, including demersal fish (about 66 percent), pelagic fish (about 18 percent), cephalopods (about 7 percent), and crustaceans (about 9 percent). During the 1985-1986 period, about 20 major species accounted for 92 percent of the total biomass of the resource populations, and about 80 species accounted for the other 8 percent. The commercial utilization of the living resources in the ecosystem dates back several centuries. With the introduction of bottom trawl vessels in the early twentieth century, many stocks began to be intensively exploited by Chinese, Korean, and Japanese fishermen, and some economically important species such as the red seabream declined in abundance in the 1920s and 1930s (Xia 1960). The stocks remained fairly stable during World War II (Liu 1979). However, due to a remarkable increase in fishing effort and its expansion to the entire Yellow Sea, nearly all the major stocks were fully fished by the mid-1970s and, by the 1980s, the resources in the ecosystem were being over-fished (Figure 6-4). Aquaculture is a major use of the coastal waters of the Yellow Sea. Mariculture is commonly practiced in all coastal areas of China and Korea, and major species of mariculture include scallop, oysters, clams, mussels and seaweed. The total yield of mariculture in 1997 was about 4.7 million t representing about 40 percent of the total fishery yield in the Yellow Sea.

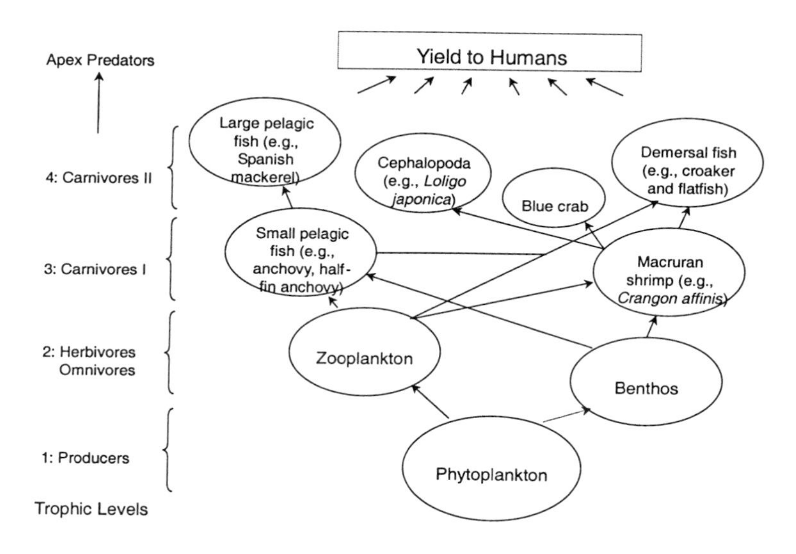

Figure 6-3. A simplified version of the Yellow Sea food web and trophic structure based on the main resource populations in 1985-1986 (Tang 1993).

CHANGING STATES OF THE ECOSYSTEM

Five indices can be used to measure changing ecosystem states: (1) biodiversity, (2) stability, (3) yields, (4) productivity, and (5) resilience (Sherman and Solow 1992). When we used these indices to assess the changing states of the Yellow Sea ecosystem, many changes in productivity, biomass yield, species composition and shifts in dominance have been found. Overexploitation is the principal source of changing states of the ecosystem, but perturbations to the natural environment should be considered important secondary drivers of change in species composition and biomass yields, at least for pelagic species and shellfish.

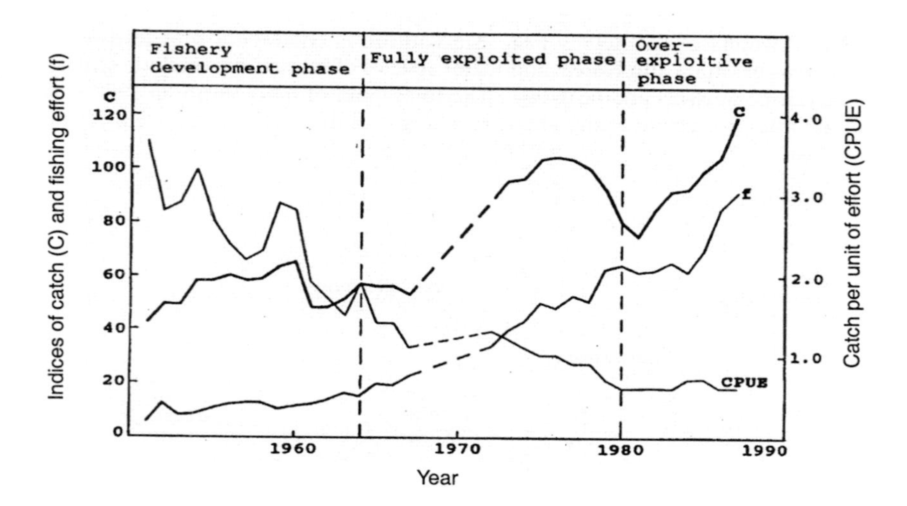

Figure 6-4. Trends in catch and effort for Yellow Sea fisheries, 1950-1980s (generalized history of the fisheries in the Yellow Sea).

Changes in Productivity

Variability in primary productivity in season and area in the Yellow Sea have been observed. Primary productivity in May 1992 varied from 12 to 425 mg $C \cdot m^{-2} \cdot d^{-1}$, with the higher values found in the northeastern part of the sea. In September, primary productivity varied from 65 to 927 mg $C \cdot m^{-2} \cdot d^{-1}$, with most of the higher values appearing at the southern part (Wu *et al.* 1995). Choi *et al.* (1995) reported that nanoplankton contributed greatly. Annual variation of primary productivity in the sea has also been observed. As shown in Table 1, primary productivity in the Bohai Sea varies in season and area, and decreased noticeably from 1982 to 1998. Over the past 40 years, an obvious declining trend of phytoplankton biomass has been found, and it seems to be linked with nutrient changes (Figure 6-5). Zooplankton is an important component of the communities in the sea. The dominant species, including *Calanus sinicus, Euphausia pacifica, Sagitta crassa* and *Themisto gracilipes,* are all important food for pelagic and demersal fish and invertebrates. The biomass of zooplankton in the Yellow Sea is lower than that in adjacent areas, ranging from 5 to 50 mg $\cdot$ m^{-3} in the north to 25 to 100 mg $\cdot$ m^{-3} in the south, because of the influence of the warm current. The annual biomass of zooplankton in the Yellow Sea has decreased noticeably since 1959 (Figure 6-6); this is similar to the declining trend found in the East China Sea (Gu *et al.* 1990). However, zooplankton biomass increased in the Bohai Sea in 1998, possibly due to the decline of anchovy stock, the most abundant species before 1998 (Figure 6-7).

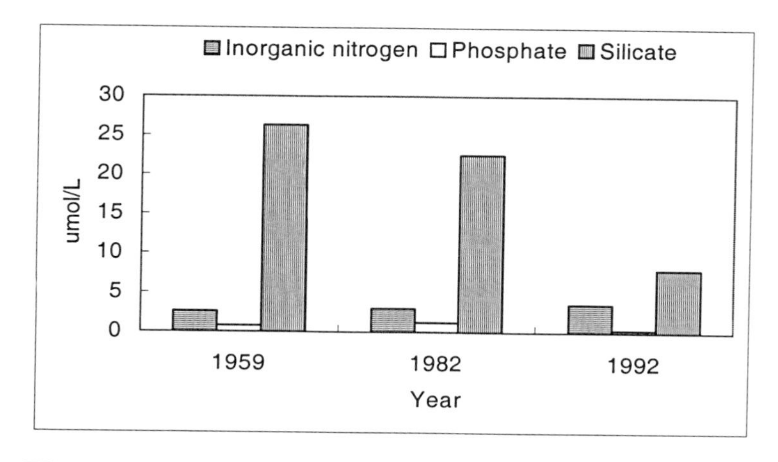

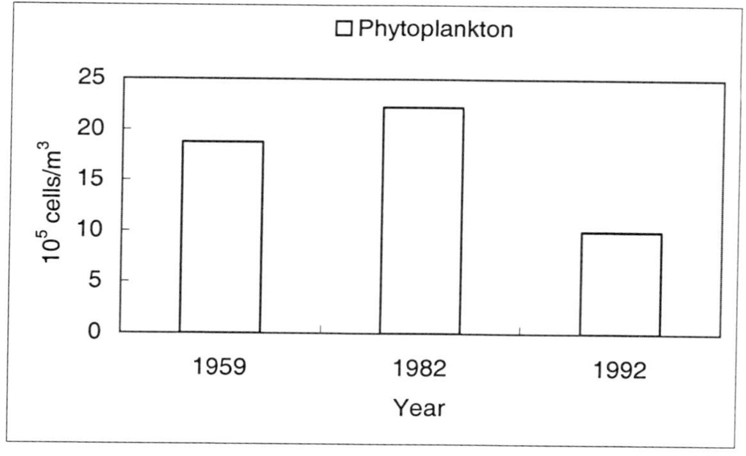

Figure 6-5. Changes of nutrients and phytoplankton in the Bohai Sea.

In the same period, the macrobenthic biomass in the sea was relatively stable — about 21 mg · m^{-2} in 1959, 24 mg · m^{-2} in 1975, and 22 mg · m^{-2} in 1992 (Tang 1993; Lee *et al.* 1998).

Major pollutants entering the Yellow Sea are organic material, oil, heavy metals and pesticides, and the pollutants mainly come with the wastewaters of industries, cities along the coast, and mariculture areas (Zhou *et al.* 1995). These pollutants are affecting valuable and vulnerable resources including habitats for most commercially important species, aquaculture harvesting sites, reserves,

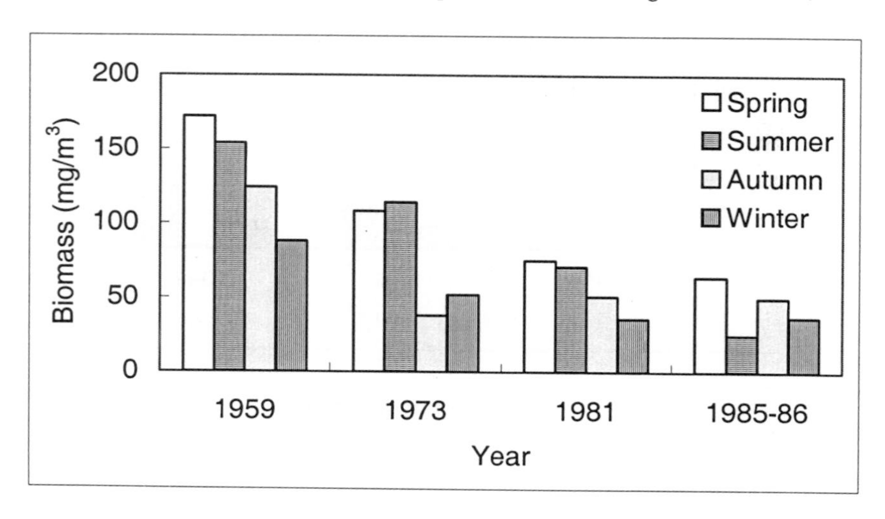

Figure 6-6. Changes in zooplankton biomass in the Yellow Sea.

recreational beaches, and potential tourism sites, especially in the shallow coastal water areas near river mouth, ports and coastal cities. After the 1970s, the frequency of occurrence of red tides gradually increased. The abnormal multiplication of red tide organisms is often associated with eutrophication. These events have been influencing the natural productivity in the ecosystem (She 1999).

Ecosystem Effects of Overexploitation

The Yellow Sea is one of the most intensively exploited areas in the world. With a remarkable increase in fishing effort and the expansion of that effort to the entire Yellow Sea, great declines in the biomass yields of resource populations in the ecosystem have been demonstrated by many studies (Xia 1978; Liu 1979; Chikuni 1985; Tang 1989, 1993; Zhang and Kim 1999).

Small yellow croaker and hairtail were formerly the important commercial demersal species in the Yellow Sea, with catches in 1957 reaching a peak of about 200,000 and 64,000 t, respectively. However, as young fish were heavily exploited in the overwintering grounds and spawning stocks were intensively fished in their spawning grounds, the biomass of these two species has declined sharply since the mid-1960s (Figure 6-8). The Yellow Sea hairtail stock became a non-target species in the 1970s, with a biomass estimated to be only 1/30 of

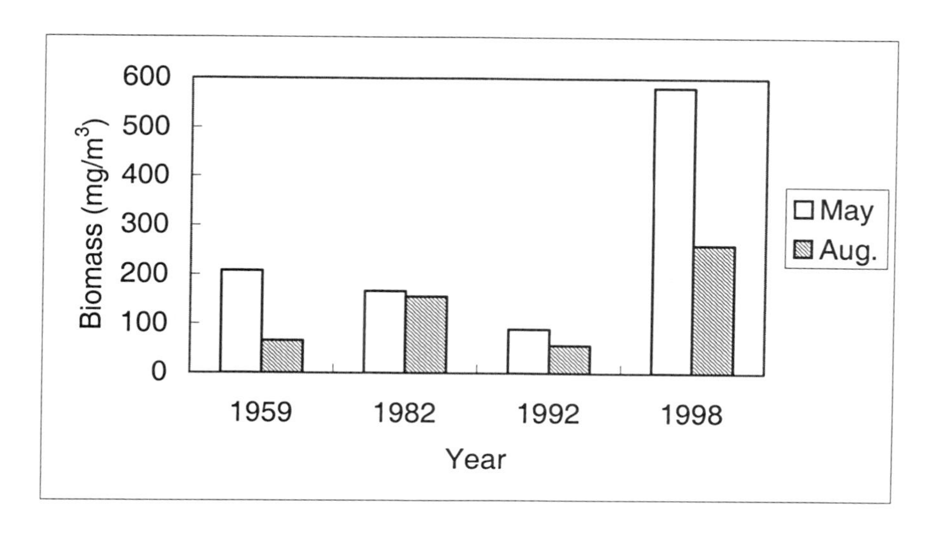

Figure 6-7. Changes in zooplankton biomass in the Bohai Sea.

Previous levels (Lin 1985). This decline in biomass was accompanied by a substantial reduction in its distribution, an increase in growth rate, earlier maturation, and a decrease in the mean age and body length of the spawning stock (Mio and Shinohara 1975; Lee 1977; Otaki *et al.* 1978; Zhao *et al.* 1990). After the resources of small yellow croaker off the Jiangsu coast in the southern Yellow Sea (about 33°N, 122°E) were depleted, the biomass yield of large yellow croaker increased, with the annual catch ranging from 40,000 to 50,000 t during the period from 1965 to 1975. Because the overwintering stock was heavily exploited in the late 1970s, the biomass decreased sharply in the early 1980s and spawning stock size declined to about one-sixth of that in the 1960s. The stock has not recovered since that overfishing. Conversely, the biomass yield of butterfish increased when small and large yellow croaker off Jiangsu decreased. There was no evidence of climatic or large-scale environmental change during this period; thus, the major cause of the fluctuations in biomass and shifts in species dominance in this area appears to be overexploitation. It is evident that overexploitation can be of sufficient magnitude to result in a species "flip." A biomass flip occurs when the population of a dominant species rapidly drops to a very low level and another species becomes dominant (Sherman 1989).

Flatfish and Pacific cod are important boreal species whose distribution and migration are related to the movement of the Yellow Sea Cold Water Mass. The catches of flatfish and Pacific cod reached peaks of 30,000 t in 1959 and 80,000 t in 1972, respectively. As fishing efforts increased, young fish were heavily harvested,

Table 1. Annual variation in season and area of primary productivity in the Bohai Sea.

Season	Winter	Spring	Summer	Autumn	Mean
1982-83	207	208	537	297	312
1992-93	127	162	419	154	216
1998		82	129	60	(90)
Area	**Laizhou Bay**	**Bohai Bay**	**Liaodong Bay**	**Central Part**	**Mean**
1982-83	412	162	325	394	312
1992-93	535	90	96	186	216
1998	(76)	(90)	(96)	(89)	(90)

Note: Unit = mg C $\cdot$ m^{-2} $\cdot$ d^{-1}.

and the biomass yields of both species have decreased since the mid-1970s (Figure 6-8).

Overfishing has also caused a decline in biomass of several other demersal species, including sea robin, red seabream, *Miichthys miiuy*, and *Nibea albiflora*. However, under the same fishing pressure, the biomass yields of other exploited resources, including cephalopods, skate, white croaker, and daggertooth pike-conger, appear to be fairly stable (Figure 6-8). This stability may be a result of their scattered distributions.

Pacific herring, chub mackerel, Spanish mackerel, and butterfish are the major, larger sized pelagic stocks in the sea. The annual catch from 1953 to 1988 fluctuated widely, ranging from 30,000 to 300,000 t per year. The causes of these fluctuations are more complicated. There may be two patterns of population dynamics. The variability is particularly significant for Pacific herring and chub mackerel stocks, whereas Spanish mackerel and butterfish stocks appear to be relatively constant (Figure 6-9).

Commercial utilization of Spanish mackerel stock began in the early 1960s, when both the catch and abundance of small yellow croaker and airtail decreased. In 1964, the catch reached 20,000 t. Since then, yields of this stock have steadily increased. Catch has peaked at 140,000 t in recent years, although the stock has borne excessive fishing predation. The reason for this is not clear. Spanish mackerel feed on anchovy. As mentioned above, anchovy, a major prey for carnivorous species in the ecosystem, is very abundant in the Yellow Sea. Perhaps this abundance is caused by an unusual combination of natural and anthropogenic conditions.

Due to overfishing, harvestable living resources in the ecosystem declined in quality. About 79 percent of the biomass yield in 1986 consisted of fish and

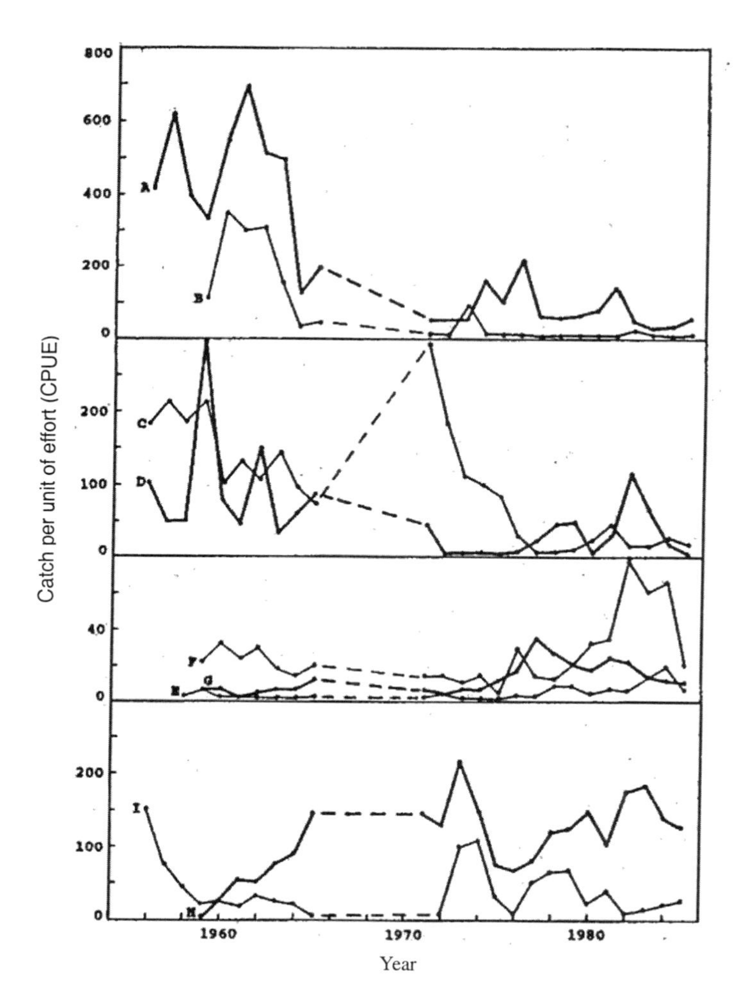

Figure 6-8. Catch per unit of effort (CPUE), expressed in kg caught per haul by paired trawlers, for (A) small yellow croaker, (B) largehead hairtail, (C) flatfish, (D) Pacific cod, (E) white croaker, (F) skate, (G) daggertooth pike-conger, (H) cephalopods, and (I) fleshy prawn

invertebrates smaller than 20cm. Their mean standard length was only 11cm, with a mean weight of 20g, compared to a mean standard length in the 1950s-1960s of 20cm. The biomass of less-valuable species increased by about 23 percent between the 1950s and 1980s.

With a mean weight of 20g, compared to a mean standard length in the 1950s and 1960s of 20 cm. The biomass of less valuable species increased by about 23 percent between the 1950s and 1980s.

Species Shifts in Dominance and Their Causes

Over the past 50 years, dramatic shifts in the dominant species of resource populations in the ecosystem have been observed — from small yellow croaker and hairtail in the 1950s and early 1960s to Pacific herring and chub mackerel in the 1970s. Small-sized, fast-growing, short-lived, low-valued species, such as

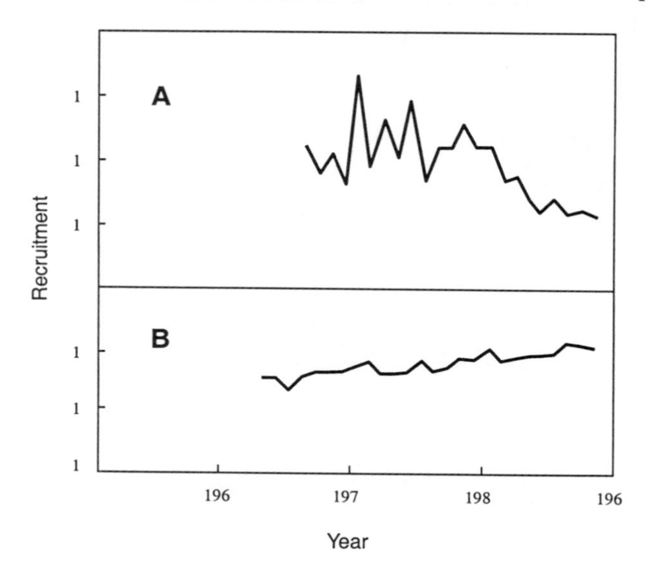

Figure 6-9. Recruitment of Pacific herring (A) and Spanish mackerel (B) in the Yellow Sea.

Japanese anchovy, half-fin anchovy, and scaled sardine, increased markedly in abundance during the 1980s and assumed a prominent position in the ecosystem resources and food web thereafter (Figure 6-10). Larger, higher trophic-level, commercially important demersal species were replaced by smaller, lower trophic level, pelagic, less-valuable species. The major resource populations in 1958-59 were small yellow croaker, flatfish, cod, hairtail, skate, sea robin, and angler, which accounted for 71 percent of the total biomass yield. Planktophagic species represented 11 percent, benthophagic species, 46 percent, and ichthyophagic species, 43 percent; by 1985-86, the major exploitable resources had shifted to Japanese anchovy, half-fin anchovy, squid, sea snail, flatfish, small yellow croaker, and scaled sardine. Of these, 59 percent were planktophagic species, 26 percent were benthophagic species, and 16 percent were ichthyophagic species (Figure 6-11). The trophic levels in 1959 and 1985 were estimated to be 3.8 and 3.2, respectively. Thus it appears that the external stress has affected the self-regulatory mechanism of the ecosystem. Recent surveys have indicated that the abundance of Japanese anchovy is declining, while the biomass of sandlance is increasing and the stock of small yellow croaker shows a recovery trend.

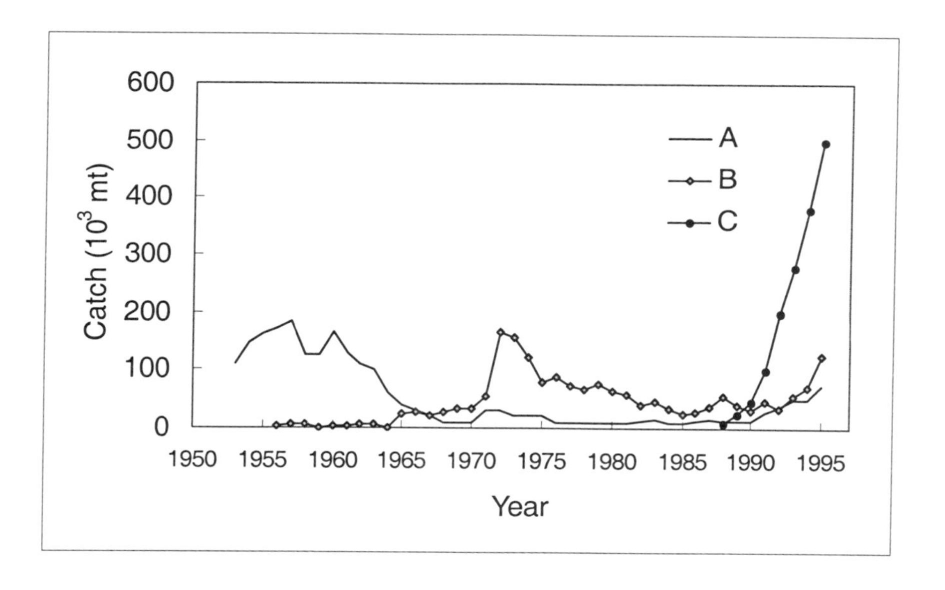

Figure 6-10. Annual catch of dominant species: (A) small yellow croaker and hairtail, (B) Pacific herring and Japanese mackerel, and (C) anchovy and half-fin anchovy.

Although we believe that the cause of changes in quantity and quality of the biomass yield and species shifts in dominance are attributable principally to human predation, this is not the case for all of species. Fluctuation in recruitment of penaeid shrimp, which is a commercially important crustacean distributed in the Bohai Sea and Yellow Sea, provides a good example. Fluctuation in recruitment was related both to environment and spawning size and it should be noted that the relative importance of the two factors varied from year to year *et al.* 1989). Pelagic species are generally sensitive to environmental changes and the fluctuations in recruitment of Pacific herring, for example, are very large. This species in the Yellow Sea has a long history of exploitation, and the fishery is full of drama. In the last century, the commercial fishery experienced three peaks (in about 1900, 1938, and 1972), followed by a period of little or no catch. Environmental conditions such as rainfall, wind and daylight are supposed to be major factors affecting fluctuations in recruitment. Long-term changes in abundance may be correlated with the 36-yr cycle of dryness/wetness oscillation in eastern China (Tang 1981; Figure 6-12).

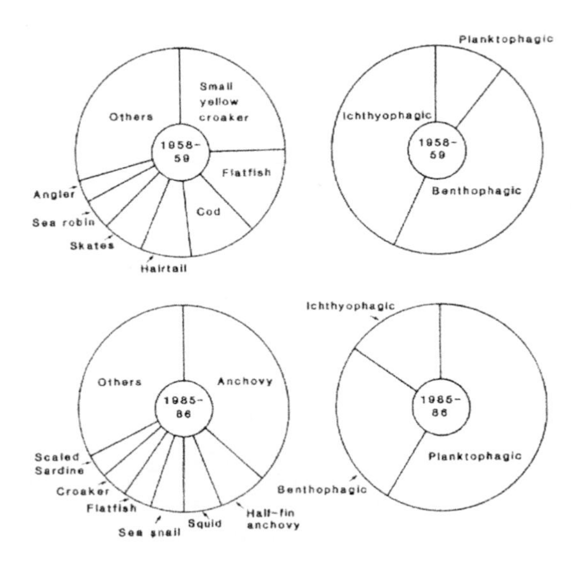

Figure 6-11. Proportion of major species and various feeding habits in total catch based on research vessel bottom trawl surveys of the Yellow Sea in 1958-1959 and 1985-1986.

There may be two types of species shifts in the ecosystem resources: systematic replacement and ecological replacement. Systematic replacement occurs when one dominant species declines in abundance by overexploitation and another competitive species uses surplus food and vacant space to increase its abundance. Ecological replacement occurs when minor changes in the natural environment affect stock abundance, especially of pelagic species. The data, based on catch, indicate that warm and temperate species tend to increase in abundance during warm years (e.g., half-fin anchovy and cuttlefish (*Sepiella maindroni*) increased in the 1980s), while boreal species (e.g., Pacific herring) tend to increase during cold years, such as the 1970s (Figure 13). Thus, natural factors may have an important effect on long-term changes in dominant species of various ecotypes.

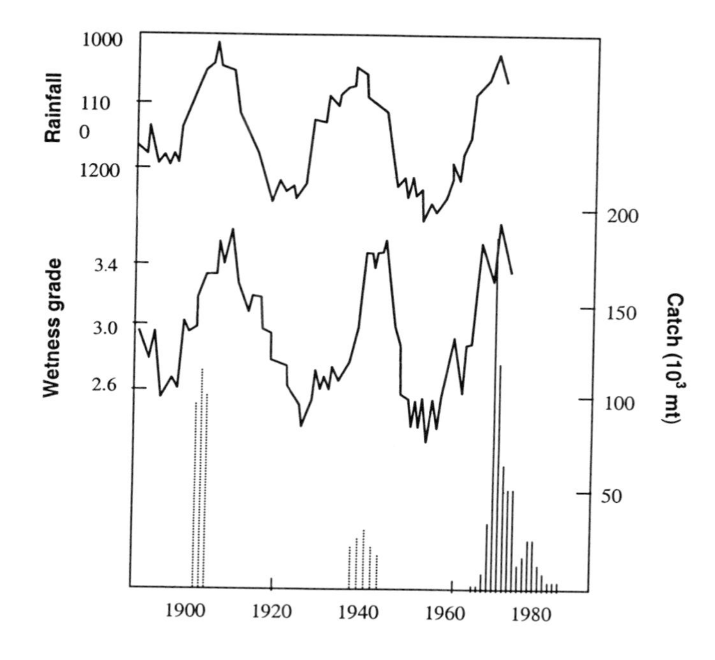

Figure 6-12. Relationship between the fluctuations in herring abundance of the Yellow Sea and the 36-yr cycle of wetness oscillation in eastern China.

MITIGATION ACTIONS

Establish a Monitoring, Assessment and Process-Oriented Studies Program

An essential component of an effective ecosystem management is the inclusion of a scientifically based strategy to monitor and assess the changing states and health of the ecosystem by tracking key biological and environmental parameters (Sherman and Laughlin 1992; Sherman 1995). Under this requirement grant has been provided by the Global Environmental Facility (GEF) for supporting a

Yellow Sea LME Project to be initiated in 2003. This proposal is a highly worthwhile activity bringing together the scientists and marine specialists from China and Korea to solve common marine resource problems within the Yellow Sea LME. Four major components were developed for the project based on the areas of intervention identified. The first component, "Regional Strategies for Sustainable Management of Fisheries and Mariculture," addresses the need for sustainable fisheries management and fisheries recovery plans agreed upon on a regional basis. The second component, "Effective Regional Initiatives for Biodiversity Protection,"

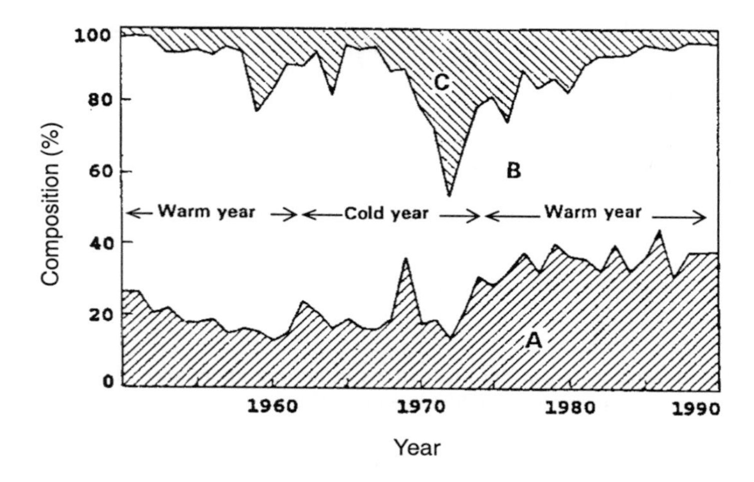

Figure 6-13. Trends in relative abundance of various ecotypes comprising the annual catch in the Yellow Sea, and long-term changes in environmental conditions. (A) Warm-water species; (B) Temperate-water species; (C) Boreal species.

addresses the need for coordinated regional action to preserve globally significant biodiversity. The third component, "Actions to Reduce Stress to the Ecosystem, Improve Water Quality and Protect Human Health," addresses the YSLME as a marine ecosystem, and develops management practices based on an understanding of ecosystem behavior, the very basis for the Large Marine Ecosystem concept. The fourth component, "Development of Regional Institutional and Capacity Building," focuses the intervention on the required national and regional institutional and capacity building and strengthening, on the preparation of investment portfolios, and on coordination of preparation of the project. The long-term objective of the project is to ensure environmentally sustainable management and use of the Yellow Sea LME and its watershed by reducing development stress and promoting sustainable development of the ecosystem. Sustainable use and development of marine resources must be achieved alongside a densely populated, heavily urbanized and industrialized region contiguous with the semi-enclosed shelf sea (Project Brief of the Yellow Sea LME, 2000).

In order to understand the interactions among the important biological, chemical and physical characteristics of the Yellow Sea LME, and further increase the predictive capability of resource managers, a comprehensive process-oriented study of ecosystem services and goods should be considered also. Economics is an important component in the ecosystem sustainability studies as shown in the following diagram:

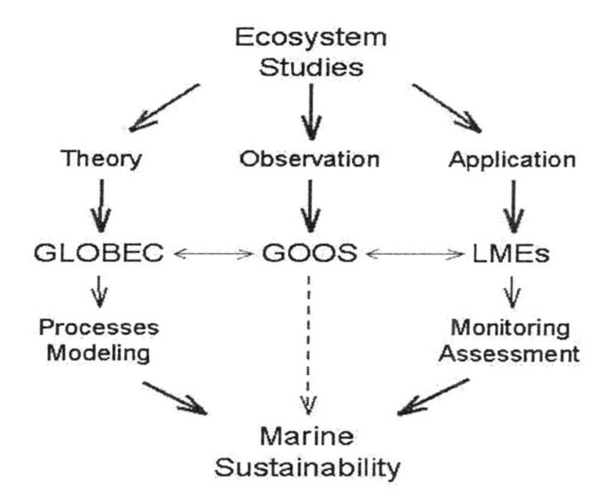

The ongoing program is China-GLOBEC II, entitled "Ecosystem Dynamics and Sustainable Utilization of Living Resources in the East China Sea and the Yellow Sea" (Tang 2000). This program has been approved as a Programme of the National Key Basic Research and Development Plan in the People's Republic of China, with a funding of 4.5 million U.S. dollars for the period of 1999-2004. Multidisciplinary and comprehensive studies in field work are carried out aiming at the following key scientific questions: energy flow and conversion of key resource species, dynamics of key zooplankton populations, cycling and regeneration of biogenic elements, ecological effects of key physical processes, pelagic and benthic coupling, and the microbial loop contribution to the main food web. The program goals are to identify key processes of ecosystem dynamics, improve predictive and modeling capabilities, provide scientific underpinning for sustainable utilization of marine ecosystem and rational management system of fisheries and other marine life in the East China Sea and the Yellow Sea.

Promote an Ecosystem Resource Recovery Plan

There are many ways to recover the resources in a perturbed LME, such as reducing excessive fishing mortality, controlling the point sources of pollution and improving the understanding of the effects of physical perturbations. In 1995, China started completely closing fishing in the Yellow Sea and East China Sea for 2-3 months in summer. This fishing ban has effectively protected the juveniles so that the catches and quality have obviously increased and improved. In addition to these efforts, artificial enhancement in the Yellow Sea should be encouraged. One proposed action is to introduce juveniles to enhance biomass yields. Since 1984, several enhancement experiments of economic-yield species have been in practice. The released penaeid shrimps in the Bohai Sea, the north Yellow Sea and southern waters off Shandong Peninsula have achieved remarkable social, ecological and economic benefit. Meanwhile the seeded shellfish such as scallop, abalone, and arkshell were also successful. The *Patinopecten yesoensis* was seeded around Haiyang Island in the north Yellow Sea with a high recapture rate of 30 percent during 1989-1991. The jellyfish released in the Bohai Sea and Yellow Sea were also encouraging, with a recapture rate of 0.07-2.56 percent. These successful activities not only promote the development of artificial enhancement programs in the sea, but also bring hope for recovery of ecosystem resources. Therefore, as an effective resource recovery strategy, artificial enhancement practices should be expanded to an LME level in the entire Yellow Sea.

REFERENCES:

Cheng, C. 1959. Notes on the economic fish fauna of the Yellow Sea and the East China Sea. Oceanol. Limnol. Sin. 2 (1): 53-60. in Chinese.

Chikuni, S. 1985. The fish resources of the Northwest Pacific. FAO Fish. Tech. Paper 266. FAO, Rome.

Choi, J.K., J.H. Noh, K.S. Shin and K.H. Hong. 1995. The early autumn distribution of chlorophyll-a and primary productivity in the Yellow Sea, 1992. The Yellow Sea. 1: 68-80.

Dong, Z. 1978. The geographical distribution of cephalopods in Chinese waters. Oceanol. Limnol. Sin. 9(1): 108-116. in Chinese with English abstract.

Gu, X. *et al.*, eds. 1990. Marine fishery environment of China. Zhejiang Science and Technology Press. in Chinese.

Guan, B. 1963. A preliminary study of the temperature variations and the characteristics of the circulation of the cold water mass of the Yellow Sea. Oceanol. Limnol. Sin. 5(4): 255-284. in Chinese.

Ho, C., Y. Wang, Z. Lei and S. Xu. 1959. A preliminary study of the formation of the Yellow Sea cold water mass and its properties. Oceanol. Limnol. Sin. 2:11-15. in Chinese.

Kumpf, H., K. Steidinger and K. Sherman, eds. 1999. The Gulf of Mexico Large Marine Ecosystem. Cambridge, MA: Blackwell Science.

Lee, J. 1977. Estimation of the age composition and survival rate of the yellow

croaker in the Yellow Sea and East China Sea. Bull. Fish. Res. Dev. Agency 16: 7-13. in Korean.
Lee, Y. C., Y.S. Qin, and R.Y. Liu. 1998. The Yellow Sea Atlas. OSTI, Korea and IOCAS, China.
Lin, J. 1985. Hairtail. Agriculture Publishing House, Beijing. In Chinese.
Liu, J. 1959. Economics of micrurus crustacean fauna of the Yellow Sea and the East China Sea. Oceanol. Limnol. Sin. 2(1): 35-42. in Chinese.
Liu, J. 1963. Zoogeographical studies on the micrurus crustacean fauna of the Yellow Sea and the East China Sea. Oceanol. Limnol. Sin. 5(3):230-244. In Chinese.
Liu, X. 1979. Status of fishery resources in the Bohai and Yellow Seas. Mar. fish. Res. Paper 26: 1-17. In Chinese.
Mio, S., and F. Shinohara. 1975. The study on the annual fluctuation of growth and maturity of principal demersal fish in the East China and the Yellow Sea. Bull. Seikai Reg. Fish. Res. Lab. 47: 51-95.
Otaki, H. and S. Shojima. 1978. On the reduction of distributional area of the yellow croaker resulting from decrease abundance. Bull. Seikai Reg. Fish. Res. Lab. 51: 111-121.
She, J. 1999. Pollution in the Yellow Sea LME: monitoring, research, and ecological effects. In K. Sherman and Q. Tang (eds.), Large Marine Ecosystem of Pacific Rim. Cambridge, MA: Blackwell Science. 419-426.
Sherman, K. 1989. Biomass flips in large marine ecosystems. In K. Sherman and L. M. Alexander (eds.), Biomass yields and geography of large marine ecosystems. AAAS Selected Symposium 111. Boulder, CO: Westview Press. 327-331.
Sherman, K. 1995. Achieving regional cooperation in the management of marine ecosystem: the use the large marine ecosystem approach. Ocean & Coastal Management, 29(1-3): 165-185.
Sherman, K., L.M. Alexander and B.D. Gold, eds. 1990. Large Marine Ecosystems: Patterns, Processes, and Yields. Washington, DC: AAAS Press.
Sherman, K., L.M. Alexander and B.D. Gold, eds. 1991. Food Chains, Yields, Models, and Management of Large Marine Ecosystems. Boulder CO: Westview Press.
Sherman, K., L.M. Alexander and B.D. Gold, eds. 1993. Large Marine Ecosystems: Stress, Mitigation, and Sustainability. Washington, DC: AAAS Press.
Sherman, K. and L.M. Alexander, eds. 1986. Variability and management of large marine ecosystems. AAAS Selected Symposium 99. Boulder CO: Westview Press.
Sherman, K. and L.M. Alexander, eds. 1989. Biomass Yields and Geography of Large Marine Ecosystems. AAAS Selected Symposium 111. Boulder CO: Westview Press.
Sherman, K. and T. Laughlin. 1992. Large marine ecosystems monitoring workshop report. U. S. Dept of Commerce, NOAA Tech. Mem. NMFS-F/NEC-93.
Sherman, K., and A.R. Solow. 1992. The changing states and health of a Large Marine Ecosystem. CM/L38 Session V.
Sherman, K., and Q. Tang, eds. 1999. The Large Marine Ecosystems (LMEs) of the

Pacific Rim. Cambridge, MA: Blackwell Science.
Sherman, K., N.A. Jaworski and T.J. Smayda, eds. 1996. The Northeast Shelf Ecosystem: Assessment, Sustainability, and Management. Cambridge, MA: Blackwell Science.
Sherman, K., E. Okemwa, and M.J. Ntiba, eds. 1998. Large Marine Ecosystems of the Indian Ocean. Cambridge, MA: Blackwell Science.
Tang, Q. 1981. A preliminary study on the causes of fluctuations on year class size of Pacific herring in the Yellow Sea. Trans. Oceanol. Limnol. 2: 37-45. In Chinese.
Tang, Q. 1988. Ecological dominance and diversity of fishery resources in the Yellow Sea. J. Chinese Academy of Fishery Science. 1(1): 47-58. In Chinese with English abstract.
Tang, Q. 1989. Changes in the biomass of the Yellow Sea ecosystem. In K. Sherman and L. M. Alexander (eds.), Biomass yields and geography of large marine ecosystem. AAAS Selected Symposium 111. Boulder, CO: Westview Press, pp. 7-35.
Tang, Q. 1993. Effects of long-term physical and biological perturbations on the contemporary biomass yields of the Yellow Sea ecosystem. In K. Sherman, L. M. Alexander, and B. D. Gold (eds.), Large Marine Ecosystems: stress, mitigation, and sustainability. Washington, DC: AAAS Press, pp. 79-83.
Tang, Q. 2000. The new age of the China-GLOBEC study. PICES Press, 8(1): 28-29.
Tang, Q., J. Deng and J. Zhu. 1989. A family of Ricker SRR curves of prawn under different environmental conditions and its enhancement potential in the Bohai Sea. Can. Spec. Publ. Fish. Aquat. Sci. 108: 335-339.
Wu, Y., Y. Guo and Y. Zhang. 1995. Distributional characteristics of chlorophyll-a and primary productivity in the Yellow Sea. The Yellow Sea. 1: 81-92.
Xia, S. 1960. Fisheries of the Bohai Sea, Yellow Sea and East China Sea. Mar. fish. Res. Pap. 2: 73-94. In Chinese.
Xia, S. 1978. An analysis of changes in fisheries resources of the Bohai Sea, Yellow Sea, and East China Sea. Mar. fish. Res. Paper 25: 1-13. In Chinese.
Zhang, C. I. and S. Kim. 1999. Living marine resources of the Yellow Sea ecosystem in Korean waters: status and perspectives. In K. Sherman and Q. Tang (eds.), Large Marine Ecosystems of the Pacific Rim. Cambridge, MA: Blackwell Science. 163-178.
Zhao, C., *et al.*, eds. 1990. Marine Fishery Resources of China. Zhejiang Science and Technology Press. In Chinese.
Zhou, M., J. Zou, Y. Wu, T. Yan and J. Li. 1995. Marine pollution and its control in the Yellow Sea and Bohai Sea. The Yellow Sea. 1: 9-16.

Large Marine Ecosystems of the World
G. Hempel and K. Sherman (Editors)

7

The Baltic Sea

Bengt-Owe Jansson

ABSTRACT

The enclosed, brackish Large Marine Ecosystem of the Baltic is rather a "Sea of Surprises" than a system near steady state. Formed after the last glaciated period, influenced by alternating fresh and saline conditions, and populated by a depauperate assembly of organisms, the Baltic is, nonetheless, a productive ecosystem, although one of changes. A permanent halocline, a water residence time of 25-30 years, lack of tide, and polar conditions in the northern parts make the sea sensitive to wastes from a human society of some 85 million people living in a drainage basin area four times larger than the Sea itself. Large pulses of North Sea water cause deeper areas to flip between oxic and anoxic conditions. Increasing nutrient and decreasing water input transfer dominance of perennial seaweeds to annuals and change hard bottoms to soft bottoms through increased sedimentation. Toxic substances and heavy metals have affected the stocks of sea eagles and seals. Institutional evolution, inspired by the international HELCOM organization (Baltic Marine Environment Protection Commission – The Helsinki Commission), has resulted in banning of hazardous substances and in programs for decreasing emissions of nutrients and wastes from industries and the transportation sector. Three stages of the recent Baltic Sea are recognized: Stage I , the oligotrophic stage, 1900-1940; Stage II, the eutrophic stage, 1940-1980; and Stage III, the stage of transition, after 1980.

INTRODUCTION

Compared to most other Large Marine Ecosystems, the Baltic is very young. Over a period spannning some 20 000 years, it has evolved from a freshwater Baltic Ice Lake, through a marine Yoldia Sea, a freshwater Ancylus Lake, and a brackish Littorina Sea to today´s less salty state, (Winterhalter *et al.* 1981). The Baltic is a 'Sea of Surprises,' forced by runoff from land by stochastic, larger inflows from the Atlantic, and by waste discharges from a drainage basin that is four times larger than the sea (Figure 7-1). There is an extensive literature on the environmental conditions of the Baltic Sea, including several condensed overviews (e.g. Elmgren 1978, 1989; Jansson 1980; Kullenberg 1983; Jansson and Velner 1995; Jansson and Dahlberg 1999; Karjalainen 1999, Kautsky and Kautsky 2000, Nehring 2001).

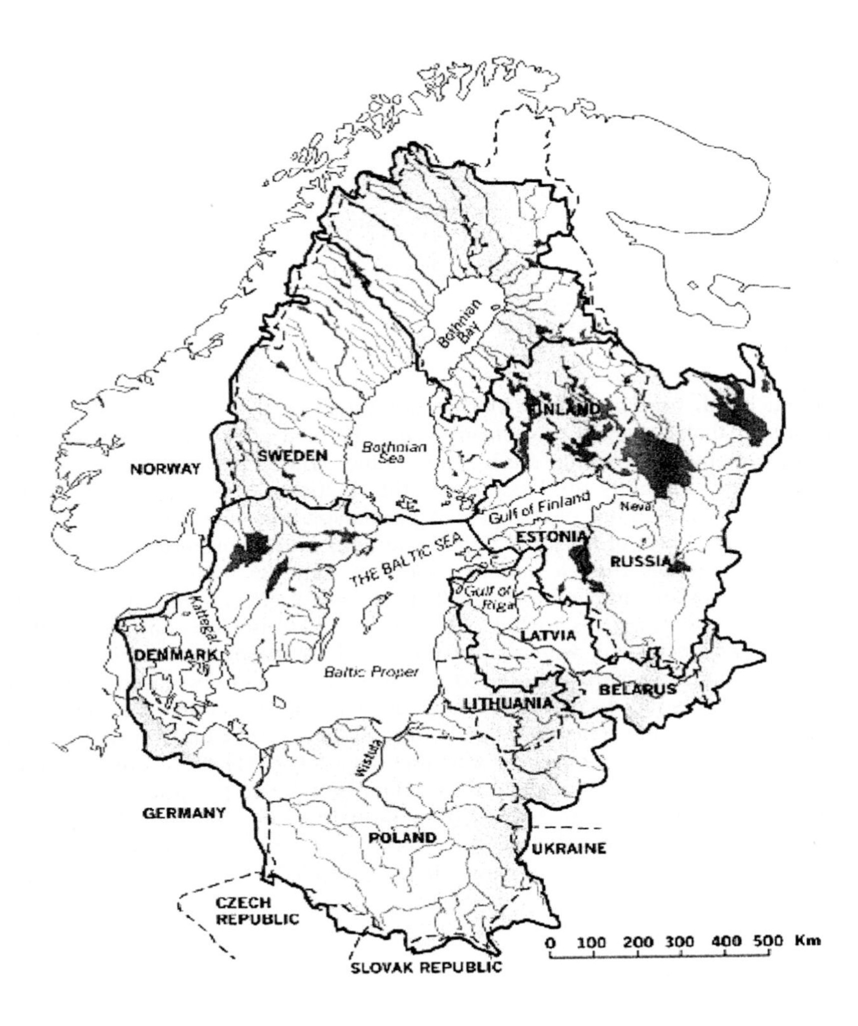

Figure 7-1 The Baltic Sea and drainage basins

THE BALTIC ECOSYSTEM

The semi-enclosed Baltic Sea is divided into several basins, the main ones being the Baltic Proper, the Gulf of Bothnia (comprising the Bothnian Sea and the Bothnian Bay), the Gulf of Finland and the Gulf of Riga (Figure 7-1 and Table 7-1).

Table 7-1. Some oceanographic data on the Baltic Sea (Voipio 1981, Håkansson *et al.* 1996). [1] between Åland and Sweden, [2] Archipelago Sea, [3] Åland Sea.

Parameters	Baltic Sea	Bothnian Bay	Bothnian Sea
Area	415,000 km^2	36 800	66 000
Area, Baltic Proper	267,000 km^2		
Length (N-S)	1300 km		
Width (W-E)	1200 km		
Volume	21 580 km^3	1490	4340
Mean depth, m	58	40	66
Maximum depth, m	459	148	293
Sill depth, m	18[1], 8[2]	25	70[3], 18
Runoff, km^{-3} yr^{-1}	446	98	91
Water residence time, yr	25-30		
Tides	2-3 cm		
Drainage area	1, 641,650 km^2		

The connections to the North Sea consist of the Kattegat and the Belt Sea. The northern parts of the basin consist of bedrock, sculptured by the movements of the inland-ice, while the southern shores consist of glacio-fluvial material and are low and sandy. The rocky shores of Sweden and Finland form extensive archipelagos, which are important areas of marine production. The large volume, the shallow sills to the North Sea, and the shallow mean depth (Table 7-1) make the Baltic an effective trap for sediments and matter emanating from the human activities in the watershed.

Hydrodynamic Processes

Measurements of the total freshwater inflow to the Baltic since the early 20th century show large variability (Figure 7-2A) with the occurrence of wet periods such as the 1920s and 1981-1990 (Bergström and Carlsson 1994). The mean annual river input of ca. 450 km^2 generates an outflow of low salinity surface water and a corresponding near bottom inflow of saline water from the North Sea over the sills between Sweden and Denmark (Öresund) and Denmark and Germany (Darss). This follows the bottom topography anti-clockwise around the island of Gotland, and is forced westwards by the shallow Archipelago Sea between Finland and Sweden and finally joins the south-flowing coastal current off the Swedish coast. A primary halocline is created at 50-70 m depth in the Baltic Proper, between a surface layer of low and locally very stable salinity (Figure 7-4) and a cold bottom layer of more variable

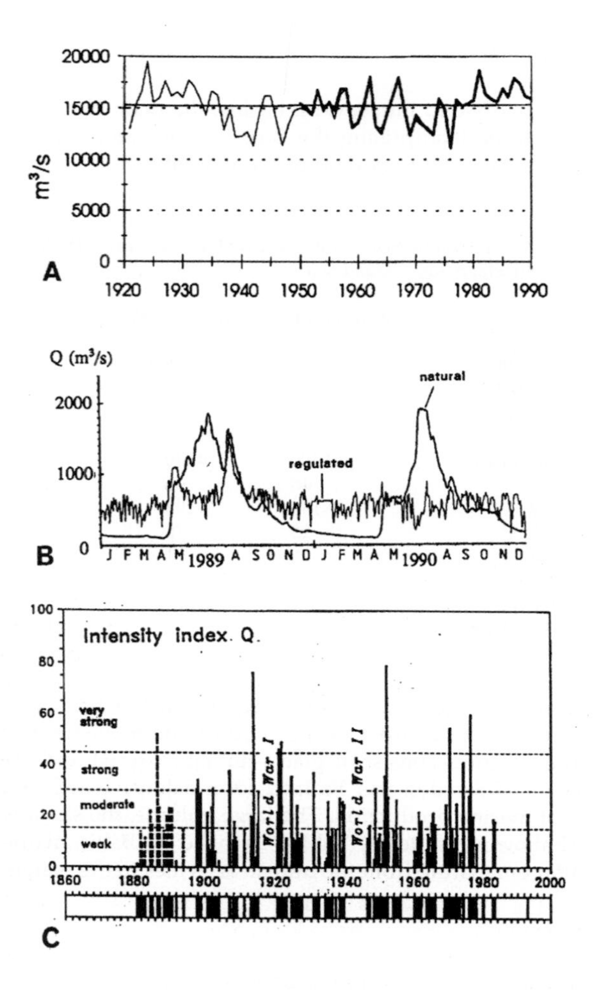

Figure 7-2, A, B and C. River runoff and inflows from the North Sea to the Baltic. A. Annual input of freshwater during the period 1921-1990. The bold curve shows mean values for 1950-1990. B. Effects of river regulation on the natural flow (reconstructed) of the lower part of the river Luleälven. (A and B from Bergström and Carlsson 1994). C. Major inflows of highly saline and oxygenated water into the Baltic Sea between 1880 and 1994, characterized by an intensity index. They seem to occur in clusters with a striking decrease in frequency and intensity after the mid-seventies (Matthäus 1995).

salinity in the 10-15 PSU range, effectively obstructing the vertical advection between the bottom water and the surface layer (Stigebrandt and Wulff 1987).

The south to north salinity gradient in the surface layer of the Baltic is to a large extent maintained by pulse-like larger inflows of water from the North Sea, generated by a specific distribution of high and low atmospheric pressure over the Baltic and the North Sea (Schinke and Matthäus 1998), resulting in stormy, westerly winds for at least ten days. The inflowing water brings not only salt, that reinforces the primary halocline, but also fish larvae, plankton and oxygen to the often stagnant, deeper bottoms. The overall oxygen consumption rate of the deep water, has been calculated to correspond to an annual carbon input of 50 g C m^{-2} yr^{-1} (Rahm 1987). The stochastic inflows come in clusters (Figure 7-2C) of highly variable volume (Matthäus1995). After the longest stagnation period on record, starting in 1977, a major inflow in 1992 carried some 300 km^3 of water of high salt and oxygen content into the Baltic, oxygenating the bottom waters. The whole Baltic Proper was oxygenated from the surface to the bottom. The salinity of the bottom water has decreased successively since 1976 due to decreased frequency and volume of inflows.

Much of the imported oxygen was consumed in oxidizing hydrogen sulphide, and the rest was soon consumed by the decomposition processes in the seabed returning the bottoms to an anaerobic state with formation of hydrogen sulphide. Figure 7-3 shows the distribution of bottom water with less than 2 ml oxygen per liter and areas of hydrogen sulphide in August 1998. The successful spawning of cod is partly regulated by the oxygenated periods in the deep water but it is also controlled by biotic factors like spawning biomass and population pressure (Figure 7-3C).

The level of intrusion of the inflowing water is determined mainly by its salinity, but also temperature has an influence. In the surface layer sharp gradient fronts in sea surface temperature were monitored by satellite imagery and shown to be an important mechanism for transporting water and substances over long distances (Kahru *et al.* 1995). Those fronts are generated mainly along coasts with straight and uniform topography, sites for upwelling, coastal jets and eddies. Typical areas are the eastern coasts of Bothnian Sea and the north-western coasts of the Gulf of Finland. During summer a thermocline at 15-20 m depth is developed in the Baltic proper, the formation of which is important for the start of a strong vernal plankton bloom.

Its large volume and shallow entrance straits give the Baltic Sea a water residence time of between 25 and 30 years. The more shallow Gulf of Bothnia has an intensive exchange of surface water with the Baltic Proper, far higher than the land runoff (Wulff and Stigebrandt 1989). The water renewal time is 4-5 years and the land runoff generates a slow, southbound current along the western coast of the Bothnian Sea (Håkansson *et al.* 1996). The deep water of the Gulf of Bothnia is derived from the surface water layer of the Baltic Proper. The halocline, being less persistent, saves the Gulf from the severe stagnation problems of the Baltic proper.

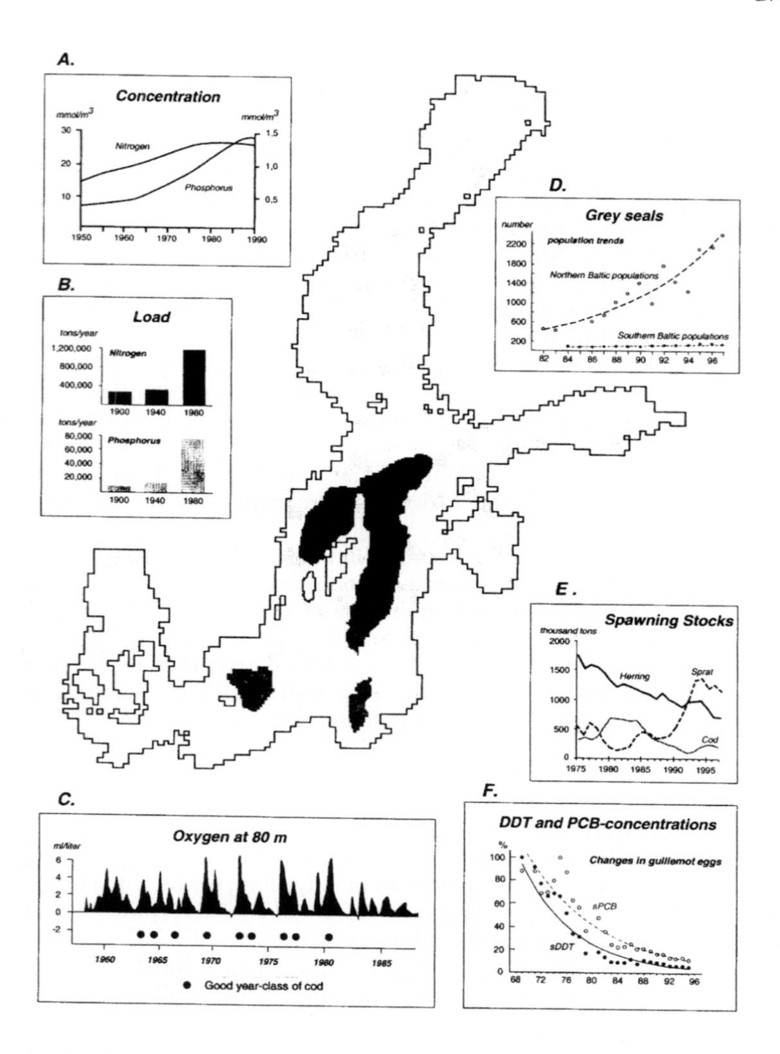

Figure 7-3. Map of the computer-processed distribution of oxygen (grey) and hydrogen sulphide (black) in the Baltic Sea in August 1998. Note the stagnant conditions in the classical Bornholm Deep, the Gdansk Deep (at the Polish coast) and in the eastern and western Gotland Basin (courtesy F. Wulff, Stockholm Univ.). A. Winter mean concentrations of phosphorus and nitrogen for the whole water column in the Baltic Proper (redrawn from a simulation in Wulff *et al.* 1990. B. Total input of nitrogen and phosphorus since 1900 (from

Jansson and Dahlberg 1999). C. Oxygen concentrations at 80 m depth in the Bornholm Deep and successful spawning of cod (from Bernes 1989). D. Number of grey seals in the Baltic since the early 1980s. The reason for the continuing depression of the southern population is still unknown (Bignert 1998). E. Spawning stocks of cod, herring and sprat, showing the critical low values for the cod population due to low recruitment and overfishing (ICES, C.M. 1998). F. DDT and PCB concentrations in the fish-eating guillemots from Gotland, Central Baltic (Bignert 1998).

The Baltic Sea can be characterized as a fiord-like estuary but a nearly closed system. This also holds for the heat balance (Omstedt and Rutgersson 2000). The climatic gradient from south to north causes up to 6 months of ice cover in the north, giving a productive season in the northern Gulf of Bothnia of 4-5 months, compared to 8-9 months in the far south with its short and intermittent spells of sea ice cover.

The semi-enclosed condition of the total basin means a near lack of tide, a few cm at most. Wind stands out as the important forcing of the vertical advection processes. Boundary mixing, through breaking of internal Kelvin waves and upwelling and downwelling, is important. Besides areas indicated in Kahru *et al.* (1995), the Polish coast is an area of frequent Ekman-transport (Siegel *et al.* 1999). Accompanying coastal jets can attain impressive magnitudes. During downwelling in the Askö-Landsort area Shaffer (1975) measured a southbound coastal jet, transporting 0.5 million m^3 of water per second. This means that the contained nutrients may be mixed over the whole northern Baltic Proper in a week.

Organism Assembly

The invasion of aquatic organisms during the late phase of the last glaciation was mainly by marine species, hardy enough to form surviving populations in the diluted medium. Figure 7-4 shows some characteristic marine organisms at their present innermost limit for persisting populations. Only a few habitat-building seaweeds penetrate far. Of seagrasses, only Zostera grows in small quantities in the Baltic Proper. Large animal taxa are missing due to their limited osmoregulation: elasmobranchs, echinoderms - except a few in the Danish Straits - radiolarians. Freshwater organisms are even fewer: luxuriant green algae such as Cladophora, a few species of pond weed, *Potamogeton spp.*, several snails and fish including pike, perch, roach, bream. The phytoplankton community is represented by a few percent of the world species total. Of zooplankton some 10 species make up most of the biomass and production. About 30 immigrants have established themselves in the system, including several recent ones (Leppäkoski 1994; Jansson 1984; Elmgren and Hill 1997). The low-diversity assembly is characterized by small size and high numbers of individuals.

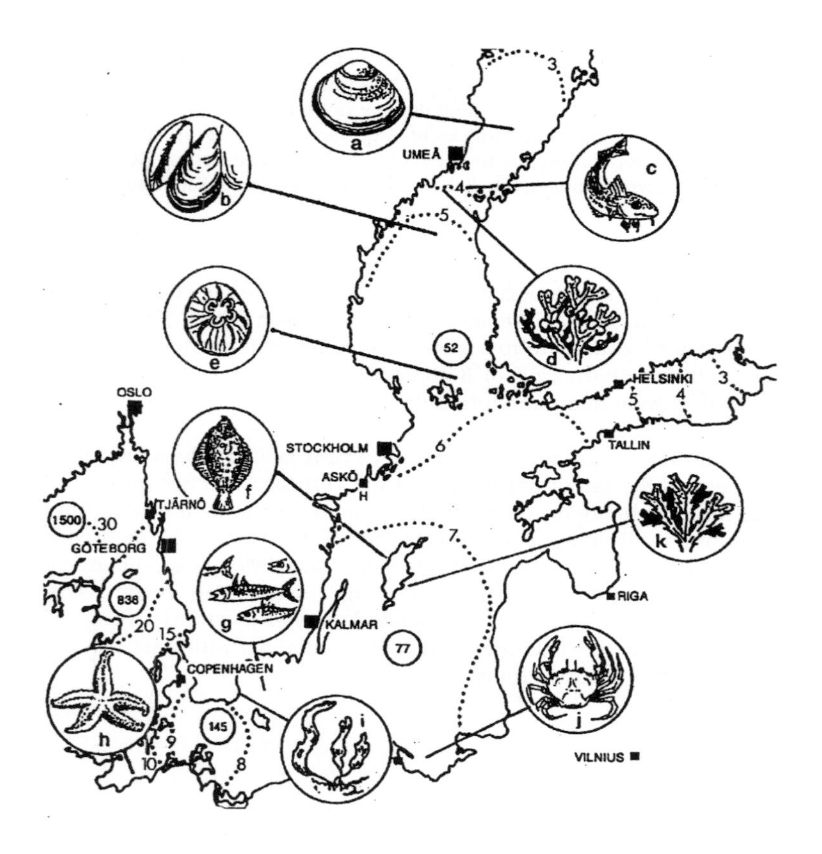

Figure 7-4. The Baltic Sea fauna and flora. Typical marine species are shown at their inner locations in some quantities. Dotted lines indicate approximate gradient of surface salinities. Figures in circles are numbers of macrofaunal species (Zenkevitch 1963). (a) *Macoma balthica*,(b) *Mytilus edulis*, (c) *Gadus morhua*, (d) *Fucus vesiculosus*, (e) *Aurelia aurita*, (f) *Pleuronectes platessa*, (g) *Scomber scombrus*, (h) *Asterias rubens*, (i) *Laminaria saccharina*, (j) *Carcinus maenas*, (k) *Fucus serratus*.

A Large-scale Ecosystem

With the physical system as a backbone, the organism assemblies are loosely tied to each other and to the physical-chemical variables forming a dynamic, complex system, which is summarized in the strongly aggregated model of Figure 7-5. The flora and associated habitats of the coast consist of hard bottoms with luxuriant seaweed and shallow soft bottoms with often extensive reed and pondweed areas. Both bottom types harbour a diverse fauna of invertebrates and fish, maintained by solar energy and flows of nutrients from land runoff and from offshore. As in the pelagic system, not all the organic matter synthesized by the plants is consumed. There is a surplus, which sinks to the bottom and forms the base of a soft bottom system of microbes and animals, decomposing the organic matter and producing biomass and products such as inorganic nutrients. There is a migration of organisms back and forth between the three subsystems in the processes of spawning, foraging and hibernation, often with both diel and annual patterns, determined by the light cycle. In the anoxic basins, oxygen is a limiting factor due to slow water exchange and accumulation of organic matter from the other systems. Phosphorus, bound in all oxygenated sediments, is released during periods of anoxia, while nitrate-nitrogen is reduced to nitrogen gas and exported to the atmosphere in large quantities (Shaffer and Rönner 1984). When a non-anoxic period follows, significant amounts of dissolved phosphate remain in the water phase, in contrast to normal conditions in freshwater. This is explained by the lower ratio of iron to phosphorus in marine systems (Gunnars and Blomqvist 1997). The nutrient pools are connected by the coastal dynamics and by the slower flows from and to the North Sea.

The coastal region, with its complex hydrodynamics and morphology, often has a lower water quality in sheltered areas, due to accumulated decomposing, organic material and oxygen deficiency. Pulses in primary production are found both in seaweed and phytoplankton communities: a heavy spring bloom, a summer maximum with strong internal cycling, and a weaker autumn bloom. Nearly half of the organic material synthesized during the spring bloom, most of it heavy diatoms, settles on the bottoms and provides about half of the annual food supply for the bottom community (Elmgren 1978). The Gulf of Bothnia shows typical polar dynamics, where spring and summer blooms have merged into one annual pulse in the middle of the short light season after the break up of the ice cover. The pelagic primary production of the Baltic Proper and the weighted average for the total Baltic have been estimated at 160 and 134 $gCm^{-2}yr^{-1}$ respectively (Elmgren 1984). Using a mass balance approach, Sandberg *et al.* (2000) studied the carbon flow in the foodwebs of Baltic Proper, Bothnian Sea and Bothnian Bay. They found a surplus of organic matter in the pelagic zones of 45, 25 and 18 $g\ Cm^{-2}\ yr^{-1}$, respectively. Except for the higher productivity of the Baltic Proper the average trophic level was lower for demersal fish in the Bothnian Sea and higher for the macrofauna in the Bothnian Bay. The lower primary production in the Bothnian Bay is compensated for by bacterioplankton using riverine organic carbon (Kuparinen *et al.* 1996). This further stresses the importance of the microbial loop, previously demonstrated for the Baltic Sea by Hagström *et al.* (1979).

Eutrophication

The most striking changes in the Baltic ecosystem since World War II are due to eutrophication from increased input of nutrient elements, primarily phosphorus and nitrogen. Society´s fossil-fuel based exploitation of natural resources has disrupted the global biogeochemical cycles, especially that of nitrogen (Jickells 1995), leading to overshoots of the capacities of basic, recycling processes (Table 7-2).

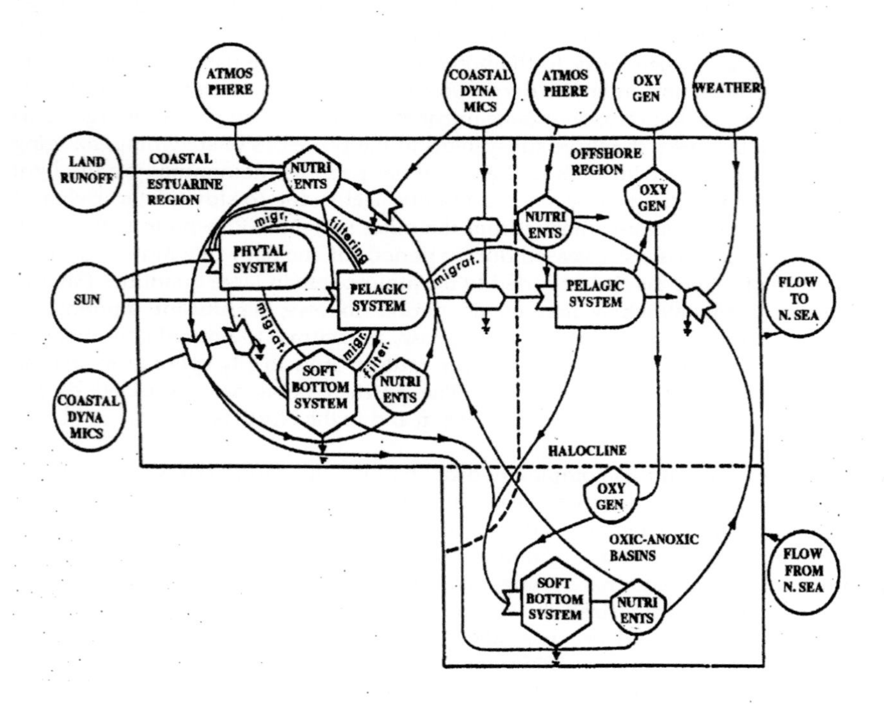

Figure 7-5. A total model of the Baltic Sea. The diffuse, functional boundary between the coastal/estuarine region and the offshore region constitutes the area of up/downwelling and coastal jets and is, approximately, where the primary halocline intersects the bottom. The functional boundary between the offshore region and the oxic-anoxic basins is represented by the persistent halocline, checking the exchange of particles and dissolved matter such as nutrients. Circles represent major forcing functions; bullets symbolize production systems (where production exceeds consumption); hexagons are consumption symbols and "birdhouses" signify pools of matter. Broad arrows and two-way arrows symbolize processes driven by both variable states (from Jansson 1981).

The loads of nitrogen and phosphorus have increased by 4 and 8 times, respectively since 1900 (Larsson *et al.*1985 and Figure 7-3B). The concentrations in the water column in the Baltic Proper show a levelling off after the 1980s (Figure 7-3A). According to Elmgren (1989) the primary production has increased by 30-70 percent, zooplankton by about 25 percent and sedimentation by 70-190 percent since the turn of the century. This increase of particles has decreased the transparency of the surface water by 2.5-3 m since 1914 (Sandén and Håkansson 1996).

In the Baltic proper both the increase in nutrients and an N/P ratio lower than 16:1, usual in oceanic waters, have clear effects on the phytoplankton community. The former dominance of diatoms, especially in the spring bloom has switched to the dinoflagellates and increased blooms of the cyanobacterian *Nodularia spumigena* (Kahru *et al.* 1994). This nitrogen-fixing organism appears in summer blooms in the nitrogen starved surface waters at N/P ratios below 7/1 (Niemi 1979). In calm and warm weather it forms floating masses, often covering a large part of the Baltic Proper, fixing annually on the order of 130 000 tons of nitrogen (Table 7-2). This is an old phenomenon, starting soon after the transition from the Ancylus Lake to the Littorina Sea some 7000 years ago at the formation of a halocline between the nutrient poor ice-generated lake water and the intruding, more nutrient-rich seawater (Bianchi *et a*l. 2000). *Nodularia* is sometimes toxic but toxic algal blooms in the Baltic are rare compared to the Swedish west coast. An exception is the dinoflagellate *Prorocentrum minimum*, which has recently become established in the Baltic Proper and central Gulf of Finland (Hajdu *et al.* 2000).

Table 7-2. Estimated annual major inputs of nitrogen and phosphorus to the Baltic Sea (Stålnacke 1996).

Agencies	Period	N, tons yr^{-1}	P, tons per yr^{-1}
Riverine	1980-1993	830 000	41 000
Coastal point sources	1990	100 000	13 000
Atmospheric deposition	1985-1989	300 000	5 000
Nitrogen fixation	1980	130 000	
Total		1 360 000	595 000

The decrease in diatoms is correlated with a decreasing ratio of dissolved silica and nitrogen, DSi:N, since the 1960s (Humborg *et al.* 2000). The increasing number of hydroelectric dams in the rivers, especially in the northern part of the drainage area, have probably trapped DSi in thriving phytoplankton and caught the inflowing mineral particles.

The Bothnian Sea is an effective sink for nitrogen and phosphorus. It imports phosphorus from the Baltic Proper and exports silica. The high N:P ratios in the silica-enriched and phosphorus-limited Bothnian Bay decrease through the Bothnian Sea to the Baltic Proper. The Gulf of Finland has the highest winter concentrations of nutrients, causing high primary production in spring and summer (Wulff *et al.* 1990). By a biogeochemical model Savchuk (2000) showed that reduction in land runoff nitrogen to the eastern Gulf would reduce local eutrophication and export to the open Gulf, whereas phosphorus reduction would increase assimilation of phosphorus imported from the west.

Changes in the composition and distribution of flora are conspicuous. The green border of the freshwater alga *Cladophora glomerata* on the rocky shores has spread seawards in the archipelagos (Waern and Pekkari 1973). The filamentous brown alga *Pilayella littoralis* increased explosively in the mid-1970s, and now covers most hard surfaces, including other seaweeds, from the surface down to some 20 meters during summer (Jansson and Kautsky 1976). Aquatic higher plants such as pondweed, *Potamogeton spp*., and reed, *Phragmites communis,* form wet "savannahs" along coasts with freshwater outlets. Due to the decreasing light penetration the bladderwrack, *Fucus vesiculosus* , shows a decreased depth range in the outer archipelagos from 11 to 8 meters, corresponding to a loss of half its biomass (Kautsky *et al.* 1986). The decline of *F. vesiculosus* in the outer Finnish archipelagos over areas of some 3 500 km^2 during 1970s and early 1980s still defies clear explanation (Kangas *et al.* 1982, Rönnberg *et al.* 1985). The decline was followed by a slow recovery in part of the area (Kangas and Niemi 1985). A complicated propagation pattern, low dispersal capacity and competition for space with the fast-growing annuals *Cladophora, Pilayella* and *Ceramium tenuicorne* (Serrão *et al.*1999) make the recovery slow, though recognizable, in areas of improved water quality. Drifting algal masses of filamentous and palmate algae accumulate on the bottoms, affecting sediment-water fluxes and bottom communities (Bonsdorff *et al.* 1997).

The estimated total primary production for the Baltic Proper, the Bothnian Sea and the Bothnian Bay of 165, 113 and 28 $gCm^{-2}\ yr^{-1}$, respectively (Elmgren 1984), constitute the basis for the softbottom system, which accordingly shows the effects of eutrophication.

Revisiting some 20 stations around the islands of Öland and Gotland in the Baltic, sampled by the Swedish fisheries biologist Hessle in 1920-1923, Cederwall and Elmgren (1980) found a more than fourfold increase in macrofaunal biomass at stations above the primary halocline. A similar increase was found for the central Bothnian Sea during a 40-year-period (Cederwall and Leonardsson 1984).

The benthic communities below the primary halocline suffer from the non-anoxic/anoxic fluctuations and formation of hydrogen sulphide. Until the 1940s the area of intermittent oxygen deficiency amounted to some 20 000 km^2 but has since expanded, due to long and frequent stagnation periods, to some 70 000 km^3 in 1990 (Elmgren 1989; Jonsson *et al.* 1990; Bernes 1988). Since then it has not increased any further. The increase in salinity in the bottom water during the 1970s, which could be traced also in the surface waters of the northern Baltic Proper through increasingly frequent appearances of the jellyfish *Cyanea capillata* in the autumn (Jansson 1978), caused a switch of the bottom communities in the southern Baltic. The previous suspension-feeding Macoma community (an association of mussels, clams, crustaceans and bristleworms, including Arctic relicts), was replaced by an Atlantic boreal community of hydrogen sulphide resistant deposit-feeders, dominated by the bristleworm *Harmothoë sarsi* (Leppäkoski 1975).

A 5-10 fold increase of carbon in the sediments during the 1980s (Jonsson and Carman 1994) has increased the oxygen consumption of the softbottoms, previously postulated through computer simulation models by Shaffer (1979) and Stigebrandt (1983). The sensitivity of the deeper basins was demonstrated by Hallberg (1973), who through sediment records showed periods of anoxia as early as the 14th and 15th centuries, coinciding with periods of warmer climate.

Modelling the effects of global warming, including increased precipitation over the Baltic region, indicates a likely gradual salinity reduction of the Baltic Proper, which in a century might give it the same surface salinity and biota of the present Bothnian Bay (Omstedt *et al.* 2000).

Fish and fisheries

Cod, herring and sprat are the main, economically important fish species in the Baltic. The combined yield of those species was around 200 000 tonnes at the end of the 1940s (Hansson and Rudstam 1990). Modern equipment and favourable environmental conditions gave higher yields: 660 000 tonnes in 1970 and 860 000 tonnes in the mid 1980s (Thurow 1989). The total yields have since declined, but remain high (Figure 7-6). The trends of the spawning stocks (Figure 7-3E) are falling for cod and herring, whereas sprat increased strongly during the 1990s, due to good spawning conditions.

Other commercially important species are eel and salmon, both moving between freshwater and saltwater and both decreasing in landings. The eel population is dependent on the natural immigration of elvers from their spawning areas in the Sargasso Sea. Westerberg (1998) estimated the number of immigrating elvers through the Öresund at 50 million individuals per year. By rearing elvers caught in Central Baltic and larvae bought in France in a lake on the island of Gotland, Westin (1998) was able to trace the tagged fertile, migrating populations southward through the Baltic, where the eels of French origin ended up in the Belt Sea while the Baltic eels found the natural migration trail through the Öresund, indicating they learn the way during migration. The Baltic salmon population, *Salmo salar*, has suffered from the extensive damming of rivers. But the compensatory artificial rearing and release of smolt by the power companies have been very successful. Unfortunately, by being based on selected breeders from a few spawning habitats only, this supported breeding has eroded the genetic diversity, and the resistance towards environmental stress has decreased for the total population. Disease and overfishing has lowered the wild population to 10-15 percent of the total standing stock, the rest being artificially reared.

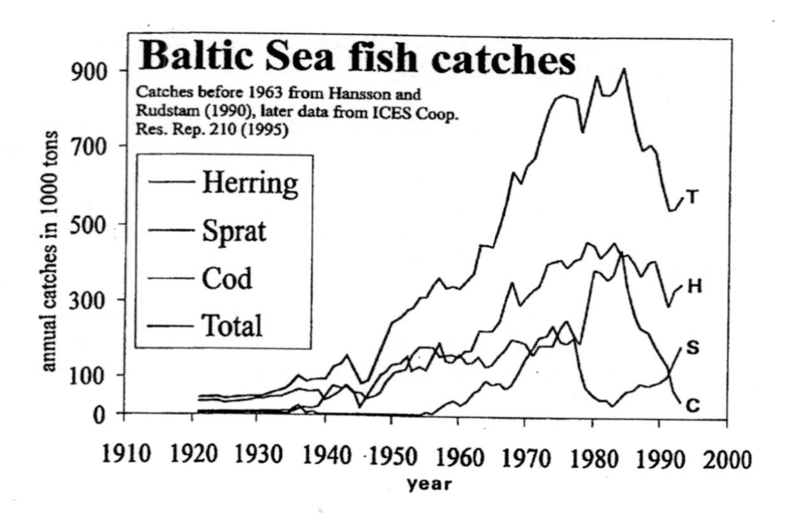

Figure 7-6. Baltic Sea fish catches.

The Baltic salmon has suffered substantially from a disease syndrome called M-74 (first encountered in 1974), causing up to 90 percent mortality of the fry in Swedish hatcheries. This has been connected to the low levels of the carotenoid astaxanthin (which give the fish flesh its red colour) and the vitamin thiamin (Snoeijs and Pedersen 1997, Bengtsson 1998). Thiamine, given to hatchery populations, restores their survival rate. The explanation is probably to be found in the foodweb dynamics. The natural salmon feeds mainly on sprat and herring. There is a strong correlation between the status of sprat, including weight, and M74 occurrence (Karlsson *et al.* 1999). Astaxanthin is produced by microalgae and certain crustaceans, thiamine by microalgae and bacteria. The decrease in the diatom populations and the observed poor nutrition conditions of sprat and herring would explain the M74 syndrome.

During the 1977-1992 stagnation period the mean weight-at-age of herring decreased by 50 percent (Flinkman 1998). This correlates well with the documented decrease in the population of large, neritic copepods in the northern Baltic Proper, due to a decrease in salinity because of high freshwater runoff and few saltwater intrusions (Vuorinen *et al.* 1999).

Freshwater species such as pike, perch and whitefish are economically important both in coastal commercial fisheries and the growing recreational fishery. Pike-perch, *Stitzostedion lucioperca*, is a highly priced species in the lake fisheries and has successively invaded the Baltic archipelagos as it favours turbid waters--a positive effect of the eutrophication process. A negative one is the shift in the fish communities in the shallow waters from pike and perch dominance to the less palatable cyprinids including bream (*Abramis brama*), roach (*Rutilus rutilus*) and ruffe *Gymnocephalus cernua* (Neuman and Sandström 1996, Figure 7-7). The increase of filamentous, brown algae on the hard bottoms has adversely affected the hatching frequency of the eggs of the spring-spawning herring which deposit the eggs on shallow hard bottoms in the Northern Baltic Proper. An approximately 75 percent mortality rate (Aneer 1987) can be compared to less than 10 percent in the spawning grounds on the Swedish westcoast which lack filamentous algal cover. Toxic exudates from the algae are the suspected cause.

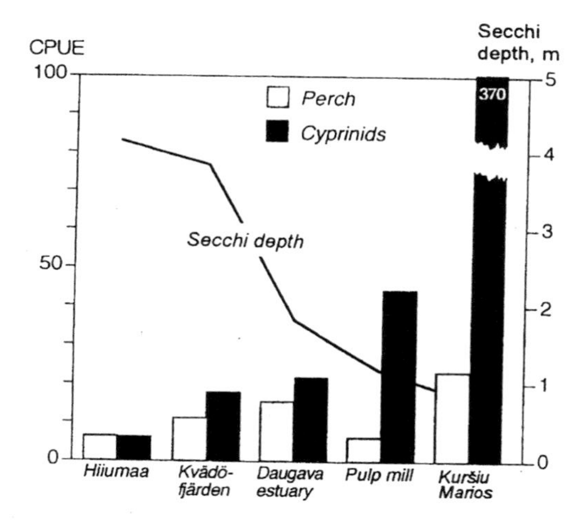

Figure 7-7. Catch-per-unit-effort (CPUE) of perch and cyprinids, related to degree of eutrophication, measured as water transparency (Secchi depth) in Baltic coastal areas: undisturbed (island of Hiiuma, Estonia; Kvädöfjärden, south Swedish coast), major estuary (Daugava, Latvia), locally polluted areas (pulp mill, Sweden; the coastal lagoon Kursiu Marios, Poland). From Neuman and Sandström (1996).

The International Baltic Sea Fishery Commission regulates the offshore fishing efforts within the Convention Area. Nevertheless overfishing is the common practice, partly due to lack of complete enforcement. Loss of spawning grounds in rivers threatens fish species such as wild salmon and sturgeon. Cod, herring, salmon and eel fisheries are at present unsustainable due to over-exploitation and impairment of the assessment of the fish stocks and for quantifying the impact of the fisheries. Introduction of seasonally and geographically closed areas should be considered, as well as areas where certain fishing practices are prohibited (HELCOM 2000).

Aquaculture in the Baltic Sea is dominated by farming of Baltic salmon. The other economically important species is the rainbow trout (*Oncorhynchus mykiss*), cultivated in large farms around Åland and in southwestern Finland with a production of some 5000 tons and with 100-200 employees in each area. Of the total nutrient loads in these areas during summer, 44 percent of the phophorus and 24 percent of the nitrogen come from the fish cultures. Increased turbidity, growth of filamentous algae and phytoplankton and fouling of fishing gear cause great concern (Kikkala 1994; Lindholm 1997).

Baltic Europe - a system of nature and society

It should be evident that, for a full understanding of the Baltic Sea, the dynamics of the drainage area, with its 85 million inhabitants in an area four times larger than that of the Sea itself, have to be considered; one has to look at the larger system of Baltic Europe (sensu Zaleski and Wojewòdka 1972). In the drainage basin, forests cover 48 percent of the area, agricultural land 20 percent and so-called unproductive land 17 percent (Sweitzer *et al.* 1996).

The increasing human population after 1800 and the surging industrialization after the two World Wars, meant an accelerating exploitation of the natural resources and increasing emissions of compounds new to nature, and therefore called contaminants. Wastes in the form of metals and chlorinated substances from forestry, agriculture and urban activities find their way through land runoff and atmospheric deposition to sediments and biota of the marine system, where they accumulate in organisms, especially those feeding on fish at the top of the foodchain, such as seabirds and seals.

Metals

By analyzing bottom sediments from different time periods Lithner *et al.* (1990) have shown the successive deposition of various metals (Figures 7-8 A-B), many of them airborn (Hallberg 1991). The anthropogenic loads of cadmium, lead and mercury to the Baltic Proper are 5 to 7 times the natural background loads, and twice as high for copper and zinc. The concentrations started to rise during the 1950s, peaked during the 1960s and 1970s, and decreased after the 1980s. Figures 7-8 C-F show how the concentrations of lead, copper and zinc in fish increased and that lead decreased during the 1980s (Bignert 1998). This might be an effect of reduced emmissions from the car traffic in Finland, Sweden, Denmark and Germany, where leaded gasoline is largely not used anymore. Mercury, the most toxic heavy metal, was a major threat to the stocks of top predators such as birds of prey and seals. Meanwhile the pollution by mercury has been drastically reduced.

the stocks of top predators such as birds of prey and seals. Meanwhile the pollution by mercury has been drastically reduced.

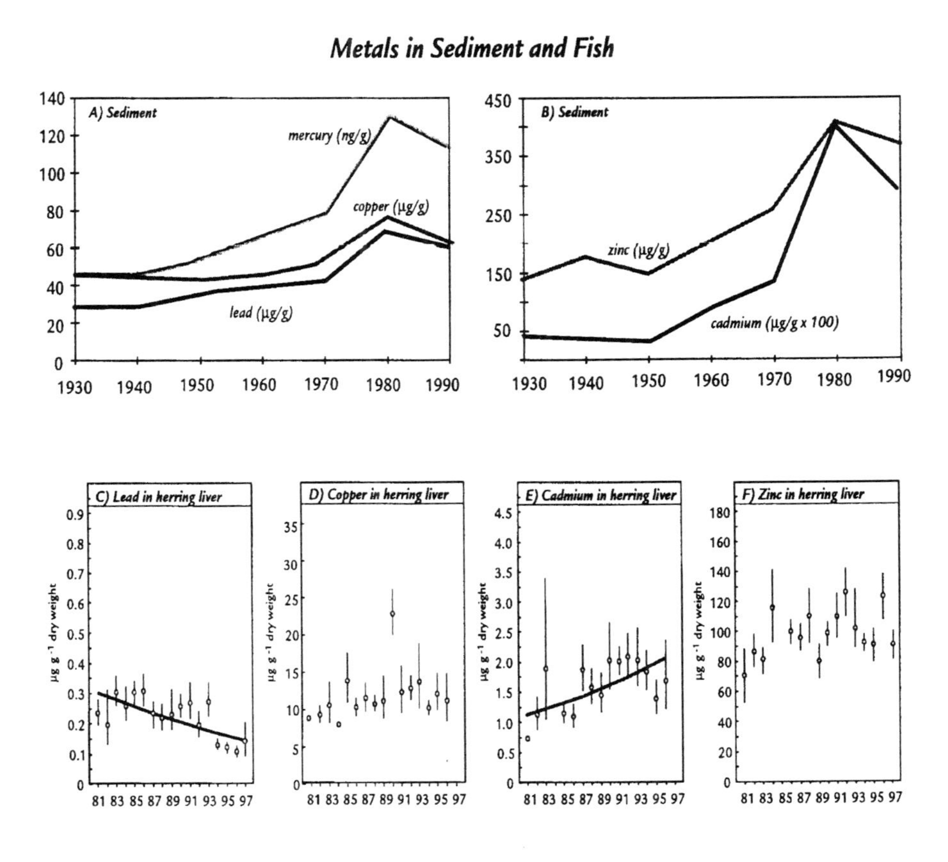

Figure 7-8. A-B Metal concentrations in sediment layers from the Baltic Proper, deposited during different time periods (redrawn from Lithner *et al.* 1990). C-F. Metal concentrations in liver of herring from the Landsort area, northern Baltic Proper (redrawn from Bignert 1998).

PCBs and DDT

Introduced during the 1930s as a component in electrical equipment, paints and plastics, PCBs were widely used during the 1960s and 1970s. The use of DDT started to extinction because of reproductive failures. After these substances were banned in the 1970s, the body burdens of both PCBs and DDT are going down, e.g. in the fish-eating guillemots (Figure 7-3F) and in herring, and the threatened populations are now increasing. The eagle population has grown from some 50 pairs of eagles in the 1950s, to around 250 pairs. Of the 100,000 grey seals at the beginning of the twentieth century, hunting took a large share and PCBs the rest, except for a few hundred in the 1970s, which have now started multiplying again (Figure 7-3F). Thanks to strong institutional measures and hard scientific and voluntary work the PCB/DDT-problem was tackled successfully. Unfortunately many hundreds of potentially hazardous chemicals continue to be discharged to the Baltic Sea. In some countries pregnant women are still advised not to eat fat fish such as herring and salmon from the Baltic Sea.

Effects of managing the landscape-seascape level

The hierarchial structure of living systems is crucial, but mostly neglected in the exploitation of natural resources. Jansson and Hammer (1999) stress the importance of patches and pulses when matching ecological and cultural diversity in a sustainable development of forestry and agriculture in the archipelagos. The shift from traditional to recreational fisheries, with accompanying common rights, neglects the traditional ecological knowledge, built on the natural patterns of pulses and patchy distributions at the seascape scale. They recommend the intermediate scale of years to a century as the critical scale for integrating natural and cultural systems.

Exploitation of wetlands has strongly affected the water quality of the Baltic Sea. Jansson et al. (1998) estimated that wetlands, amounting to some 8 percent of the total Baltic drainage area, annually retain nitrogen corresponding to 5-13 percent of the natural and anthropogenic emissions reaching the Baltic. Some 90 percent of the wetlands in Poland have been reclaimed (Rydlöv *et al.* 1991). The catastrophic floodings of the rivers Oder and Vistula in July 1997, when river water in the order of one week´s total freshwater inflow to the Baltic entered via the Polish coast, exemplify the negative effects of decreasing the buffering capacity of wetlands. A detailed study, however, by the Warnemünde Institute (IOW) did not show any significant effect of the 1997 flooding on nutrient levels and primary production in the Oder Bucht.

Institutional evolution for meeting sustainable development

Man-made capital is increasingly replaced by natural capital and the (so far) free services of the ecosystems such as continuous production of groundwater, clean air, fresh food, building material, and the recycling of wastes are increasingly eroded. The total ecosystem area needed for maintaining the 29 cities with populations of more than 250 000 inhabitants around the Baltic amounts to 0.75-1.5 times the whole drainage area, according to Folke *et al.* 1997. Folke *et al.* (1991) estimated that up to 85 percent of the total primary production in the Baltic was needed to support the commercial fish catches.

For the future development of Baltic Europe, the prime ministers ordered the planning organizations of the countries to formulate a "Vision on Strategies around the Baltic Sea 2010" (VASAB). To minimize the environmental effects of the enlarged network of transportation on land and at sea, a Joint Comprehensive Action Program (JCP) was adopted. The planned measures range from investment in waste-treatment facilities to institutional strengthening, human resource development, to development of policies and regulations, public awareness campaigns and environmental education (Helsinki Commmission 1993). To reduce the nutrient load to the Baltic Sea to the level of the 1930s, a reduction of about 65 percent for nitrogen and 80 percent of phosphorus has been estimated (Stockholm Environmental Institute 1990). As a start, the load of both nitrogen and phosphorus should be reduced by half between 1987 and 1995 (Ministerial Declaration 1988). A total of 132 "hot spots" of industrial, municipal and agricultural discharges were identified and are supervised by the JPC.

So far only Poland and Russia are considered to be close to the 50 percent nutrient reduction target for the agricultural sector, mostly due to the economic recession and the resulting decline in fertilizer use during the 1990s (Stockholm Environmental Institute 1990). The Agenda 21 for the Baltic Sea Region (Kristoferson and Stålvant 1996) has produced scenarios for the major sectors of resource exploitation.

Concerning toxic substances, a 50 percent reduction target for many persistent, organic pollutants and heavy metals has been decided (Ministerial Declaration 1988). A complete success seems far off as 50 000 to 100 000 substances are in use and some 1000 new ones are introduced every year in industry and households. Nevertheless, the success with the PCB/DDT problem should be a strong stimulus. Although established time-tables for the restoration of the Baltic Sea have not been met, most countries report significant improvements in, e.g. hygienic condition of the beaches, due to the operation of new or improved sewage treatment plants. Public awareness has increased and programs formulated for raising it further (Partanen-Hertell *et al.* 1999). The present state of the sustainability of Baltic Europe and the importance of stakeholder participation has been explored by Jansson and Stålvant (2001).

The Fourth Periodic Assessment (HELCOM 2000) summarized the present environmental state of the Baltic Sea. Since the 1980s total loads of phosphorus and organic matter have decreased. The atmospheric nitrogen loads also have decreased, perhaps because emissions from agriculture and car traffic are lower. PCB-emissions still occur through dumping of old material and incineration. Cadmium concentrations in organisms from the central Baltic Proper are increasing in spite of restrictions instituted, and no explanations have been found so far. Some 88 percent of the marine and coastal biotopes of the Baltic are affected by human activities. Several top carnivores suffer from pollution. The most threatened areas and species are in need of active protection (HELCOM 2000).

The three stages of the Baltic Sea

Summarizing the trends of the environmental variables in the system since 1900, three stages can be defined as in Table 7-3.

Table 7- 3. Developmental stages in the Baltic Proper system during the 1900s. The changes in the fish community reflect complex couplings, involving predation, overfishing and anoxic periods.
[1] Sandén and Håkansson, 1996. [2] Elmgren 1989.

VARIABLE	< 1940	1940 - 1980	1980 - 2000
Nutrients	low	increasing	leveling off
Secchi disc, meters[1]	11	8	
Pelagic prim. prod.[2] ($gCm^{-2}yr^{-1}$)	79-103	134	160
Seaweeds perennial	luxuriant	decreasing	leveling off
Seaweeds annual	low	increasing	stable
Herring, sprat	common	decreasing	increasing
Cod	common	increasing	decreasing
Salmon wild	common	decreasing	close to extinct
Salmon reared	-	increasing	leveling off
Metals in sediments	low	increasing	decreasing

Stage I covers the period to 1940, when the impact of the World War era starts. It is characterized by clear water, low primary production, luxuriant perennial, brown seaweed vegetation free from epiphytes and moderate green algal vegetation accompanied by large fish stocks, but only intermediate catches of the major, commercial fish species, including wild salmon. This is the oligotrophic period.

Stage II represents the period 1940-1980 with accelerating, industrial activities. Nutrients are increasing , macrozoobenthos is increasing except in areas of oxygen deficiency, *Fucus* vegetation is decreasing and annual green algae are spreading. Landings of herring, sprat and cod are increasing. Accumulating heavy metals and body burdens of DDTs and PCBs are causing high mortality of fish-eating top consumers. The Baltic has become eutrophic. Environmental awareness is increasing.

Stage III 1980 to present. Primary production and nutrients show leveling off with signs of silica depletion, and some recovery of the *Fucus* vegetation. The fish community shows displacement due to the decline of cod, and wild salmon is replaced by artificially reared salmon. Sea eagle and seal populations are recovering. This is the period of HELCOM and of economic transition. A shift to the western, industrialized, large-scale agriculture in Poland and in the eastern Baltic States could drive the Baltic Sea deeper into the eutrophic stage and periods of anoxia. On the other hand, a decrease in the nutrient loadings could instead increase the water transparency, re-establish the large seaweeds and increase the production of cod and salmonid fish.

ACKNOWLEDGEMENTS

I thank Ragnar Elmgren, Gotthilf Hempel and Dietwart Nehring for constructive comments on the paper and Per Jansson for computer artwork.

REFERENCES

Aneer, G. 1987. High natural mortality of Baltic herring (*Clupea harengus*) eggs caused by algal exudates? Mar. Biol. 94:163-169

Bengtsson, B.-E. 1998. Present status of the M74 Syndrome - Report from the Swedish FiRe-project. Proc. 1st Baltic Mar. Sci. Conf., Rönne, Bornholm, Denmark.ICES

Bergström, S and B. Carlsson. 1994. River runoff to the Baltic Sea:195-1990. Ambio Vol.23 No.4-5

Bernes, C. 1988. Monitor 1988. Sweden´s marine environment - ecosystems under pressure. Nat. Swedish Env. Protection Board, Stockholm, Sweden

Bernes, C. 1989. Monitor 1989. Climate and the natural environment. Nat. Swedish Env. Protection Board, Stockholm, Sweden

Bianchi, T.S., P. Westman, T. Andrén, C. Rolff and R. Elmgren. 2000. Cyanobacterial blooms in the Baltic Sea: natural or human induced? Limnol. Oceanogr. 45:716-726

Bignert, A., ed. 1998. Swedish Marine Monitoring Data. Swedish Museum of Natural History, Stockholm, Sweden

Bonsdorff, E., E. Blomqvist, J. Mattila and A. Norkko. 1997. Coastal eutrophication - causes, consequences and perspectives in the archipelago areas of the Northern Baltic Sea. Estuar. Coastal Shelf Sci. 44:63-72

Cederwall, H. and R. Elmgren. 1980. Biomass increase of benthic macrofauna demonstrates eutrophication of the Baltic Sea. Ophelia, Suppl. 1:287-304

Cederwall, H. and K. Leonardsson. 1984. Monitoring of softbottom fauna in the Gulf of Bothnia. Report for 1983. Naturvårdsverket, Rapport SNV-PM 69 (in Swedish)

Elmgren, R. 1978. Structure and dynamics of Baltic benthos communities, with particular reference to the relationship between macro- and meiofauna. Kieler Meeresforschungen Sonderheft 4, 1

Elmgren, R. 1984. Trophic dynamics in the enclosed, brackish Baltic Sea. Rapp.P. v. Reun. Cons. Int. Explor. Mer 183:152-169

Elmgren, R. 1989. Man´s impact on the ecosystem of the Baltic Sea: energy flows today and at the turn of the century. Ambio 18: 326-332

Elmgren, R. and C. Hill. 1997. Ecosystem function at low biodiversity - the Baltic example. In Ormond, R.F.G, Gage, J.D. and M.V. Angel, eds. Marine biodiversity. Patterns and processes, Cambridge Univ. Press. 319-336

Flinkman, J., Aro, E., Vuorinen. I. and M. Viitasalo. 1998. Changes in northern Baltic zooplankton and herring nutrition from 1980s to 1990s: top-down and bottom-up processes at work. Mar. Ecol. Prog. Ser. 165:127-136

Folke, C., M. Hammer and A.M. Jansson. 1991. Life-support value of ecosystems: a case study of the Baltic Sea region. Ecological Economics 3:123-137

Folke, C., Å. Jansson, J. Larsson and R. Costanza. 1997. Ecosystem appropriation by cities. Ambio 26:167-172

Gunnars, A. and S. Blomqvist. 1997. Phosphate exchange across sediment-water interface when shifting from anoxic to oxic conditions - an experimental comparison of freshwater and brackish-marine systems. Biogeochemistry 37:203-226

Hagström. Å., U. Larsson, P. Hörstedt and S. Normark. 1979. Frequency of dividing cells, a new approach to the determination of bacterial growth rate in aquatic environments. Appl. Environ. Microbiol. 37:805-812

Hajdu, S., L. Edler, I.Olenina and B. Witek, 2000. Spreading and establishment of the potentially toxic dinoflagellate Prorocentrum minimum in the Baltic Sea. Internat. Rev. Hydrobiol. 85:561-575

Håkansson, B., P. Alenius and L. Brydsten. 1996. Physical environment in the Gulf of Bothnia. Ambio Spec. Rep. 8:5-12

Hallberg, R. O. 1973. Palaeoredox conditions in the Eastern Gotland Basin during the last 400 years. Merentutkimuslait. Julk./ Havforskningsinst. Skr. 238:3-16

Hallberg, R.O. 1991. Environmental implications of metal distribution in Baltic Sea sediments. Ambio 20:309-316

Hansson, S. and L.G. Rudstam. 1990. Eutrophication and Baltic fish communities. Ambio 19: 123-125

HELCOM. 2000. Forth Periodic Assessment, in print

Helsinki Commission. 1993. The Baltic Sea Joint Comprehensive Action Programme. Baltic Sea Environmental Proceedings 48

Humborg, C., D.J. Conley, L. Rahm, F. Wulff, A. Cociasu and V. Ittekot. 2000. Silicon retention in river basins: Far-reaching effects on biogeochemistry and aquatic food webs in coastal marine environments. Ambio 29: 45-50

ICES, C.M. 1998. Report of Baltic Fisheries Assessment Working Group

Jansson, A-M. and N. Kautsky. 1976. Quantitative survey of hardbottom communities in a Baltic archipelago. In Keegan, B.F., O'Ceidigh, P.O. and P.J.S. Boaden: Biology of benthic organisms. 11th Europeam Symposium on Marine Biology, Galway, Oct.1976, 359-366. Pergamon Presss, Oxford

Jansson, Å., C. Folke and S. Langaas. 1998. Quantifying the nitrogen retention capacity of natural wetlands in the large-scale drainage basin of the Baltic Sea. Landscape Ecology 13:249-262

Jansson, A-M and M. Hammer. 1999. Patches and pulses as fundamental characteristics for matching ecological and cultural diversity: the Baltic Sea archipelago. Biodiversity and Conservation 8:71-84

Jansson, B.-O. 1972. Ecosystem approach to the Baltic problem. Ecol. Res. Comm. NFR, Bull 16. 82p

Jansson, B.-O. 1978. The Baltic - a systems analysis of a semi-enclosed sea. In Charnock, H. and G. Deacon, eds. Advances in oceanography, Plenum Press, New York. 131-183

Jansson, B.-O. 1980. Natural systems of the Baltic Sea. Ambio 9:128-136

Jansson, B.-O. 1981. Coastal lagoons: proceedings of the International Symposium on Coastal Lagoons [Montreuil]: Gauthier-Villars, [1982]

Jansson, B.-O. 1984. Baltic Sea ecosystem analysis: critical areas for future research. Limnologica (Berlin) 15:237-252

Jansson, B.-O. and H. Velner. 1995. The Baltic: the sea of surprises. In L. H. Gunderson, C. S. Holling, and S. S. Light, eds. Barriers and Bridges to the Renewal of Ecosystems and Institutions. Columbia University Press, New York, New York, USA. 292-372

Jansson, B.O. and K. Dahlberg. 1999. The environmental status of the Baltic Sea in the 1940s, today and in the future. Ambio 28: 312-319

Jansson, B.O. and C-E Stålvant. 2001. The Baltic Basin Case Study (BBCS) - Towards a sustainable Baltic Europe. Continental Shelf Research 21 (18-19):1999-2019

Jickells, T. 1995. Human impacts on the nitrogen cycle. In: Jäger, J,. Liberatore, A. and Grundlach, K., eds. Global Environmental Change and Sustainable Development in Europe. European Commission, Luxembourg: Office for official publication of the European Communities. 243p

Jonsson. P. and R. Carman. 1994. Changes in the deposition of organic matter and nutrients in the Baltic Sea during the twentieth century. Mar. Poll. Bull. 28:417-426

Jonsson, P., R. Carman, and F. Wulff. 1990. Laminated sediments in the Baltic - a tool for evaluating mass balances. Ambio 19:152-158

Kahru, M., B. Håkansson and O. Rud. 1995. Distributions of the sea-surface temperature fronts in the Baltic Sea as derived from satellite imagery. Continental Shelf Research 15:663-679

Kahru, M., U. Horstmann and O. Rud. 1994. Satellite detection of increased cyanobacteria blooms in the Baltic Sea:natural fluctuation or ecosystem change? Ambio 223:469-472

Kangas, P. and Å. Niemi. 1985. Observations of recolonization by the bladder-wrack, Fucus vesiculosus, on the southern coast of Finland. Aqua Fennica 15:133-141

Kangas, P., H. Autio, G. Hällfors, H. Luther, A. Niemi and H. Salemaa. 1982. A general model of the decline of Fucus vesiculosus at Tvärminne, south coast of Finland, in 1977-1981. Acta Bot. Fennica 118:1-27

Karjalainen, M. 1999. Effect of nutrient loading on the development of the state of the Baltic Sea - an overview. Walter and Andrée de Nottbeck Foundation Scientific Reports 17. 35p

Karlsson, L., E. Ikonen, A. Mitans and S. Hansson. 1999. The diet of salmon (Salmo salar) in the Baltic Sea and connections with the M74 Syndrome. Ambio 28:37-42

Kautsky, L. and N. Kautsky. 2000. The Baltic Sea, including Bothnian Sea and Bothnian Bay. In C.R.C Sheppard, ed: Seas at the Millennium: An environmental evaluation. Elsevier Science Ltd.

Kautsky, N., H. Kautsky, U. Kautsky and M. Waern. 1986. Decreased depth penetration of *Fucus vesiculosus* since the 1940s indicates eutrophication of the Baltic Sea. Mar. Ecol. Progr. Ser. 28:1-8

Kikkala, T. 1994. Skärgårdshavets belastningssituation och miljötillstånd. Lantbrukets och fiskodlingars belastning I kust- och skärgårdsvatten. (in Swedish) In: Blomqvist, E.A., ed. Nordiska ministerrådets skärgårdssamarbete Stockholm-Åland-Åboland, Rapport 1994: 4, 30-35

Kristoferson, L. and C-E. Stålvant. 1996. Creating an Agenda 21 for the Baltic Sea region. Main Report. Stockholm Environment Institute, SEI, Stockholm, Sweden. 59p

Kullenberg, G. 1982. Mixing in the Baltic Sea and implications for the environmental conditions. In: Nihoul, J.C.J., ed. Hydrodynamics of semi-enclosed seas. Elsevier Oceanography Series 34: 399-412.

Kullenberg, G., 1983. The Baltic Sea. In: Ketchum, B.H. Estuaries and enclosed seas. 26. Ecosystems of the world. Elsevier, Amsterdam.

Kuparinen, J., Leonardsson, K., Mattila, J. and J. Wikner, 1996. Food web structure and function in the Gulf of Bothnia, the Baltic Sea. Ambio Spe. Rep. 8:13-21

Larsson, U., Elmgren, R. and F. Wulff, 1985. Eutrophication and the Baltic Sea: causes and consequences. Ambio 14:9-14

Leppäkoski, E. 1975. Macrobenthic fauna as indicator of oceanization in the Southern Baltic. Merentutkimuslait. Julk./Havsforskningsinst. Skr. 239:280-288

Leppäkoski, E. 1994. The Baltic and the Black Sea - seriously thretatened by non-indigenous species? In: Proceeding of the Conference & Workshop Non-indigenous Estuarine and Marine Organisms (NEMO), Seattle, Washington, April 1993. U. S. Dept. of Commerce. 37-44

Lindholm, T. 1997. Vad händer i Skärgårdshavet om ingenting drastiskt görs? (in Swedish) Skärgård 1, 20:5-10

Lithner G., H. Borg, U. Grimås, A. Göthberg, G. Neumann, H. Wrådhe. 1990. Estimating the load of metals to the Baltic Sea. Ambio. Spec. Rep. (7 Sept.):7-9

Lithner, G. 1994. Biologiska effekter av metaller i limniska system. Vatten 50:64-69

Matthäus, W. 1995. Natural variability and human impacts reflected in long-term changes in the Baltic deep water conditions - a brief review. German Journal of Hydrography 47:47-65

Ministerial Declaration. 1988. The Ministerial Declaration on the Protection of the Marine Environment of the Baltic Sea Area. HELCOM, Helsinki, Finland

Nehring, D. 2001. The Baltic Sea: an example of how to protect marine ecosystems. Oceanologia 43:5-22

Neuman, E. and O. Sandström. 1996. Fish monitoring as a tool for assessing the health of Baltic coastal ecosystems. Bull. Sea Fish. Inst. 3:1-11

Niemi, Å., 1979. Blue-green algal blooms and the N:P ratio in the Baltic Sea. Acta Bot. Fenn.110:57-61

Omstedt, A., B. Gustafsson, J. Rodhe J. and G. Walin. 2000. Use of Baltic Sea modelling to investigate the water cycle and the heat balance in GCM and regional climate models. Clim. Res. 15:95-108

Omstedt, A. and A. Rutgersson. 2000. Closing the water and heat cycles of the Baltic Sea. Meteorologische Zeitschrift 9:59-66

Partanen-Hertell, M., P. Harju-Autti, K. Kreft-Burman and D. Pemberton. 1999. Raising environmental awareness in the Baltic Sea area. The Finnish Environment. Helsinki. 327 http://www.vyh.fi/eng/orginfo/publica/electro/fe_327/fe_327/.htm

Rahm, L.1987. Oxygen consumption in the Baltic Proper. Limnol Oceanogr. 32:973-978

Rönnberg, O., Letho, J. and I. Haahtela. 1985. Changes in the occurrence of *Fucus vesiculosus L.* in the Archipelago Sea, SW Finland. Ann. Bot. Fenn. 22:231-244
Rydlöv, M., H. Hasslöf, K. Sundblad, K. Robertson, and H.B. Wittgren. 1991. Wet lands – vital ecosystems for nature and society in the Baltic Sea region. World Wide Fund for Nature, WWF, Report to the HELCOM ad hoc Level Task Force
Sandberg, J., Elmgren, R., & Wulff, F. 2000. Carbon flows in Baltic Sea food webs - a re-evaluation using a mass balance approach. Journal of Marine Systems, 25, 249-260
Sandén, P. and B. Håkansson. 1996. Long-term trends in Secchi depth in the Baltic Sea. Limnol. Oceanogr. 41:346-351
Savchuk, O.P. 2000. Studies of the assimilation capacity and effects on nutrient load reductions in the eastern Gulf of Finland with a biogeochemical model. Boreal Env. Res. 5:147-163
Schinke, H. and W. Matthäus. 1998. On the causes of major Baltic inflows - an analysis of long time series. Continental Shelf Research 18:67-97
Serrão, E., S.H. Brawley, J. Hedman, L. Kautsky and G. Samuelsson. 1999. Reproductive success of *Fucus vesiculosus* *(Phaeophyceae)* in the Baltic Sea. J. Phycol. 35:254-269
Shaffer, G. 1975. Baltic coastal dynamics project - the fall down-welling regime off Askö. Contr. Askö Lab. Univ. Stockholom 7:1-16
Shaffer, G. 1979. On the phosphorus and oxygen dynamics of the Baltic Sea. Contr. Askölab., Univ. of Stockholm. 26:1-90
Shaffer, G. and U. Rönner. 1984. Denitrification of the Baltic proper deep water. Deep-Sea Research 31:197-220
Siegel H., M. Gerth and A. Mutzke. 1999. Dynamics of the Oder River plume in the Southern Baltic Sea: satellite data and numerical modelling. Continental Shelf Research 19:1143-1159
Snoeijs, P. and M. Pedersén. 1997. The significance of microalgae in astaxanthin and thiamine dynamics in the lower trophic levels of the Baltic salmon food web. Project description manuscript
Stålnacke, P. 1996. Nutrient loads to the Baltic Sea. Linköping studies in Art and Science 146. Diss. Linköping Univ. S-591 83, Linköping , Sweden
Stigebrandt, A. 1983. A model for the exchange of water and salt between the Baltic and the Skagerrak. J. Phys. Oceanogr. 13:411-427
Sweitzer, J., S.A. Langaas, and C. Folke. 1996. Land cover and population density in the Baltic Sea drainage basin:a GIS database. Ambio 25:191-198
Stigebrandt, A. and F. Wulff. 1987. A model for the exchange of water and salt between the Baltic and the Skagerrack. J. Mar. Res. 45:729-59
Stockholm Environmental Institute (SEI). 1990. "Forward to 1950: policy considerations for the Baltic Environment". Ambio Special Report 7:21-24
Thurow, F. 1989. Fishery resources of the Baltic Region. In Westing, A.H. (ed.): Comprehensive security for the Baltic. An environmental approach. PRIO, SAGE Publ., London. 54-61
Voipio, A.,ed. 1981. The Baltic Sea, Elsevier Oceanographic Series 30, 418p.

Vuorinen, I., J. Hänninen, M. Viitasalo, U. Helminen and H. Kuosa. 1999. Proportion of copepod biomass declines together with increasing salinities of the Baltic Sea. ICES J. Mar. Sci. 55:767-774

Waern, M. and S. Pekkari. 1973. Outflow studies. Nutrients and their influence on the algae in the Stockholm Archipelago during 1970. Oikos 15:155-163

Westerberg, H. 1998. Oceanographic aspects of the recruitment of eels to the Baltic Sea. Bull. Fr. Pêche Piscic. 349:177-185

Westin, L. 1998. The spawning migration of the European silver eel (*Anguilla anguilla L.*) with particular reference to stocked eel in the Baltic. Fisheries Research. 38: 257-270

Winterhalter, B., T. Flodén, H. Ignatius, S. Axberg and L. Niemistö. 1981. Geology of the Baltic Sea. In Voipio, A. (ed): The Baltic Sea, Elsevier Oceanographic Series 30:1-122

Wulff, F. and A. Stigebrandt. 1989. A time-dependent model for nutrients in the Baltic Sea. Global Biogeochemical Cycles 3:63-78

Wulff, F., A. Stigebrandt and L. Rahm. 1990. Nutrient dynamics of the Baltic Sea, Ambio 19:126-133

Zaleski, J. and C. Wojewòdka. 1972. Europa Baltyka (Baltic Europe). Ossolineum, Warszawa. (In Polish, summary in English, Russian and German).

Zenkevitch, L. 1963. Biology of the seas of the USSR. John Wiley and Sons Inc. London, New York. 955p

Large Marine Ecosystems of the World
G. Hempel and K. Sherman (Editors)

8

Overfishing Drives a Trophic Cascade in the Black Sea[1]

Georgi M. Daskalov

ABSTRACT

During the last decades environmental conditions have deteriorated in the Black Sea. Population explosions of phytoplankton and jellyfish have become frequent and several fish stocks have collapsed. In this study literature sources and long-term data are explored in order to find empirical evidence for ecosystem effects of fishing. Inverse trends of decreasing predators, increasing planktivorous fish, decreasing zooplankton and increasing phytoplankton biomass are revealed. Increased phytoplankton biomass provoked decreasing transparency and nutrient content in surface water. A massive development of jellyfish over the 1970s and the 1980's had a great impact on consumption and decreasing of the zooplankton. The dividing point in these changes occurred in the early 1970s when industrial fishing started and stocks of pelagic predators (bonito, mackerel, bluefish, dolphins) were severely depleted. The trophic cascade mechanism is invoked to explain the observed changes. According to this hypothesis, reduction in apex predators decreases consumer control and leads to higher abundance of planktivorous fish. The increased consumption by planktivorous fish causes a consecutive decline in zooplankton biomass that reduces the grazing pressure on the phytoplankton and allows its standing crop to increase. The effects of fishing and eutrophication are explored using a dynamic mass-balance model. A balanced model is built using 15 ecological groups including bacteria, phytoplankton, zooplankton, protozoan, ctenophores, medusae, chaetognaths, fishes, and dolphins. Ecosystem dynamics is simulated over 30 years assuming alternative scenarios of increasing fishing pressure and eutrophication. The changes in simulated biomass are similar in direction and magnitude to the observed data from the long-term monitoring. The cascading pattern is explained by the removal of predators and the effect of the removal on trophic interactions, while including the eutrophication effect that leads to biomass increase in all groups. The present study demonstrates that the combination of uncontrolled fishery and eutrophication can cause important

---

[1] Article first appeared in *Marine Ecology Progress Series*, 225:53-63 (January 2002) and is presented here with the permission of MEPS and the author.

alterations in the structure and dynamics of a large marine ecosystem. These findings may provide insights for ecosystem management, suggesting that conserving and restoring natural stocks of fish and marine mammals can contribute much to sustaining viable marine ecosystems.

INTRODUCTION

During recent decades, environmental conditions have deteriorated in the Black Sea. Population explosions of phytoplankton and jellyfish have become more frequent and several fish stocks have collapsed (Caddy and Griffiths 1990, Zaitsev 1993, Prodanov *et al.* 1997). Shifts in seawater quality and fisheries landings were accompanied by changes in species diversity, and structure of marine communities (Gomoiu 1985, Zaitsev and Mamaev 1997). Stocks of top predators were severely depleted during the 1950s and 1960s and the stocks of small planktivorous fish subsequently increased. By the late 1980s, an already unfavourable ecological situation was exacerbated by the coincident invasion of an exotic ctenophore, *Mnemiopsis leidyi*, and there was a severe collapse in the fisheries.

The complex dynamics of bottom-up (resource limitation) and top-down (consumer control) interactions set regulation in natural communities (McQueen *et al.* 1986, Hunter and Price 1992, Verity and Smetacek 1996). According to the bottom-up approach, organisms at on each trophic level are resource limited and their population abundance depends mainly on the availability of their food (or of nutrients for plants). The top-down view focuses on the predator control of prey dynamics. The top-predators only experience resource limitation but they regulate the abundance of their prey. At each successive lower trophic level, populations are either resource or predator regulated (Fretwell 1977). Thus, the impact of top-predators may propagate down the food web influencing lower trophic levels and ultimately regulating the primary production through a trophic cascade (Carpenter *et al.* 1985). Although the bottom-up approach is traditional in oceanography, the top-down view has been favoured in terrestrial ecology (Hairston *et al.* 1960, Fretwell 1977, Oksanen *et al.* 1981) and limnology (Hrbaček *et al.* 1961, Carpenter et al. 1985, Northcote 1988).

Most of the work on top-down effects and trophic cascades has been carried out in freshwater ecosystems, mainly relatively small lakes (e.g. Carpenter and Kitchell 1988, Persson *et al.* 1992, Carpenter and Kitchell 1993). Examples of top-down effects can be found in studies of marine littoral communities (Paine 1992, Menge *et al.* 1994, Estes and Duggins 1995). Many of these works are based on results from ecosystem experiments. The scientific knowledge on trophic interactions has been applied in lake management using so-called biomanipulation (Shapiro *et al.* 1975, Hansson *et al.* 1998). Attempts have been made to characterize explain large marine systems exploited by fisheries, as being regulated by top-down or bottom-up processes (e.g. Skud 1982, Kozlow 1983, Rudsdam *et al.* 1994, Parsons 1996). In most cases, little evidence has been found for trophic cascades extended from apex predators down to primary producers. Cascading effects resulting from the removal of marine predators by fisheries have been discussed by Parsons (1992), Estes and Duggins (1995), Steneck (1998). A recent review by Pace *et al.* (1999) summarized scientific evidence of trophic cascades in diverse ecosystems.

Previous workers have attributed the main cause of the changes in the Black Sea to cultural eutrophication (Gomoiu 1985, Caddy 1993, Zaitsev 1993, Bologa *et al.* 1995, Zaitsev and Mamaev 1997 among others). In this paper an alternative view is developed pointing out the role of fishing in structuring the marine ecosystem. It is suggested that elimination of apex marine predators by the fisheries has altered the pelagic food web by causing inverse changes across consecutive trophic levels. The trophic cascade mechanism (Carpenter *et al.* 1985, Pace *et al.* 1999) is invoked to explain the observed patterns. According to this hypothesis, reduction in apex predators decreases consumer control and leads to higher abundance of planktivorous fish. The increased consumption by planktivorous fish causes a decline in zooplankton biomass that reduces the grazing pressure on the phytoplankton and allows its standing crop to increase.

I first analyse patterns in time series indexing consecutive trophic levels from the apex predators to phytoplankton. Mass-balance dynamic modelling is then used to simulate imitate the observed ecosystem structural changes.

MATERIALS AND METHODS

Several literature sources and data series were reviewed in order to find empirical evidence for ecosystem effects of fishing. Biological time-series were collected from different sources (Table 8-1) and explored for trends and correlation. Non-linear trends were modelled using non-parametric local weighted regression ('loess', Cleveland *et al.* 1992).

The effects of fishing and eutrophication on the Black Sea ecosystem were investigated using a dynamic mass-balance model Ecopath with Ecosim (Pauly *et al.* 2000). A balanced model of the 1960s' pelagic food web was fitted using mean biomasses, production and consumption rates, and diets of 15 ecological groups (Tables 8-2, 8-3). Input data for the mass-balance model were compiled from data sources listed in Table 8-1. Biomass is expressed in carbon weight ($gC{\cdot}m^{-2}{\cdot}year^{-1}$) using conversion coefficients from Shushkina *et al.* (1983) and rates are given on an annual basis. Three fish categories represent the dominant species in each category. These are anchovy, sprat and horse mackerel in the planktivorous fish category; mackerel, bonito and bluefish represent the pelagic predators category; and whiting and spiny dogfish represent the demersal fish category.

Table 8- 1. Sources of data used in long-term data analyses and mass-balance modeling

Data	Source
Phosphate content	Bryantsev et al. (1985)
Phytoplankton	Petipa et al. (1970), Kondratieva (1979), Prodanov et al. (1997), Velikova (1998)
Zooplankton	Petipa et al. (1970), Grese (1979b), Shushkina et al. (1983), Prodanov et al. (1997)
Protozoan	Shushkina et al. (1983), Shushkina & Vinogradov (1991)
Noctiluca scintillans	Grese (1979a), Simonov et al. (1992)
Pleurobrachia pileus	Grese (1979b), Simonov et al. (1992)
Aurelia aurita	Mironov (1971), Grese (1979b), Shushkina & Musaeva (1983), Prodanov et al. (1997)
Sagitta settosa	Grese (1979b), Mashtakova (1985), Shushkina & Vinogradov (1991)
Fish larvae	Tkatcheva & Benko (1979), Dehknik (1979)
Fish	Tkatcheva & Benko (1979), Ivanov & Beverton (1985), Shul'man & Urdenko (1989), Prodanov et al. (1997), Daskalov (1998)
Dolphins	Vodyanitzkiy (1954), Sirotenko et al. (1979), Özturk (1996)

Table 8-2. Input values and results from the mass-balance model of the pelagic food web in the 1960s. Alternative entries for the "Fishing & Eutrophication 2" scenario are in bold. Flows are in gC·m^{-2}·year^{-1} and rates are on annual basis.

Groups	Biomass	P/B	Q/B	Harvest	EE	Fishing mortality
Small phytoplankton	0.20	526.30			0.98	
Large phytoplankton	0.68	223.00			0.67	
Protozoan	0.17	160.00	584.00		0.93	
Small zooplankton	0.20	65.48	420.00		0.98	
Large zooplankton	0.46	34.40	312.86		0.29	
Noctiluca scintillans	0.09	7.30	36.20		0.00	
Pleurobrachia pileus	0.02	10.95/**20.00**	29.20/**100.00**		0.02/**0.17**	
Aurelia aurita	0.03	10.95/**20.00**	29.20/**100.00**		0.00	
Sagitta setosa	0.08	36.50/**40.00**	73.00/**110.00**		0.29/**0.26**	
Fish larvae	0.01	5.00/**5.20**	20.00		0.53/**0.98**	
Planktivorous fish	0.19	1.53/**2.00**	10.99/**20.00**	0.02	0.98/**0.77**	0.13
Demersal fish	0.05	0.63	1.50	0.00	0.98	0.02
Pelagic piscivores	0.02	0.55	5.00	0.01	0.99	0.29
Dolphins	0.01	0.35	19.00	0.00	0.57	0.20
Detritus	82.53	-	-		0.67/**0.68**	

Table 8- 3. Diet composition as a part of the total consumption. Data are derived from Vodyanitzkiy (1954), Petipa *et al*. (1970), Mironov (1971) Grese (1979a), Grese (1979b), Shushkina *et al*. (1983), Shushkina & Musaeva (1983), Ivanov and Beverton (1985)

Prey	Predator											
	Protozoan	Small zooplankton	Large zooplankton	Noctiluca scintillans	Pleurobrachia pileus	Aurelia aurita	Sagitta setosa	Fish larvae	Planktivorous fish	Demersal fish	Pelagic piscivores	Dolphins
Small phytoplankton	0.220	0.450	0.300	0.170	0	0	0	0	0	0	0	0
Large phytoplankton	0.110	0.200	0.500	0.200	0	0	0	0	0	0	0	0
Protozoan	0.050	0.100	0.080	0.150	0	0	0	0	0	0	0	0
Small zooplankton	0	0.04	0.04	0	0.495	0.500	0.495	0.660	0.070	0	0	0
Large zooplankton	0	0	0	0	0.505	0.444	0.405	0.340	0.819	0	0	0
Noctiluca scintillans	0	0	0	0	0	0	0	0	0	0	0	0
Pleurobrachia pileus	0	0	0	0	0	0.005	0	0	0	0	0	0
Aurelia aurita	0	0	0	0	0	0	0	0	0	0	0	0
Sagitta setosa	0	0	0	0	0	0.050	0.1	0	0.100	0	0	0
Fish larvae	0	0	0	0	0	0.001	0	0	0.006	0	0	0
Planktivorous fish	0	0	0	0	0	0	0	0	0.005	0.977	0.595	0.960
Demersal fish	0	0	0	0	0	0	0	0	0	0.023	0	0.035
Pelagic piscivores	0	0	0	0	0	0	0	0	0	0	0.405	0.005
Dolphins	0	0	0	0	0	0	0	0	0	0	0	0
Detritus	0.620	0.210	0.080	0.480	0	0	0	0	0	0	0	0

Table 8-4. Major trends and events in the Black Sea ecosystem and fisheries before and after 1970.

prior to 1970	After 1970
Abundant stocks and operating fishery on dolphins	Low dolphins stocks, fishery stopped
Relative abundance of large pelagic predator fishes (tuna, swordfish, large bonito) in the catches	No large pelagic predator species reported in the catches in the Northern Black Sea
Regular migration and abundance of bonito and bluefish in the Black Sea	Collapse of bonito and bluefish stocks and fisheries followed by their partial recovery in the Southern Black Sea.
Regular migration and abundance of the mackerel *Scomber scombrus* in the Black Sea	Disappearance of the mackerel *Scomber scombrus* from the Northern part of the sea
Stock and fishery of the large variety of horse mackerel	No catches of the large variety of horse mackerel
Moderate standing stocks of small pelagics which are mostly important as food for predator species	Increase in the small pelagics stocks that become a base of the industrial fishery
High diversity of exploited fishes with dominance of large valuable species	Low diversity of exploited fishes with dominance of small pelagics
Relatively low abundance of gelatinous plankton	Blooms of the gelatinous plankton including invaders *Mnemiopsis* and *Beroe*
Relatively high zooplankton biomass and moderate phytoplankton biomass	Relatively low zooplankton biomass, abundant phytoplankton stock producing frequent and intense blooms, plankton structural changes
High water transparency.	Decrease in water transparency causing a dramatising reduction of the red algae *Phyllophora* in the Northwestern Black Sea
High bottom oxygen content on the shelf.	Decrease in bottom oxygen content causing hypoxia and benthic communities' degradation
High phosphate and silicate contents in surface water	Decrease in phosphate and silicate contents in surface water

The Ecopath model assumes a mass-balance over a certain period (usually a year). The model is parameterized based on two general equations (Christensen *et al.* 2000). The first equation describes the production rate, of each ecological group over the period, as a sum of the rates of fishery catch, predation mortality, biomass accumulation, net migration (emigration – immigration), and other mortality due to causes such as disease and old age.

Production = fishery catches + predation mortality + biomass accumulation + net migration + other mortality

The second equation expresses the energy balance of the group:

Consumption = production + respiration + unassimilated food

The formal expression of these equations, as well as an explicit form of the system of linear equations describing the modelled ecosystem are given by Christensen *et al.* (2000) and are available at: www.ecopath.org.

The main input parameters are the biomass (B), production/biomass (P/B), consumption/biomass (Q/B) and the ecotrophic efficiency (EE) of all groups, but if one of the parameters is unknown for a group, then the model can estimate it. The ecotrophic efficiency (EE) is the fraction of the production of each group that is used in the system. The relative production (P/B) is assumed to be equivalent to the total mortality (Z)(Allen 1971), which is the sum of all mortality due to fishing, predation, diseases, etc. Additionally, catches, assimilation, migration and biomass accumulation rates, as well as diet composition for all groups (Tables 8-2, 8-3) are required as inputs. A system of n- linear equations for n- groups is solved in order to quantify the biomass flows in the system and to estimate the missing parameters. The detritus group is formed as a model output from the 'flows to detritus' of all living groups, consisting of the non-assimilated fraction of the food and the losses due to "other mortality."

In the temporal dynamic model Ecosim (Walters *et al.* 1997) a set of coupled differential equations is derived from the Ecopath linear equations in the form:

$$dB_i/dt = g_i \sum C_{ji} - \sum C_{ij} - I_i - (M_i + F_i + E_i)B_i$$

where dB_i/dt is the biomass growth rate of the group i during the time interval dt, g_i is the growth efficiency, M_i, F_i, I_i and E_i are respectively the natural mortality, fishing mortality, immigration and emigration rates, and C_{ji} C_{ij} are consumption rates. $\sum C_{ji}$ expresses the total consumption by the group i, and $\sum C_{ij}$ is the predation on the i-group by all its predators. The consumption rates are calculated by assuming that prey biomasses consist of vulnerable and invulnerable components and the transfer rate (v_{ij}) between these components determines if the control is preponderantly top-down (Lotka-Voltera dynamics) or bottom-up (resource limitation). The vulnerability transfer rates (v_{ij}) are scaled in such a manner that the user can specify values from 0 for bottom-up control, to 1 for top-down control. In this study a top-down control was assumed and all vulnerability rates were set to 0.9 except for *Noctiluca* and fish larvae where "mixed control" v_{ij} = 0.5 was assumed.

The Ecosim model was used to simulate the temporal evolution in biomass of the groups included in response to the changes in the fishing intensity and eutrophication. In order to run Ecosim simulations, fishing mortality patterns over time must be specified for all the fishable components. Allowing an increase in primary production, as a forcing function, can simulate the effect of eutrophication.

RESULTS

Key Trends and Events Relating to Top-Down Effects

Time series of piscivorous and planktivorous fish, zooplankton and phytoplankton, and phosphate content in surface water are plotted on Figure 8-1. Inverse trends can be interpreted in the light of the trophic cascade hypothesis. The dividing point in the observed changes occurs in the early 1970s, when the stocks of pelagic predators were severely depleted. Spearman rank correlation is estimated for pairs of logarithmically transformed original series shown on Figure 8-1. The highly significant negative correlation between time-series is mainly due to the inverse trends, but not to yearly variations. Other important events attributable to the top-down effects are summarised in Table 8-4.

In the late 1960s, predator control of small planktivorous fishes (sprat, anchovy, horse mackerel) decreased to a great extent. The large dolphin population which consumed about 500 000 Tonnes of fish (mostly sprat and anchovy) diminished about 10 fold due to overexploitation (Sirotenko *et al.* 1979, Özturk 1996). Because of the great reduction of the stocks, the dolphin fishery was stopped in Bulgaria, Romania and the ex-USSR in 1966, but continued in Turkey until 1983. Before 1970, the fishery landings were dominated by large valuable migratory species, the most abundant being the bonito *Sarda sarda* followed by the Black Sea mackerel *Scomber scombrus* and the bluefish *Pomatomus saltator.* Large migratory predators such as bluefin tuna and swordfish were also regularly reported in the catch statistics. A larger morph of the horse mackerel *Trachurus mediterraneus* was caught in considerable quantities during the 1950s and 1960s. By the late 1960s all of these important fisheries had collapsed, mainly due to heavy unregulated fishing. The large variety of horse mackerel disappeared from the Black Sea and the mackerel from its main area of distribution in the northern and western Black Sea and never recovered. The bonito and bluefish stocks were severely depleted (Figure 8-1a). In the early 1970s the stocks of planktivorous fishes increased considerably and became a target for the industrial fishery. In the mid-1980s the total catch reached near one million tonnes, about 65 percent of which was anchovy and about 20 percent consisted of sprat and the smaller variety of horse mackerel (Prodanov *et al.* 1997). The consumption of zooplankton obviously increased; moreover, during the 1970s and 1980s the jellyfish biomass increased considerably. A dramatic increase of the large scyphozoan *Rhizostoma pulmo* was observed in the early 1970s (Zaitzev and Mamaev 1997) and another species: *Aurelia aurita* became dominant in the early 1980s, reaching more than 1 $kg{\cdot}m^{-2}$ of biomass (Shushkina and Musaeva 1983). By the late 1980s *Aurelia* was replaced as the dominant species by the exotic ctenophore *Mnemiopsis leydyi* which occurred in similar quantities (~2 $kg{\cdot}m^{-2}$ or $700{\cdot}10^{-6}$ tonnes in the sea, Shushkina and Vinogradov 1991). As a consequence the zooplankton biomass decreased about 2 fold compared to the 1960s (Figure 8-1b). This reduced the grazing pressure on phytoplankton and the standing crop doubled during the 1980s (Figure 8-1c). The increase of the phytoplankton biomass resulted in increased pumping of nutrients and their reduction in the surface layer since 1975 (Figure 8-1d). The frequent phytoplankton blooms and the bulk of unutilized algal biomass produced a shift in the water quality to a state characterized by low transparency and high production of detritus causing oxygen depletion and hypoxia near the bottom. Benthic mortality under low oxygen conditions acted as feedback amplifying the situation. The mortality of the stocks of mussels and other benthic filter-feeders, allowed the increase of the unutilized detritus thus causing oxygen depletion.

Most of these events have been interpreted in the light of the bottom-up approach, referring to anthropogenic eutrophication as a main causative factor. There is no doubt about the impact of eutrophication on many of the processes in the sea, but community variation in response to the changing environment can also be influenced by top-down forces. Industrial fishing has become the other powerful anthropogenic factor altering natural populations. In Table 8-5, the hypothetical cascade mechanism is sketched, showing how depletion of the top predators can change the dominance structure of the pelagic food web, leading to alternation of resource limitation and consumer control of successive trophic levels.

Dynamic mass-balance modelling

A mass-balance model of the pelagic food web in the 1960s was constructed (Table 8-2) and from the observed patterns of biomass, an attempt was made to predict the dynamics over the next decades. Results from the balanced (Ecopath) model and network analysis are discussed elsewhere (Daskalov 2000). Here I concentrate on the temporal dynamic model.

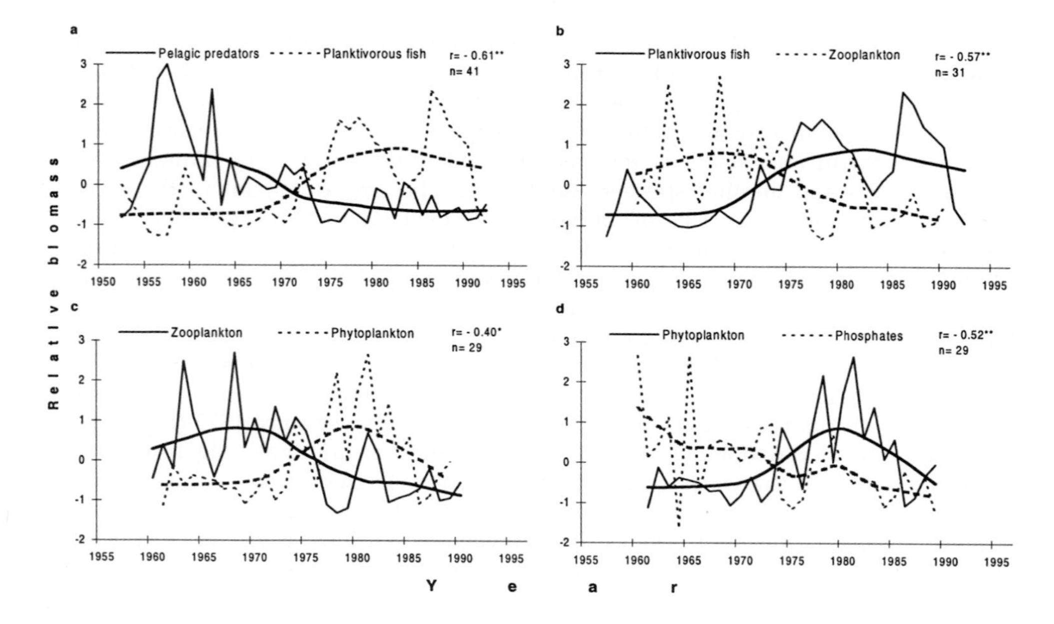

Figure 8-1. Inverse trends across consecutive trophic levels. Light curves give original data (zero mean unit variance values), bold curves give non-linear trends smoothed by loess. a. Pelagic predator fish (bonito, mackerel, bluefish) catch vs. planktivorous fish (sprat, horse mackerel) biomass, b. Planktivorous fish biomass vs. zooplankton biomass, c. Zooplankton biomass vs. phytoplankton biomass, d. Phytoplankton biomass vs. phosphate content in surface water. Spearman rank correlation is estimated for pairs of logarithmically transformed original series given on each panel. Correlation coefficients (r) are significant at $^{**}P<0.01$ and $^{*}P<0.05$ levels.

Several modeling simulations were run in order to explore the effects of fishing and eutrophication (Figure 8-2). First, the effects of fishing alone were simulated by assuming a ca 3-fold increase in fishing mortality over a 30-year period ((Figure 8-2a). The fishery forcing on the piscivorous fish and dolphins was handled in such a manner that the changes in biomass predicted by the model correspond to the observed changes. There is no information on the fishing mortality of those predatory groups, but bearing in mind the increasing capacity of the fisheries directed on them (Özturk 1996, Prodanov *et al.* 1997), the assumed values seem the lowest possible. The increase in fishing mortality of planktivorous and demersal fish corresponds to estimates by Prodanov *et al.* (1997) and Daskalov (1998).

Most of these events have been interpreted in the light of the bottom-up approach referring to the cultural eutrophication as a main causative factor. There is no doubt about the impact of the eutrophication on many of the processes in the sea, but community variation in response to the changing environment can also be influenced by top-down forces. Industrial fishing has become the other powerful anthropogenic factor altering natural populations. In Table 8-5, the hypothetical cascading mechanism is sketched, showing how depletion of the top-predators can change the dominance structure of the pelagic food web, leading to alternation of resource limitation and consumer control of successive trophic levels.

Table 8-5. Change in dominance across the pelagic food chain since the depletion of top-predators after 1970.

Trophic level	prior to 1970	After 1970
Top-predators	resource limited	-
Planktivores	consumer controlled	resource limited
Zooplankton	resource limited	consumer controlled
Phytoplankton	consumer controlled	resource limited

Several modelling simulations were run in order to explore the effects of fishing and eutrophication (Figure 8-2). First, the effects of fishing alone were simulated by assuming an approximately three-fold increase in fishing mortality over 30 years (Figure 8-2a). The fishery forcing on the piscivorous fish and dolphins was handled in such a manner that the changes in biomass predicted by the model correspond to the observed changes. There is no information on the fishing mortality of those predatory groups, but having in mind the increasing capacity of the fisheries directed on them (Özturk 1996, Prodanov *et al.* 1997) the assumed values seem the lowest possible. The increase in fishing mortality of planktivorous and demersal fish correspond to estimations by Prodanov *et al.* (1997) and Daskalov (1998).

The results indicate clear cascade patterns of all the ecosystem components (Figure 8-2a). Intensive fishing leads to about a two-fold decrease of the apex predators (large fish and dolphins). As a result of the reduced predation pressure, the biomass of the planktivorous fish increases. The increase of the demersal fish biomass is due partly to the decrease of dolphins as predators but is largely affected by the increase

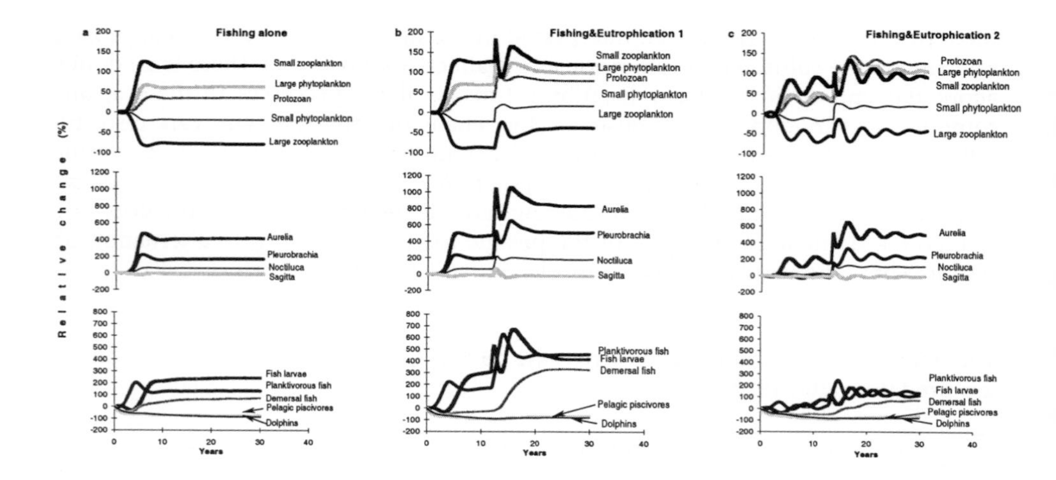

Figure 8-2. Results from the temporal dynamic model Ecosim: a. Changes due to fishing mortality forcing: "Fishing alone", b. Changes due to fishing mortality and eutrophication forcing: "Fishing & Eutrophication 1", c. Same as b. but using higher P/B and Q/B given in Table 2: "Fishing & Eutrophication 2"

of their main prey: the planktivorous fish. The parallel increase of the invertebrate zooplanktivores *Aurelia* and *Pleurobrachia* seems unexpected, keeping in mind that these species compete with fish for zooplankton food. The critical assumption leading to these results is that planktivorous fish feed in a size-selective manner and consume a dominant portion of large zooplankton in their diet (Table 8-3). There is some uncertainty about the food-selectivity of invertebrate zooplanktivores. The local populations of *Aurelia, Pleurobrachia* and *Sagitta* are regarded as unselective feeders (e.g. Mironov 1971, Sushkina and Musaeva 1983). However new studies of jellyfish feeding report controversial results of food selection (e.g. positive selection toward smaller prey (Costello and Colin 1994, Graham & Kroutil 2001), or toward larger prey (Båmstedt *et al.* 1994, Suchman and Sullivan 1998). The (assumed) opportunistic feeding resulted in about equal proportions of small and large prey in the diets of invertebrate zooplanktivores (Table 8-3). Model analyses show that results are sensitive to the relative proportions of small vs. large prey in the diets of planktivores. The structure of the zooplanktivory has its effect on the biomass evolution of the small and large zooplankton. The elimination of large zooplankton by the planktivorous fish results in its reduction. The small zooplankton biomass increases on a competitive basis fuelling the population growth of the invertebrate zooplanktivores. Because of its high *P/B, Sagitta* is an important competitor of the other zooplanktivores, but as a food for fish and *Aurelia* it is controlled from above and its biomass does not change significantly. *Aurelia* has no predators and consumes only a negligible portion of *Pleurobrachia* (Table 8-3). The increase in biomass of *Aurelia* and *Pleurobrachia* can be explained by the increase in its food, the small zooplankton. As a result the growing jellyfish populations consume a bigger quantity of large zooplankton that leads to its further decrease. The positive effect on fish larvae can be explained in the same manner, as their food consists of about 60 percent small zooplankton (Table 3). Biomass of protozoans and *Noctiluca* follow the increase in their main foods: detritus and phytoplankton. The changes in the zooplankton induce reciprocal changes in the phytoplankton that lead to an increase in the large and a decrease in the small phytoplankton.

The changes predicted by the "Fishing alone" scenario fit well with regard to the observed changes between the 1960s and 80s (Figure 8-3). In all groups the direction of the changes is the same as in the observed data (Figure 8-3a) and there is less than 0.5 percent difference in the magnitude of the change (Figure 8-3b). The model predicts a smaller than the observed decrease in zooplankton and increase in phytoplankton. As seen in Figure 8-2 changes in overall zoo- and phytoplankton biomasses are mainly due to changes in the large sized fractions of these groups. The changes in small zoo- and phytoplankton are in opposite direction. Thus the model predicts changes in the plankton size structure tending toward an increase of the small over the large zooplankton and, inversely, increasing the large over the small phytoplankton. Unfortunately, this prediction cannot be tested directly because of the lack of published size-structure data from the 1980s. However there is some evidence that such changes in the size structure might occur. Velikova (1998) reported a positive trend of the phytoplankton cell volume from 1956-1990, showing nearly a four-fold increase of the cell volume in the 1980s compared to the 1960s. Recent research on the species and stage structure of the copepod populations in relation to the dynamics of the sprat stock has shown a negative correlation between sprat biomass and copepod size indicating the effect of the size-selective consumption by the fish (Daskalov and Mihneva unpublished data).

That eutrophication was a major factor in the Black Sea during the 1970s and 1980s cannot be ignored when attempting to explain the ecosystem changes. Alternative scenarios assuming a 50 percent increase in primary production of the two phytoplankton groups due to eutrophication forcing were run in order to simulate the approximately two-fold increase in the total phytoplankton biomass. The first run was made using the same baseline parameters as in the "Fishing alone" scenarioand the second one was run using the higher P/B and Q/B coefficients for planktivores listed in Table 8-2.

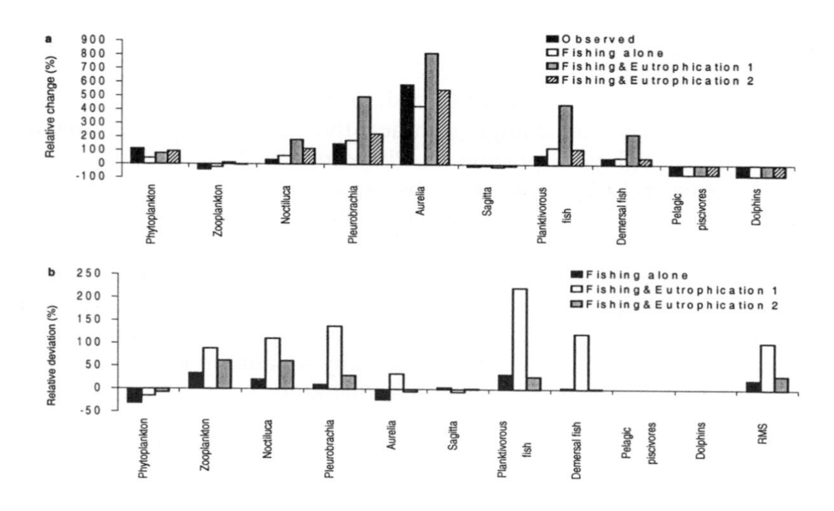

Figure 8-3a. Percent change between the 1960s and 1980s, in observed and simulated biomasses, assuming increasing fishing (Fishing alone), and increasing both fishing and eutrophication (Fishing & Eutrophication 1 and 2). **8-3b.** Relative deviation of simulated from the observed biomasses. RMS is the square root of the mean of the squares (root-mean-square) of the deviations of all groups.

The inclusion of the eutrophication does not alter radically the direction of the change in biomass of the groups (Figures 8-2 and 8-3). The primary productivity forcing by 50 percent led to a several-fold increase in apex predator biomass in the "Fishing and Eutrophication 1" scenario. An increase was not observed in the sea during the 1980s higher fishing mortality. Coefficients were applied on the dolphins and piscivores groups in order to fit their biomasses to the observed values. As a result, stronger direct and indirect effects on lower trophic levels were observed and the biomasses of planktivorous and demersal fish, and invertebrate zooplanktivores increased substantially beyond the observed values and those simulated by the "Fishing alone" option (Figure 8-3). After some initial perturbation the small zooplankton biomass stayed at the same level, while the large zooplankton biomass increased compared to the "Fishing alone" option (Figure 8-2b). As a result the total zooplankton increased compared to the level in the 1960s which contradicts the empirical trend (Figure 8-3a). Both small and large phytoplankton increased (Figure 8-2b) and the total biomass approached the observed level (Figure 8-3a). However, the changes simulated by this scenario, as a whole, differed substantially from the observed trends (Figure 8-3b).

In the "Fishing & Eutrophication 2" scenario, higher baseline mortality (P/B) coefficients were assumed to compensate for the effects of the increased primary productivity. The new P/B and Q/B coefficients used (Table 8-2) fall into the range of possible values found in the literature (Table 8-1). The fluctuations observed in the modelled trends (Figure 8-2c) are due to Lotka-Voltera effects. The results from this scenario are close to the "Fishing alone" and to the observed trends (Figure 8-3). This scenario gives the closest fit to the observed change in phytoplankton biomass (Figure 8-3b). The change in zooplankton biomass is not successfully modelled, as only a negligible decrease in total biomass is observed (Figure 8-3a). As in the previous two scenarios the decrease in total zooplankton biomass is due to the decrease in the large zooplankton. It is also possible that the small zooplankton fraction is under- presented in the observed data, due to the use of a larger mesh size (Shushkina *et al.* 1983) so that the trend in the observed total zooplankton is representative of for the change in the large zooplankton fraction. The relative deviation from the observed data in most of the groups is smaller than with the previous two scenarios. The root-mean-square (RMS) of the deviation of all groups, being a measure of the overall deviation, is close to the RMS of "Fishing alone" and significantly smaller than RMS of the "Fishing+Eutrophication 1" scenario (Figure 8-3b).

DISCUSSION

Both overfishing and cultural eutrophication are responsible for the observed changes in the Black Sea ecosystem. However, the results of this study show that top-down effects are more important determinants of the ecosystem structure.

Using the modelling experiments, there are two possible explanations—strong or weak top-down effects in the Black Sea ecosystem. According to the "strong" explanation the fishery eliminated top-predators during the 1960s. That led to a release of predator control on planktivores and their outburst during the 1980s, followed by cascading zooplankton depression and phytoplankton increase. This explains the ecosystem change solely by the action of top-down forces. The "weak" explanation includes the bottom-up influence due to the eutrophication as well. It interprets the change as an alteration of respective top-down and bottom-up controls down the food web (Table 8-5). Phytoplankton and planktivores were mostly resource controlled after 1970. Their population abundance was driven by the rise in productivity, while the zooplankton, being consumer controlled, stayed depressed. The first explanation does not seem very likely. It assumes biomass trends independent from the rise in productivity. However, it can not be tested because productivity did increase after 1970. Both explanations assume, however, the top-down influence through a trophic cascade as a factor for a radical structural and functional change in the whole pelagic ecosystem.

The analyses of the structural dynamics of the system revealed possible direct and indirect responses of different groups to top-down and bottom-up forcing. The results indicated the pivotal role of zooplanktivory in the system. Size selective feeding of small pelagic fish is of crucial importance for biomass dynamics of the intermediate and lower trophic levels. It is responsible indirectly for the trends in jellyfishes, which are a primary factor in the consumption of zooplankton. Further, predominant consumption of large zooplankton by fish has pronounced effects on size-structure and biomass trends in zoo- and phytoplankton. Total zoo- and phytoplankton biomasses are determined mainly by the dynamics of their large-size fractions which are more sensitive to top-down effects.

Observed data and modelling results both indicate that changes can be explained by the mechanism of trophic cascade. Although such results could be expected, because of using Lotka-Volterra modelling (as in Koslow 1983, Silvert 1993, 1994), here they are confirmed by the long-term data on four trophic levels; that is seldom the case in other large marine ecosystems (e.g. Koslow 1983, Rudsdam *et al.* 1994, Reid *et al.* 2000).

A question arises: are top-down effects more pronounced in the Black Sea than in other large marine ecosystems? It should be noted that the data on which this study is based, were collected during more than three decades, independently by different institutions on the eastern and western parts of the sea. These data show consistent trends, but during this time they have mainly been interpreted based on the bottom-up approach. Verity and Smetacek (1996) presented a detailed discussion of the causes of underestimating the top-down view in oceanography and the need for changing the paradigm in studying pelagic ecosystems. Another issue that seems to be overlooked by most marine scientists is the effect of industrial fishing. In the case of the Black Sea, natural patterns of abundance and behaviour of predatory fish and dolphins, described in antiquity and persisting during millennia, were suddenly destroyed by uncontrolled exploitation. Certainly, the Black Sea is a quite unique basin and some of its characteristics such its isolation from the ocean, anoxic deep layer, and relatively low taxonomic diversity contribute to its high sensitivity to human-induced disturbance. However the relatively few studies in oceanography and fishery science exploring top-down and trophic cascade effects do not allow generalizsations to be made about the relative importance of the bottom-up and top-down processes in different parts of the ocean at present. It may be that top-down effects are merely easier to detect by means of common observational and experimental approaches in relatively simple ecosystems like lakes, estuaries, enclosures or low-diversity marine systems such as the Black Sea and the

Arctic/sub-arctic ocean (Skjoldal *et al.* 1992, Shiomoto *et al.* 1997). An increasing number of studies on trophic interactions and the effects of fisheries on different marine ecosystems in the future are expected to restore the balance in understanding the roles of top-down and bottom-up factors in the ocean.

The present study demonstrates that uncontrolled fishing can cause important alterations in the structure and dynamics of a large marine ecosystem. Because most of the world fisheries are preferentially oriented toward valuable predatory species (Pauly *et al.* 1998), such effects can be expected in other areas also. These findings may provide insights for ecosystem management, suggesting that conserving and restoring natural stocks of fish and marine mammals through effective fisheries regulation, marine reserves, and other measures, together with water quality and nutrient control, can contribute much to sustaining viable marine ecosystems.

ACKNOWLEDGEMENTS

I thank: Philippe Cury (ORSTOM, France) for the invitation to the ICES/SCOR Symposium on "Ecosystem effects of fishing" where this study was initiated; Villy Christensen for reading an earlier draft and giving useful suggestions; Marliz Dimcheva and Helen Boyer for the English language corrections; three anonymous referees for evaluation and comments on the manuscript.

REFERENCES

Allen K.R. 1971. Relation between production and biomass. J Fish Res Board Can 28: 1573-1581

Båmstedt U., M.B. Martinusen, S. Matsakis. 1994. Trophodynamics of two scyphozoan jellyfishes, *Aurelia aurita* and *Cyanea capillata*, in western Norway. ICES J Mar Sci 51: 369-382

Bologa A.S., N. Bodeanu, A. Petran, V. Tiganus, Yu. P. Zaitzev. 1995 Major modifications of the Black Sea benthic and biotic biota in the last three decades. In Briand F., ed. Les mers tributaires de Méditerranée. Bulletin de l'Institut océanographique, Monaco, numéro spécial 15, CIESM Science Series n°1. p 85-110

Bryantsev V.A., D.Ya. Fashchuk, M.S. Finkel'stejn. 1985. Anthropogenic changes in the oceanographic characteristics of the Black Sea. In: Oceanographic and Fisheries Investigations of the Black Sea, VNIRO, Moskow. 3-18. In Russian

Caddy J.F. 1993. Toward a comparative evaluation of human impacts on fishery ecosystems of enclosed and semi-enclosed seas. Reviews in Fisheries Science, 1(1): 57-95

Caddy J.F., R.C. Griffiths. 1990. A perspective on recent fishery-related events in the Black Sea. Studies and Reviews. GFCM. 63, FAO, Rome. 43-71

Carpenter S.R., J.F. Kitchell, eds. 1988. Complex Interactions in Lake Communities. Springer Verlag, New York

Carpenter S.R., Kitchell J.F., eds. 1993. The trophic cascade in lakes. Cambridge, Univ Press, Cambridge

Carpenter S.R., J.F. Kitchell, J.R. Hodgson. 1985. Cascading trophic interactions and lake productivity. BioScience 35: 634-649

Christensen V., C.J. Walters, D. Pauly. 2000. Ecopath with Ecosim: a user's guide, October 2000 Edition, Fisheries Centre, University of British Columbia, Vancouver, Canada and ICLARM, Penang. Penang, Malaysia

Cleveland W.S., E. Grosse, W.M. Shyu. 1992. Local regression models. In Chambers J.M., T.J. Hastie (eds). Statistical models in S. Wadsworth Brooks/Cole Advanced Books & Software, Pacific Grove, CA. p 309-376

Costello J.H., S.P. Colin. 1994. Morphology, fluid motion and predation by the scyphomedusa *Aurelia aurita.* Mar Biol 121: 327-334

Daskalov G. 1998. Using abundance indices and fishing effort data to tune catch-at-age analyses of sprat *Sprattus sprattus*, whiting *Merlangius merlangus* and spiny dogfish *Squalus acanthias* in the Black Sea. Cah Opt Medit 35: 215-228

Daskalov G. 2000. Mass-balance modeling and network analysis of the Black Sea pelagic ecosystem. Izv. IRR Varna 25: 49-62

Dehknik T.V. 1979. Dynamics of abundance, survival and elimination of fish egg and larvae. In: Grese VN, ed. Productivity of the Black Sea. Naukova Dumka, Kiev. p 272-279. In Russian

Estes J.A., D.O. Duggins. 1995. Sea otters and kelp forests in Alaska: generality and variation in a community ecological paradigm. Ecol Monogr 65: 75-100

Fretwell S.D. 1977. The regulation of plant communities by food chains exploiting them. Perspectives in Biology and Medecine 20: 169-185

Gomoiu M-T. 1985. Problèmes concernant l'eutrophisation marine. Cercetari Marine 18: 59-96

Graham W.S. and R.M. Kroutil. 2001. Size-based prey selectivity and dietary shifts in the jellyfish *Aurelia aurita.* J Plank Res 23:67-74

Grese V.N. 1979a. On the bioproductive system of the Black Sea and its functional characteristics. Gidrobiologicheskiy jurnal, 15 (4): 3-9. In Russian

Grese V.N. 1979b. Zooplankton. In: Grese VN, ed. Productivity of the Black Sea. Naukova Dumka, Kiev. 143-169. In Russian

Hairston N.G., F.E. Smith, L.B. Slobodkin. 1960. Community structure, population control and competition. Am Nat 94: 421-424

Hansson L-A., H. Annadotter, E. Bergman, S.F. Hamrin, E. Jeppesen, T. Kairesalo, E. Luokkanen, P-A. Nilsson, M. Sondergaard, J. Strand. 1998. Biomanipulation as an Application of Food-Chain Theory: Constraints, Synthesis, and Recommendations for Temperate Lakes. Ecosystems 1:558-574

Hrbaček J, M. Dvořakova, V. Kořniek, L. Prochazkova. 1961. Demonstration of the effect of the fish stock on the species composition of zooplankton and the

intensity of metabolism of the whole plankton association. Verh Int Ver Limnol 14: 192-195
Hunter, M.D. and P.W. Price. 1992. Playing chutes and ladders: bottom-up and top-down forces in natural communities. Ecology. 73: 724-732
Ivanov L, R.J.H. Beverton. 1985. The fisheries resources of the Mediterranean, part two: Black Sea. Studies and Reviews. GFCM. 60, FAO, Rome
Kondratieva T.M. 1979. Role of different algal size groups in the phytoplankton production. In: Grese V.N., ed. Productivity of the Black Sea. Naukova Dumka, Kiev. 99-109. In Russian
Koslow J.A. 1983. Zooplankton community structure in the North Sea and Northeast Atlantic: Development and test of a biological model. Can J Fish Aquat Sci 40: 1912-1924
Mashtakova G.P. 1985. Long-term dynamics of plankton community in the Eastern Black Sea. In Oceanographic and fisheries investigations of the Black Sea, VNIRO, Moskow. 50-61. In Russian.
McQueen D.J., J.R. Post, E.L. Mills. 1986. Trophic relationships in freshwater pelagic ecosystems. Can J Fish Aquat Sci 43: 1571-1581
Menge B.A., E.L. Berlow, C.A. Blanchette, S.A. Navarette, S.B. Yamada. 1994. The keystone species concept: variation in interaction strength in a rocky intertidal habitat. Ecol Monogr 64: 249-286
Mironov G.N. 1971. Biomass and distribution of the jellyfish *Aurelia aurita* (L.) from trawl survey data in 1949-1962 in the Black Sea. Biologiya morya, Kiev 24: 49-69 In Russian
Northcote T.C. 1988. Fish in the structure and function of freshwater ecosystems: a "top-down" view. Can J Fish Aquat Sci 45: 361-379
Oksanen L, S.D. Fretwell, J. Arruda P. Niemela. 1981. Exploitation ecosystems in gradients of primary productivity. Am Nat 118:240-261
Özturk B., ed. 1996. Proceedings of the first international symposium on the marine mammals of the Black Sea. UNEP Istanbul
Pace M.L., J.J. Cole, S.R. Carpenter, J.F. Kitchell. 1999. Trophic cascades in diverse ecosystems. Trends Ecol Evol 14: 483-488
Paine R.T. 1992. Food-web analysis through field measurements of per capita interaction strength. Nature 355: 73-75
Parsons T.R. 1992. The removal of marine predators by fisheries and the impact of trophic structure. Marine Pollution Bulletin 25: 51-53
Parsons T.R. 1996. The impact of industrial fisheries on the trophic structure of marine ecosystems. In: G.A. Polis, K.O. Winemiller, eds. Food Webs. Integration of Patterns and Dynamics. Chapman and Hall, New York. 352-357
Pauly D., V. Christensen, J. Dalsgaard, R. Froese, F.C. Torres Jr. 1998. Fishing down marine food webs. Science 279: 860-863
Pauly D., V. Christensen, C. Walters. 2000. Ecopath, Ecosim, and Ecospace as tools for evaluating ecosystem impact of fisheries. ICES J Mar Sci 57: 697-706

Persson L., S. Diehl, L. Johansson, G. Andersson, S.F. Hamrin. 1992. Trophic interactions in temperate lake ecosystems: a test of food chain theory. Am Nat 140: 59-84

Petipa T.S., E.V. Pavlova, G.N. Mironov. 1970. The food web structure, utilization and transport of energy by trophic levels in the planktonic communities. In: J.H. Steele, ed. Marine Food Chains. Oliver & Boyd, Edinburgh, 142-167

Prodanov K., K. Mikhaylov, G. Daskalov, K. Maxim, E. Ozdamar, V. Shlyakhov, A. Chashchin, A. Arkhipov. 1997. Environmental management of fish resources in the Black Sea and their rational exploitation. Studies and Reviews. GFCM. 68, FAO, Rome

Rudsdam L.G., G. Aneer, M. Hilden. 1994. Top-down control in the pelagic Baltic ecosystem. Dana 10: 105-130

Shapiro J., V. Lammara, M. Lynch. 1975. Biomanipulation: an ecosystem approach to lake restoration. In: P.L. Brezonik, J.L. Fox, eds. Water Quality Management Through Biological Control. University of Florida. 85-96

Shiomoto A., K. Tadokoro, K. Nasawa, Y. Ishida. 1997. Trophic relation in the subarctic North Pacific ecosystem: possible feeding efect from pink salmon. Mar Ecol Prog Ser 150: 75-85

Shul'man G.E., S.Yu. Urdenko. 1989. Productivity of the Black Sea fishes. Naukova Dumka, Kiev. In Russian

Shushkina E.A., E.I. Musaeva. 1983. The role of medusae in the energy of the Black Sea plankton communities. Oceanology, Moskow, 23: 125-130. In Russian.

Shushkina E.A., Yu.I. Sorokin, L.P. Lebedeva, A.F. Pasternak, K.E. Kashevskaya. 1983. Production-destruction characteristics of the plankton community in the Northeastern Black Sea during the season 1978. In: Yu.I. Sorokin and V.I. Vedernikov, eds. Seasonal Changes of Black Sea Plankton. Nauka, Moskow, 178-201

Shushkina E.A., M.E. Vinogradov. 1991. Fluctuations of the pelagic communities in the open regions of the Black Sea and changes influenced by ctenophore *Mnemiopsis*. In Vinogradov M.E., ed. The variability of the Black Sea ecosystem: natural and anthropogenic factors. Nauka, Moscow. 248-261. In Russian

Silvert W. 1993. Size-structured models of continental shelf food webs. In V..Christensen and D. Pauly, eds. Trophic models of aquatic ecosystems. ICLARM Conf Proc 26, 40-43

Silvert W. 1994. Bloom Dynamics in Marine Food Chain Models with Migration. ICES C.M. 1994/R:2

Simonov A.I., A.I. Ryabinin, D.E. Gershanovitch, eds. 1992. Project "The USSR seas". Hydrometeorology and hydrochemistry of the USSR seas. Vol. 4: Black Sea, no. 1: Hydrometeorological conditions and oceanological bases of the biological productivity. Hydrometeoizdat, Sankt Peterbourg. In Russian

Sirotenko M.D., M.M. Danilevskiy, V.A. Shlyakhov. 1979. Dolphins. In: K.S. Tkatcheva, Yu.K. Benko, eds. 1979. Resources and raw materials in the Black Sea, AztcherNIRO, Pishtchevaya promishlenist, Moskow. 242-246 In Russian

Skjoldal J.D., H. Gjosaeter, H. Loeng . 1992. The Barents Sea ecosystem in the 1980s: ocean climate, plankton, and capelin growth. ICES Mar Sci Symp 195: 278-290
Skud B.E. 1982. Dominance in fish: the relation between environment and abundance. Science 216:144-149
Steneck R.S. 1998. Human influences on coastal ecosystems: does overfishing create trophic cascades? Trends Ecol Evol 13: 429-430
Suchman C.L., B.K. Sullivan. 1998. Vulnerability of copepod *Acartia tonsa* to predation by the *scyphomedusa Chrysaura quinquecirrha:* effect of prey size and behavior. Mar Biol 132:237-245
Tkatcheva K.S., Yu.K. Benko. (eds). 1979. Resources and raw materials in the Black Sea, AztcherNIRO, Pishtchevaya promishlenist, Moskow. In Russian
Velikova V. 1998. Long-term study of red tides in the western part of the Black Sea and their ecological modeling. In T. Wyatt *et al.*(eds). Proceedings of the VIII International conference on harmful algal blooms, Vigo
Verity P.G., V. Smetacek. 1996. Organism life cycles, predation and the structure of marine pelagic ecosystems. Mar Ecol Prog Ser 130: 277-293
Vodyanitzkiy V.A. 1954. On the problem of the biological productivity of water basins, with special reference to the Black Sea. Trudi Sevastopolskoy Biologicheskoy Stnantzii 8: 347-424 In Russian
Walters C., V. Christensen, D. Pauly. 1997. Structuring dynamic models of exploited ecosystems from trophic mass-balance assessments. Rev Fish Biol Fish 7: 139-172
Zaitsev Yu. 1993. Impact of eutrophication on the Black Sea fauna. Studies and Reviews. GFCM. 64, FAO, Rome. 63-86
Zaitsev Yu., V. Mamaev. 1997. Marine biological diversity in the Black Sea: A study of change and decline. UN Publications, New York

II
Upwelling Current LMEs

Large Marine Ecosystems of the World
G. Hempel and K. Sherman (Editors)

9

Interannual Variability Impacts on the California Current Large Marine Ecosystem

Daniel Lluch-Belda, Daniel B. Lluch-Cota and Salvador E. Lluch-Cota

ABSTRACT

Interannual variability at scales shorter than one century in the California Current System are described as detected by both physical and biological indices, and their likely effects on populations and the ecosystem are discussed. Major scales include the high frequency (< 10 years), mostly associated with ENSO events, the bi-decadal variation and the Very Low Frequency (VLF) variation, normally associated with regime changes. Existing evidence and possible impacts of longer term change expected from anthropogenic global warming effects are reviewed.

INTRODUCTION

The California Current LME is essentially the eastern boundary current of the north Pacific gyre, roughly extending some 40° latitudinally. Its most relevant feature consists of a surface current transporting Pacific subarctic water equatorward, together with Eastern North Pacific Central water entering from the west along its path, while Equatorial Pacific water penetrates through the southern limit of the system in the form of a deeper countercurrent (Lynn and Simpson 1987). Seasonally variable wind-driven upwelling incorporates cool, nutrient-rich waters alongshore; inshore, a narrow countercurrent often flows poleward during fall and winter. Four major faunal assemblages exist: Transitional, or the California Current itself, limited to the north by the Subarctic, to the west by the Central and by the Equatorial domains at the south (Figure 9-1).

The California Current is one of the best known marine ecosystems of the world, due in considerable proportion to the California Cooperative Oceanic Fisheries

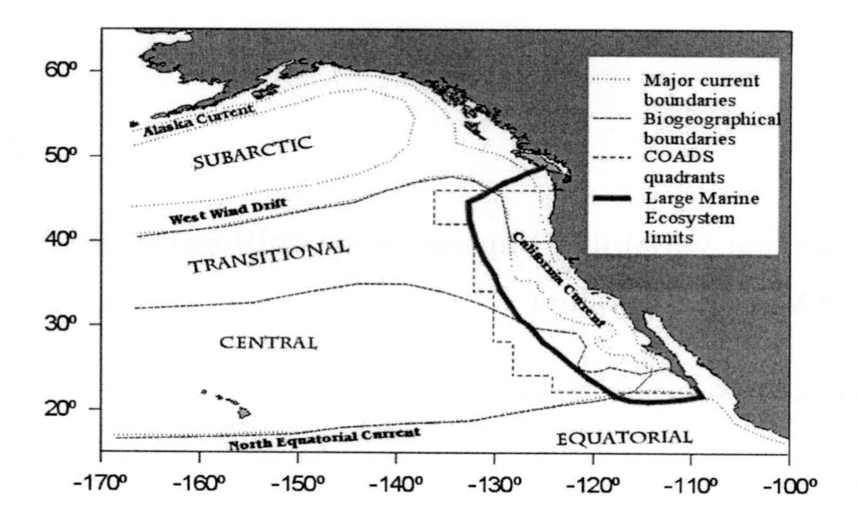

Figure 9-1. The broad geographic context of the California Current LME. Biogeographical domains from Moser *et al.* (1987), originally defined by Brinton (1962). The southern part of the California Current from Lynn and Simpson (1987). Also shown are the 76 Comprehensive Ocean-Atmosphere Data Set (COADS) database 2° x 2° quadrants later used for description of SST variation.

Investigation (CalCOFI) program, a continuous effort that has extended over 50 years. It combines a very complex array of oceanographic features, with considerable seasonal variation, as corresponds to a temperate mixing area (Lynn and Simpson 1987). Its interannual variability has been the subject of an increasing number of studies in the last decades. The main purpose of the present work is to review and discuss available information on the likely effects of the principal identified scales of < 100 years inter-year change on the ecosystem.

SCALES OF INTERANNUAL CHANGE

Figure 9-2 presents the interannual variation of sea surface temperature (SST) and sea level height (SLH) in the area, while Figure 9-3 shows the main periods of variability

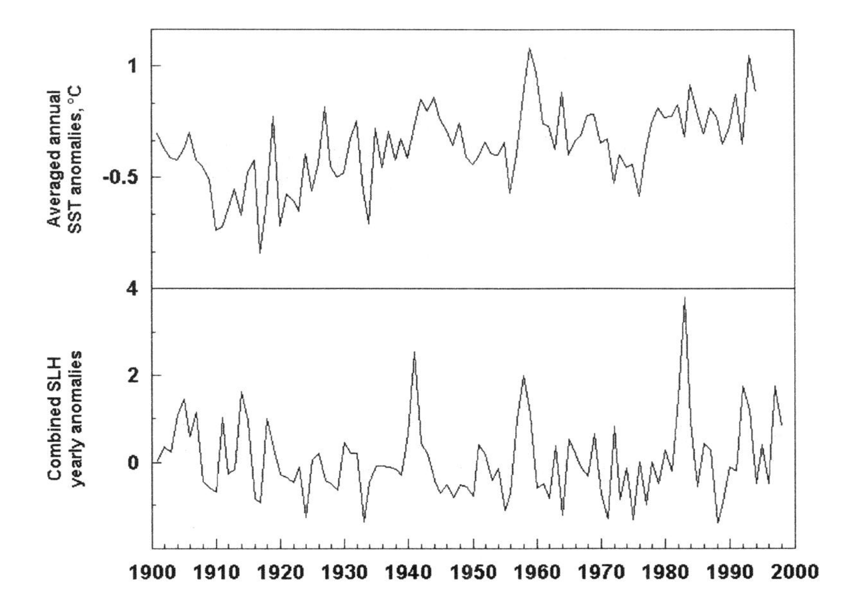

Figure 9-2. Interannual variation of sea surface temperature (SST) and sea level height (SLH). Above: monthly values of SST for each 2° x 2° COADS quadrant shown in Figure 9-1 were used to estimate the climatology for that quadrant; then the monthly anomalies (departures from the climatology) were calculated. The yearly anomaly was estimated as the averaged monthly departures. Finally, the yearly averaged anomalies from each quadrant were averaged to obtain the series. Below: the mean yearly sea level height anomalies in San Francisco and San Diego CA (COADS).

detected in the SST series by means of spectral analysis. From both figures, it is evident that physical variability in the California Current Large Marine Ecosystem (CCLME) occurs in several time scales, including high-frequency signals, decadal-scale variability and inter-decadal changes.

The high frequency (< 10 years)

Interannual variation was recognized long ago, particularly within the California Current LME. El Niño (and La Niña) events, originating in the tropics in connection with the atmospheric large-scale Southern Oscillation, were early noted by Wooster (1960) as having effects in the subtropical and temperate areas of the California Current. Nowadays, it is well established that the El Niño-

Southern Oscillation (ENSO) influence reaches the subtropical and temperate areas, and is an important source of variability in these regions (Horel and Wallace 1981, Diaz and Kiladis 1992).

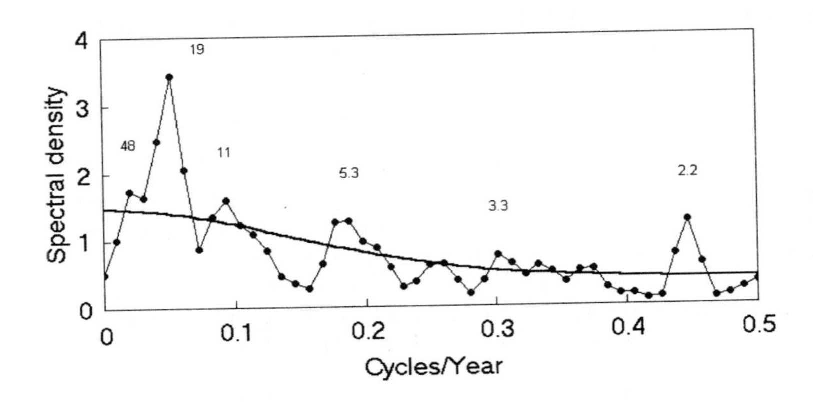

Figure 9-3. Spectral densities of the first principal component of yearly anomalies of sea surface temperature in the California Current shown in Figure 9-2 (circled curve), also included for comparison is a "red-noise" spectrum (thick curve). The anomaly series was de-meaned, de-trended, padded and tappered, then the spectral densities were estimated by smoothing the periodogram values with a 5-year term Tukey filter. The "red-noise" spectrum resulted from applying the same filter to the periodogram but with a 49-year window. Numbers indicate the main periods in years.

The spectral characterization of almost any series related to ENSO reveals two main components, one related to the stratospheric quasi-biennial oscillation (QBO, 1.5-2.9 year periods), and the other of a lower frequency (LF, 2.6-7.3 year periods)(Yasunari 1985, Xu 1992, Barnett *et al.* 1995). Both modes have been identified along the North American coast (Ware 1995, Unal and Ghil 1995).

Decadal variability (10-20 years)

Regarding longer periods, decadal-scale variability has been the subject of increasing attention as the need to understand the natural variability over this time scale becomes evident, before the proper assessment of the possible effects of global warming can be achieved. Nevertheless, decadal changes were early recognized by Hubbs (1948), who described both annual and long term persistent changes based on the presence of southern and northern fauna in the California Current.

The most documented example of decadal change is that of the mid 1970s. Its signal was early identified from the fluctuations of sardine and anchovy populations, including those of the California Current (Kawasaki 1983, Lluch-Belda *et al.* 1989, 1991). Ebbesmeyer *et al.* (1991) documented this shift using data from many time series (Kerr 1992), and others detected the signal from both atmospheric and oceanic variables (Trenberth 1990, Graham 1994, Parrish *et al.* 2000).

Most authors agree that this shift was not an isolated phenomenon, but an extreme phase of decadal-scale, natural oscillations of the climate systems (Mantua *et al.* 1997, White and Cayan 1998). Ware (1995) made an analysis of the marine climate of the northeast Pacific, including the coast of California. Among the dominant periods he noted the signal of a Bi-Decadal Oscillation (BDO, 20 years), and provided a review of mechanisms proposed to explain its origin.

One mechanism relates to astronomical phenomena, and was proposed by Royer (1989, 1993). After noting a 19-year, roughly periodic signal in records of air and sea temperatures off Alaska, he concluded that this was not predominantly wind-forced and speculated that it originated from the 18.6 year synodic period lunar cycle. However, Ware (1995) noted that the signal is dominant at mid-latitudes, where theory predicts nodal tidal forcing should not be evident (Loder and Garrett 1978, Royer 1993).

Several have proposed this variability to be ENSO-related, but authors disagree on the signal being transmitted by oceanic or atmospheric processes. For example, Graham (1994) suggested that very strong El Niño events may have extended and prolonged extra-tropical effects due to reflected Rossby waves, generated when the ENSO-related Kelvin waves reach the eastern boundary of the Pacific basin. Clarke and Lebedev (1996, 1999) noted decadal variations of the equatorial trade winds, and suggested these may remotely force variability in extratropical areas via the poleward propagation of pressure-restoring forces along the boundary (Clarke and Van Gorder 1994).

Hollowed and Wooster (1992) and Wooster and Hollowed (1995) reviewed the decadal variability in the northeast Pacific, and described a pattern of alternating warm and cold eras with an average period of 17 years. After Chelton and Davis (1982) proposed that the Alaskan and California Currents fluctuate out of phase, Hollowed and Wooster postulated that the alternating eras correspond to two environmental states in the North Pacific, and suggested that the timing and duration of the warm and cold eras are determined by the effects of the Southern Oscillation on the atmospheric circulation patterns in the northern hemisphere.

Ware (1995) pointed out that the existence of a 20-year oscillation in temperature in the eastern equatorial Pacific temperatures (Schlesinger and Ramankutty 1994) supports the idea that this scale of variability could be originated in the tropics, and could then be transmitted to higher latitudes by atmospheric teleconnections (Mysak 1986, Graham 1995). More recently, Minobe and Mantua (1999) showed the inter-decadal component in the atmospheric variability over the North Pacific to be related to the Pacific-North America pattern (PNA), a teleconnection that captures the changes of the Aleutian Low pressure cell in response to ENSO (Horel and Wallace 1981).

However, others considerate unlikely that this variability be related to ENSO (e.g. McGowan *et al.* 1998). The strongest support for this idea resulted from the study by Latif and Barnett (1994) of the North Pacific that analyzed results from a coupled model of global ocean-atmosphere general circulation, "ECHO," developed at the Max-Planck-Institute, Hamburg. The model showed a cyclic signal corresponding to the heat content of the surface layer, which results in an oscillatory behavior of simulated sea surface temperatures such that negative/positive anomalies rotate in a pattern similar to that of the North Pacific gyre but complete a cycle in about 20 years. This pattern emerges regardless of tropical variability, thus suggesting that the mechanism underlying decadal variability in the North Pacific is independent of ENSO. Ghil and Vautard (1991) and Zhang *et al.* (1997) found a cyclic component of similar period from global air and sea temperature records (respectively), and suggested that the signal likely originated outside the tropics.

Very low frequency (> 50 years)

A longer scale of inter-year change was described by Ware (1995) as very low frequency (VLF, 50-90-year period). These periods grossly correspond to the regime variation (Lluch-Belda *et al.* 1989) originally described on the basis of biological indices and later on physical indices (Mantua *et al.* 1997), and also in studies on the detection of regime shifts (Ebbesmeyer *et al.* 1991).

The regime shift during 1976-1977 marked the end of a cooling phase that had extended since the 1940s and marked the beginning of a warming trend. The previous regime shift occurred around 1941, or soon after, and represented a reversal of a very long warming trend that began about the mid 1910s (Lluch-Belda *et al.* 2001). It may be noted that the cooling period between the 1941 and 1976 regime shifts included one of the strongest brief-period warming events, the 1957-1959 event.

Seemingly, the mid 1970s shift resulted from the interaction of the Bi-Decadal Oscillation (BDO) and the VLF variability (Ware 1995, Minobe 2000); thus it is not surprising that often the two scales of variation have been confused. Though both signals probably are produced by different, yet unknown physical mechanisms, recent studies suggest that their interaction plays a major role regarding the timing and amplitude of the decadal regime shifts in the North Pacific (Minobe and Mantua 1999, Hare and Mantua 2000, Minobe 2000, Ware and Thomson 2000).

Physical effects of interannual variation

Given its high frequency, the impacts of ENSO events on the physical environment have been far more revealed than those of any other scale of interannual variation. ENSO is known to affect the physical environment of the CCS (Chelton 1981, Chelton *et al.* 1982, Wooster and Fluharty 1985, Beamish 1995, McGowan *et al.* 1998).

Longhurst (1966) analyzed the likely mechanisms resulting in the northward dispersal of the pelagic red crab (*Pleuroncodes planipes*) during the strong 1958-1960 ENSO event and concluded that four features were possibly associated with it: the Davidson Current, the intensification of the great permanent eddy south of Point Conception, the amplification of the coastal countercurrent along the Baja California coast and the overall southward drift of the main current offshore of the above features. Later, Alvariño (1976), also looking at the northward advection of pelagic red crabs, pointed out the relaxation of the California Current and the intensification of the countercurrent during 1972-1973.

Bernal (1979) found that the principal source of productivity at the northern part of the California Current, as related to macrozooplankton abundance, was southward advection and not upwelling as previously assumed. High sea level height (SLH) anomalies were found to be associated with a relaxation of the California Current southward advection (Bernal and Chelton 1984), concurrently with lower zooplankton productivity. Intensified northward advection of the California Countercurrent during ENSO events was also described by McLain and Thomas (1983). La Niña (LN) events have basically opposite effects to those of El Niño (EN) (Diaz and Kiladis 1992).

At least two types of mechanisms, oceanic and atmospheric, have been proposed to relate ENSO to the California Current. One oceanic mechanism is characterized by coastally trapped waves that originate in the tropics, propagate to higher latitudes and cause a lowering of the thermocline along the coast (Parés-Sierra and O'Brien 1989, Jacobs *et al.* 1994). In addition, Clarke and colleagues (Clarke and Van Gorden 1994, Clarke and Lebedev 1996, 1999) proposed that the same effect may merge from the poleward coastal propagation of oceanic pressure-restoring forces resulting after the weakening of the equatorial trade winds. In both cases, the outcome would be thickening of the mixed layer and increased stratification along the tropical and subtropical coasts of the northeast Pacific, which may prevent Ekman pumping from reaching the cold, subsurface waters that, under these conditions, would be deeper than normal (Wyrtki 1975, Norton *et al.* 1985, Shkedy *et al.* 1995).

The relation of ENSO to extra-tropical atmospheric variability is expressed by the so-called teleconnections. The term refers to planetary-scale, atmospheric waves forced at the tropics which may drive large changes in the atmospheric circulation patterns in distant, extra-tropical regions. Although at first the term was applied to one specific pattern (Horel and Wallace 1981, Wallace and Gutzler 1981), currently the term has been extended to describe any persistent pattern of atmospheric pressure fields, whether related to ENSO or not.

The best-documented of several teleconnections that have been proposed to relate ENSO to the California Current region (Mo and Livezey 1986, Barnston and Livezey 1987) is the Pacific-North America pattern (PNA) (Horel and Wallace 1981). This involves changes in the Aleutian Low (AL) pressure cell, such that El Niño conditions may promote an acceleration of the atmospheric cyclonic circulation, and thus of the wind-driven, poleward oceanic flow along the Gulf of Alaska (Namias and Cayan 1984, Emery and Hamilton 1985, Hamilton 1988).

Many authors have noted links between the AL and the California Current. Chelton and Davis (1982) suggested that the changes in atmospheric and oceanic circulation might be coupled, so that the Alaska gyre is strong (weak) while the California Current is weak (strong). Norton *et al.* (1985) suggested that remotely-forced atmospheric patterns may result in variations of upwelling winds along the CCS. More recently, Hollowed and Wooster (1992) and Wooster and Hollowed (1995) proposed a general scheme of the northeast Pacific, with two characteristic and opposite stages: a weakened (intensified) AL and a diminished (enhanced) flow of the Alaska gyre, accompanied by a strong (weak) flow and enhanced (diminished) upwelling in the California Current. Bakun (1996) noted that the intensification of the AL would result in an increased Ekman transport towards the coast, and probably also an increase in runoff towards the coastal areas. He suggested that these signals, qualitatively identical to those of coastal waves in terms of thermocline depth and sea level variability, could extend to the California Current during the winter expansion of the AL.

As for the bi-decadal scale of variation, a number of authors have reported it for the eastern north Pacific ocean (Ware 1995, Wooster and Hollowed 1995) as alternating multi-year warm/cool periods in temperature (both sea and air, actual and estimated), location and intensity of atmospheric pressure cells (particularly the Aleutian Low) and associated wind fields, sea level, upwelling intensity and mixed layer depth. There also seems to be a relationship between warm periods and the east-west displacement of the Aleutian Low (Emery and Hamilton 1985) and intensification/relaxation of either the Alaska gyre or the California Current (Wooster and Hollowed 1995, Brodeur *et al.* 1996).

Mantua *et al.* (1997) found evidence for reversals in the prevailing polarity of the interdecadal oscillation occurring around 1925, 1947 and 1977 at the mid-latitude North Pacific Basin and built the Pacific Decadal Oscillation index (PDO). Regarding the most recent reversal, Brodeur *et al.* (1996), Ingraham *et al.* (1998) and Ingraham (2000) reported on a surface drift simulation model (OSCURS) that showed that winter trajectories started at Ocean Station P drifted more toward the California Current before 1976 and more into the Alaska Current after 1976.

Parrish *et al.* (2000) also found several roughly bi-decadal periods resulting from changes in mid-latitude wind stress and characterized them through SST and southward wind stress associated with upwelling off San Francisco; an extensive cold period with moderate southward wind stress extending from 1908-1930, while a warm period associated with very weak southward wind stress from 1931-1947. Warm periods with intense southward wind stress occurred during 1958-1961 and 1977-1995. They concluded that surface water entering the California Current was of more subtropical origin after 1976. They also noted that after that year the surface of the central north Pacific cooled by 1°C or more while, along the North American coast and in the Gulf of Alaska, it warmed by a similar amount.

The very low frequency variation scale has been less well described, mostly because it has often been mixed with decadal scale variations, and because there are fewer completely recorded cycles. Lluch-Belda *et al.* (2001) particularly looked into this scale and concluded that regimes really correspond to long term sustained warming or cooling periods, and not to warm or cool eras as previously perceived. They further concluded that the California Current appears to be part of a subarctic/subtropical mixing area, permanently under the changing dominance of one or the other.

Thus, all the above described scales of variation would seem to have a common large scale mechanism, the predominance of either southward influence from the subarctic region or northward impact of the subtropical one: alternating states of intensified northward or southward flow at different time scales. This is illustrated by the similar trends, particularly at the decadal and very low frequency scales, of the SST anomalies of the California Current (Figures 9-2 and 9-3) and the Aleutian Low Pressure Index (Figure 9-4), recurrently linked to several scales of interannual variation.

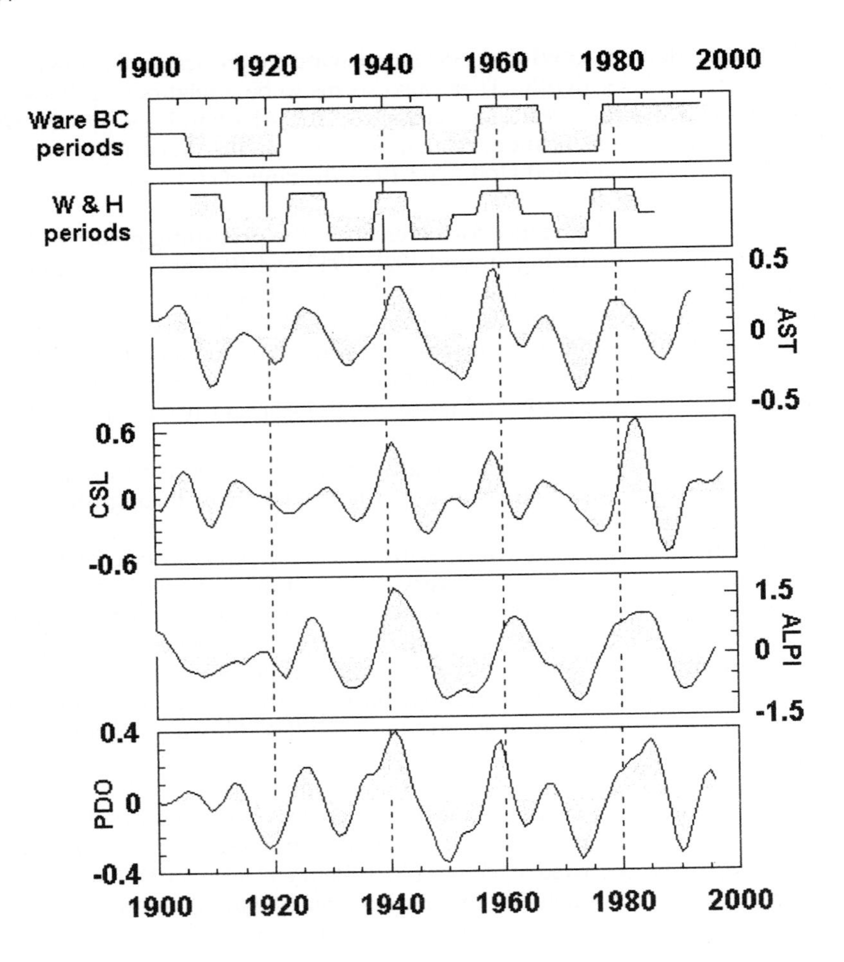

Figure 9-4. Decadal variation: Top two panels, warm and cold periods as described by Ware (1995) and Wooster and Hollowed (1995). The series in the lower four panels were obtained filtering the raw series for high (< 10 years) and low (> 30 years) frequencies by means of "Hamming" smoothing, thus leaving the 10-30 year window. AST: averaged yearly SST anomalies in the California Current; CSL: yearly standardized and averaged sea level height (SLH) anomalies in San Francisco and San Diego, CA.; ALPI: Aleutian Low Pressure Index; PDO: Pacific Decadal Oscillation (Mantua *et al.* 1997).

In short, during warming (cooling) periods the AL is stronger (weaker), the Alaska Current intensifies (relaxes), the California Current relaxes (intensifies), sea level heights increase (decrease) and the activity and size of eddies increase (decrease).

The biological consequences of interannual change

Within this scenario, ecosystem changes could be expected to show alternating trends of domination by either cold- or warm-adapted communities and their associated species, resulting from a southward progression of northern (subarctic) species, a retraction of southern (subtropical) species during cooling periods, and the opposite during warming periods.

There are, however, some complications to proving this straightforward scheme. First, from the observational standpoint we often have information based only on partial extents of the full range of variation. For instance, the CalCOFI program made an extensive and intensive latitudinal coverage of the California Current during the 1950s, but the frequency was lowered afterward; since the late 1970s the southern part (corresponding to Baja California) has been essentially untouched. During recent decades, sampling has been concentrated in the southern California Bight area (Figure 9-5). Further, some data series correspond to point locations, such as fishery landings or scale deposition rates.

Second, change occurs both at different time scales, as discussed above, and apparently at different intensities (Lluch-Belda *et al.* 2001). The interaction of scales results in further complex patterns when, for instance, an ENSO event occurs during a prolonged cooling period.

A final difficulty lies in the non-homogeneous nature of latitudinal change. In effect, there are no smooth, linearly graded transitions along the coast, but some rather abrupt changes resulting from fixed ocean features, fronts, eddies and areas of high or low productivity (Lluch-Belda 2000).

Taking into account those complications, some blurred indications of ecosystem changes may be inferred:

Primary productivity

Through both oceanic and atmospheric processes, ENSO is known to affect the ecological conditions of the marine environment along the Pacific coast of North America (Chelton 1981, Chelton *et al.* 1982, Wooster and Fluharty 1985, Beamish 1995, McGowan *et al.* 1998). Of particular significance are the effects on phytoplankton communities, because these are the basis of marine biological production (Balech 1960, Barber and Chavez 1983). In the California Current, a

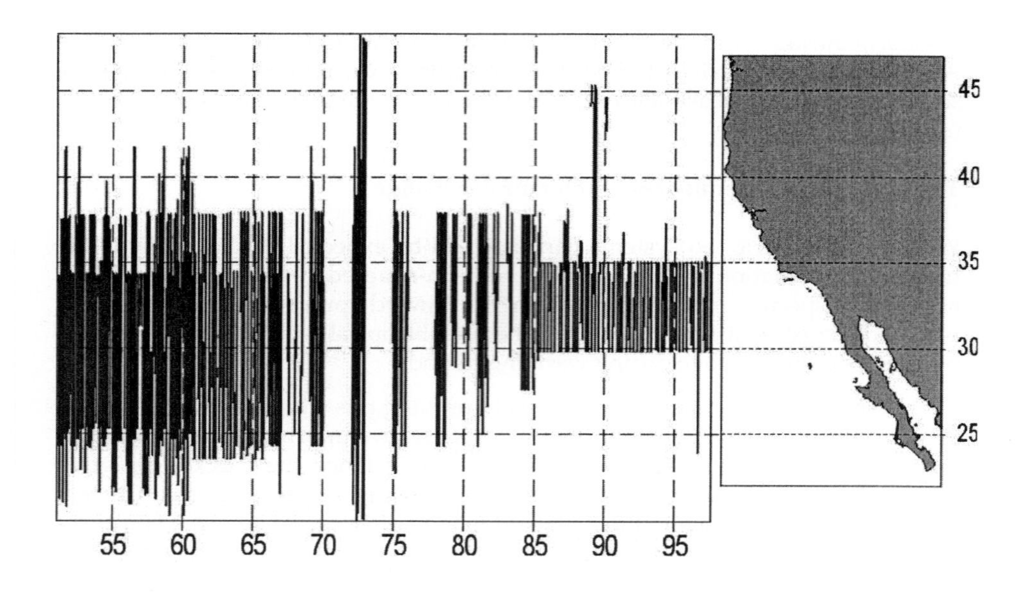

Figure 9-5. The latitudinal and temporal coverage of CalCOFI cruises. Map of California coast at right shown for reference.

region more productive than most other oceanic areas because of the existence of enrichment mechanisms, El Niño events are known to reduce biological production after partially disrupting the enrichment effect of coastal upwelling (Barber and Chavez 1986).

Several studies have analyzed satellite-derived phytoplankton biomass estimates from the California Current, particularly off the U.S. coast and in relation to the strong 1982-83 El Niño. Fiedler (1984) compared images off southern California and found lower pigment levels during March, 1983, than in April, 1982 and suggested that this reduction was associated with weakened wind-driven upwelling. Abbott and Barksdale (1991) also noted lower than average pigment concentrations during 1983 off Central California. Strub *et al.* (1990) applied empirical orthogonal function analysis to imagery of most of the California Current, including the western coast of Baja California. They showed that El Niño caused a large-scale decrease in pigment concentration starting in April 1983 and extending to 1985-1986.

Others looking at El Niño 1982-1983 effects in tropical areas also found a decrease in satellite-derived phytoplankton biomass, but with a different timing. Zuria-Jordan *et al.* (1995) analyzed data from an alongshore track extending from central Baja California to the eastern entrance to the Gulf of California. They found low pigment levels starting in September 1982 and lasting until mid-1984 at the Gulf entrance, but extending until late 1985 along the west coast of Baja California. For the tropical eastern Pacific, Fiedler (1994) found low pigment concentration along the coast of southern Mexico and Central America during 1982-1983, with values recovering to their average levels during 1984. S.E. Lluch-Cota *et al.* (1997) also examined data from the Gulf of Tehuantepec and found low levels of pigment concentrations from late 1982 until early 1984, coincident with a period of deeper than normal thermocline, warm surface temperatures, and weakened winds.

Recently, Lluch-Cota (2002) compared selected, productive areas along the Pacific coast of Central America, Mexico, and the California Current. He noted that phytoplankton biomass decreased during 1982-83 in all areas; however, in the tropical areas the decrease occurred during 1982 while, for the California Current, this was not observed until mid-1983. Though not documented for the California Current, this was noted from *in situ* studies in the Gulf of California (Lara-Lara *et al.* 1984, Valdez-Holguín and Lara-Lara 1987), which suggested that advection of warm waters during El Niño causes an initial increase of tropical phytoplankton. Lluch-Cota (2002) also noted the recovery of phytoplankton to pre-El Niño levels was larger and occurred earlier in the tropics as compared to the California Current. This result, summarized in Figure 9-6, suggests that the delayed recovery might be related to the persistence of warm conditions in the California Current due to a positive phase of the Pacific Decadal Oscillation (PDO). This points to a role of decadal-scale variability in modulating the ecological effects of El Niño.

Regarding primary productivity changes in longer time scales, Strub *et al.* (1990) analyzed the large-scale surface pigment concentration in the California Current from roughly 30°N to about 50°N during the 1979-1986 period using principal estimator patterns (PEP). The period is immersed in the warming interval that has obtained since the mid-1970s, and shows the long-term decline from positive values during 1979-1980 to negative ones during 1985-1986; however, the lowest values occur during 1983, resulting from the 1982-1983 ENSO event. This trend in thermocline deepening parallels one found by Miller (1996) that also shows two dominant signals, a decadal-scale change and the ENSO time-scale waves. There are no large-scale synoptic accounts of primary productivity prior to the very late 1970s, so no comparison with later trends is possible.

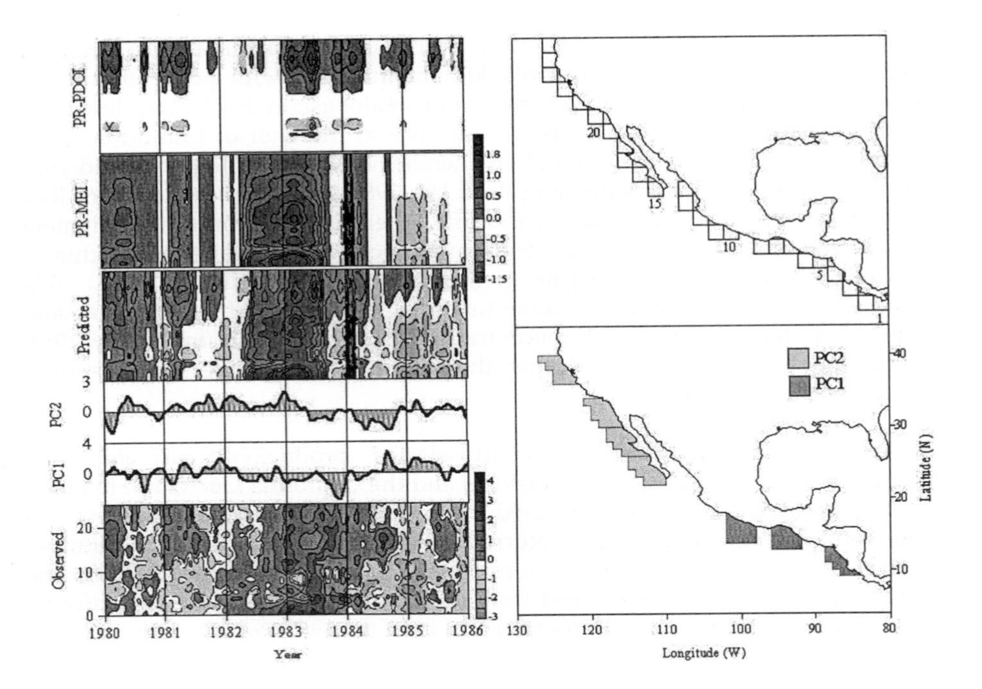

Figure 9-6. Line plots: first (PC1) and second (PC2) principal components of satellite-derived phytoplankton biomass resulting from the Principal Components analysis of a series of monthly anomalies in each of the seven areas shown in the lower map. Shades of grey in the map indicate whether the highest eigenvector (i.e., the closest relationship) was obtained from PC1 or PC2 (see Lluch-Cota 2002 for details). Contours: SST monthly anomalies in the coastal areas shown in the upper map (y-axis areas are numbered poleward), as derived from the raw (observed) data (the Reynolds dataset, see Reynolds and Smith 1994) and as estimated by multiple linear regression using the Multivariate ENSO index (MEI) and the Pacific Decadal Oscillation Index (PDO) as independent variables. PR stands for the partial regressions of each index (i.e., their contribution to the predicted values (see details in Lluch-Cota *et al.* 2001).

Polovina *et al.* (1995) found increased production after the 1976-1977 regime shift at the central north Pacific and the northeastern Pacific (Miller and Schneider 2000). On the other hand, Lluch-Cota and Teniza-Guillén (2000) analyzed the interannual variability of surface pigment concentrations along the west coast of Baja California and found no significant decline in either the long term trend (1980-1985) or during the 1982-1983 ENSO event at Punta Eugenia (about 28°N), and concluded that other local mechanisms, different from those general to the coastal area, were responsible for phytoplankton productivity at this spot, as previously proposed by Roesler and Chelton (1987).

Other primary producers include macroalgae, of which the giant kelp, *Macrocystis* is a very relevant one in the coastal area from Santa Cruz, California to central Baja California. Tegner *et al.* (1996) examined data from the Point Loma kelp forest and concluded that there is a weak but significant relation between average sea surface temperature and kelp harvest and that deleterious effects of ENSO events on kelp forests are evident. Casas-Valdés *et al.* (in press) also found effects of short-term ENSO events on kelp harvest off Baja California.

Secondary productivity

Zooplankton volume has often been used as a proxy for secondary productivity. There are reports of persistent declines off southern California (Roemich and McGowan 1995) since the 1976 regime shift. However, Brodeur *et al.* (1996) found the opposite in the Gulf of Alaska and discussed the various hypotheses linking the contrary trends in the California Current and the Gulf of Alaska. Off Baja California, in the Punta Eugenia region, Lavaniegos *et al.* (1998) found no significant changes in zooplankton volumes between the 1980s and the 1950s and 1970s.

Macrozooplankton displacement values are available since 1950 for the CalCOFI area (Figure 9-7). The series is dominated by very high values along the entire coast during 1953 and especially 1956, a rapid decline during the 1957-1959 ENSO and a slower recovery until the early 1970s, particularly at the northern areas. High volumes are found during 1978-1979 and 1985, but there is a clear diminishing trend into the 1990s (Lluch-Belda *et al.* 2001). Nevertheless, zooplankton data also seem to be very sensitive to high frequency variations.

Effects on fish populations

The small pelagic fishes, including sardines and anchovies, are dominant components of the so called "wasp-waist" ecosystems of temperate coastal upwelling systems such as the California Current (Bakun 1996). Here, the California sardine (*Sardinops sagax caerulea*) and the northern anchovy (*Engraulis mordax*) inhabit the area, showing considerable fluctuations in their relative abundance during the present century and, in synchrony with other coastal systems, their relative abundance has alternated.

It has been proposed that the California sardine has had its permanent habitat in the Sebastián Vizcaíno region, along the central west coast of the Baja California

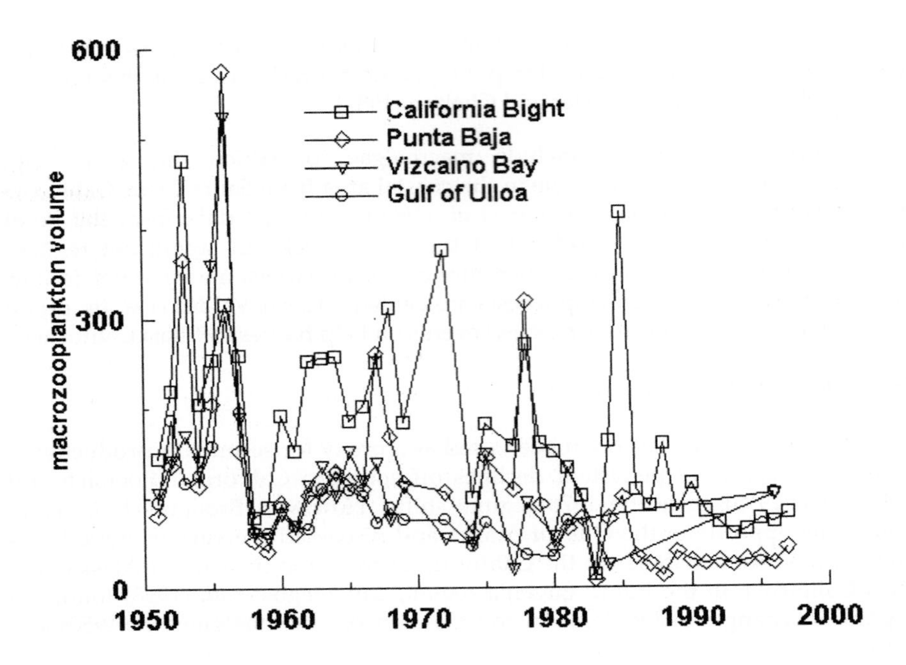

Figure 9-7. Yearly averaged macrozooplakton displacement volume from four, one-degree square coastal areas off the southern California Current: Southern California Bight (33-34°N), Punta Baja (30-31°N), Vizcaino Bay (27.5-28.5°N) and Gulf of Ulloa (25-26°N). Figure modified from Lluch-Belda *et al.* (2001).

peninsula (Lluch-Belda *et al.* 1991) at least during the last century. This has long been recognized as one of the major areas of sardine spawning (Tibby 1937, Ahlstrom 1954). On the other hand, the southern California bight area is the major secondary spawning center, where an even larger population grows during adequate periods and extends northward.

At the short-term scale the latitudinal distribution of spawning may change considerably, as shown in Figure 9-8. Several features are evident in the figure. First is the south to north annual progression of spawning, especially notable during the 1954-1956 lapse, with spawning at the northern area concentrated during the mid part of the year. Secondly, note that the spawning season extends to the early part of the year in the southern California Bight (SCB) area during the warm 1958-1960 event. The third point is the way in which spawning

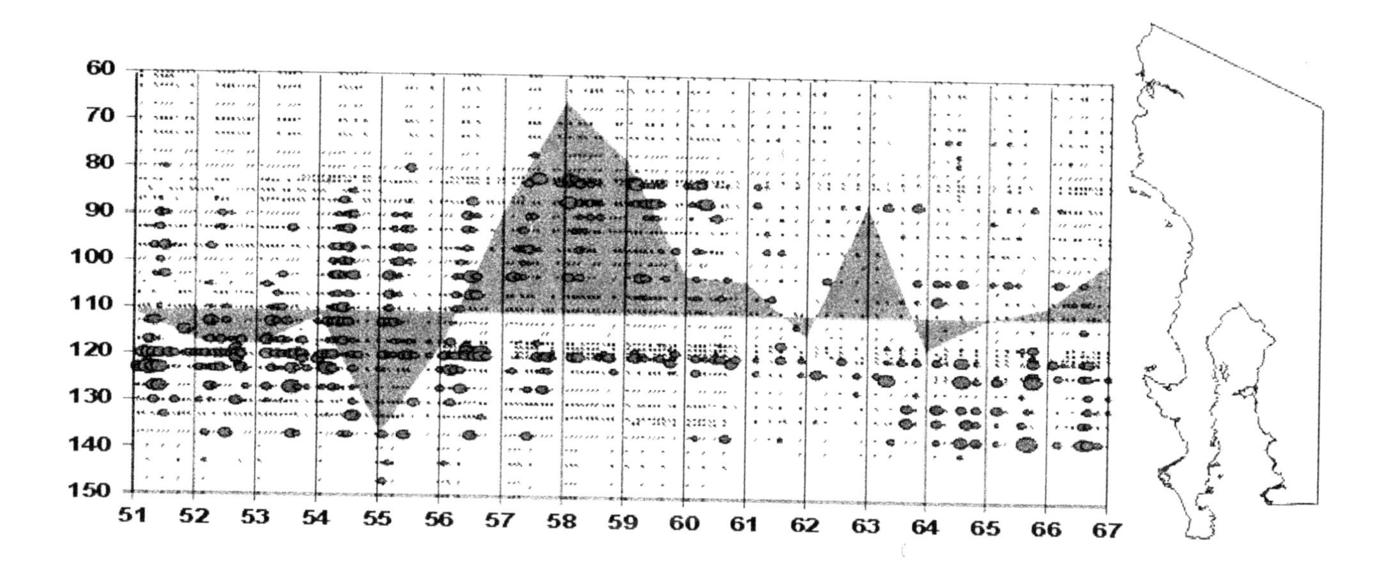

Figure 9-8. Averaged relative abundance of sardine eggs per line in the CalCOFI cruises. The radius of each of the filled circles is proportional to the log of the average. Shadowed curve in the background represents yearly averaged SST anomalies.

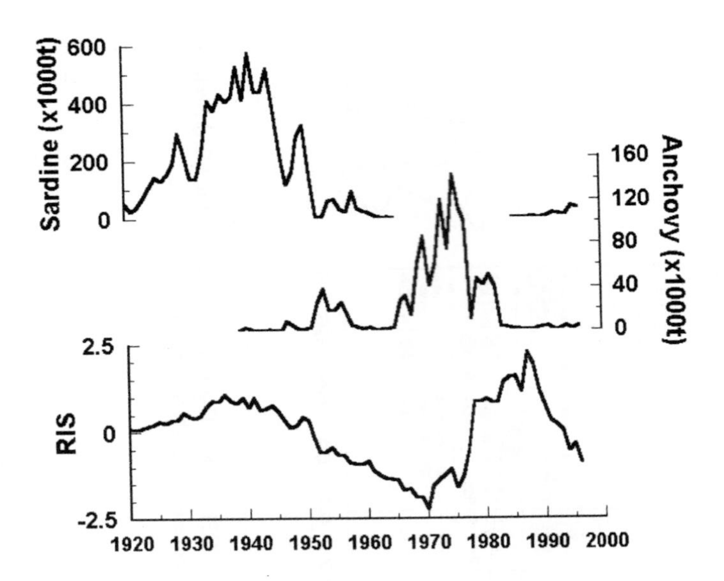

Figure 9-9. Historic sardine (upper) and anchovy (middle) catches (1920-2000) from the California Current System, and the Regime Indicator Series (RIS: lower). Catch data were obtained from Schwarzlose *et al.* (1999). RIS is a composite series reflecting synchronous variability of sardine and anchovy populations of the Japan, California, Benguela and Humboldt currents. Modified from Lluch-Cota D.B. *et al.* (1997).

ceases at the northern areas and extends southward during the cooling period of 1961 to 1967. The fourth feature to be noted is how the Sebastián Vizcaíno area shows consistent spawning during the whole interval.

At the decadal time scale, major changes in the distribution and relative latitudinal abundance of sardine populations have been revealed by catches of juveniles for the live bait fishery for tunas (Rodríguez-Sánchez *et al.* 2001), particularly during the 1930-1980 declining trend. However, it is the low frequency regime change that becomes more evident from the large scale variations of sardine abundance. Figure 9-9 shows the sardine and anchovy catches off California, together with a composite series reflecting variability of sardine and anchovy stocks worldwide. The sardine population in the northern area (north of the southern California Bight, SCB) has been increasing at a rapid rate since the mid-1970s to a considerable biomass, 1.6 million tonnes in 1999 (Calif. Dept. of Fish and Game, 2000), extending its spawning northward all the way to British Columbia. From a low of 10,000 t in 1965. During the 20th century, another episode of high sardine population abundance peaked during the 1940s; sardine and anchovy high abundance periods have alternated not only at the northeastern Pacific, but in all major abundance areas around the world (Lluch-Cota D.B. *et al.* 1997, Schwartzlose *et al.* 1999).

Such increases parallel other events such as the sardine scale deposition rates from the Santa Barbara basin varved sediments series (Soutar and Issacs 1969, Baumgartner *et al.* 1992). These outbursts of sardine abundance consist of a brief period of population growth (typically less than 40 years long), followed by an even shorter collapse; on average, such events are spaced ~ 60 years (Baumgartner *et al.* 1992, Lluch-Belda *et al.* 2001).

Regarding other fish species, while there were considerable increases of recruitment in many stocks of marine fishes right after the 1976-1977 regime shift in the eastern North Pacific (Hollowed and Wooster 1992, Beamish and Bouillon 1993), Parrish *et al.* (2000) found that this trend did not continue into the 1980s. Further, many of the fisheries in the northern part of the California Current have declined in recent years. Bottom fish and salmonid standing stocks have been at record low levels and some of the latter have shown significant decreases in their growth rates. In short, there has been a marked reduction of fish production between Vancouver Island and Oregon since the 1976-1977 regime shift (Parrish *et al.* 2000). The contrary has occurred in the Alaska Current, where fishery production has been high and the Alaska salmon stocks have increased while the California ones decreased (Mantua *et al.* 1997, Miller and Schneider 2000).

Johnson (1962) related sea temperatures with albacore availability off Oregon and Washington. Clark *et al.* (1975) related interannual variations in the percentage of albacore tuna fished north or south of San Francisco to climatic fluctuations. Biological responses of local populations, such as increased recruitment of the southeastern Alaska herring (Mysak 1986) or the survival index of Pacific mackerel (Sinclair *et al.* 1985) during ENSO years, have been also reported.

There are two different proposed components of the 1976-1977 regime shift: the change itself and the condition following it. Parrish *et al.* (2000) suggested that the principal source of the biological bonanza in the California Current had to do with the shift from one state to the other, and not to the new state itself. The possibility that change itself is the main cause of the variation regarding sardine population growth was also postulated by Lluch-Belda *et al.* (2001) .

Turning to fish fauna as a whole at the high frequency scale, note that Hubbs and Schultz (1929) described a single-season northward movement of southern fauna to the California coast during 1926 and analyzed a number of similar reports together with some temperature data. Later Hubbs (1948) described further instances of annual changes in the distribution of fauna. Radovich (1961) described the 1957-59 warm event and analyzed previous reports of faunal shifts, including those by Hubbs (1948).

At lower frequencies, Hubbs (1948) also described long-term changes in the fish fauna composition off California and suggested that there had been a warming period around the 1860s. MacCall (1996) reviewed low-frequency variability of fishes in the California Current and referred to biological variability in 50-70 year scales. He further noted that the conditions after 1976 were similar to those of 1850-1870. Lluch-Belda *et al.* (2001) found that the number of fish larvae species at selected areas along the California-Baja California coast changed both in the high and low frequencies, with a long term declining trend between 1950 and the mid 1970s.

Other faunal movements

Pelagic red crabs (*Pleuroncodes planipes*) are usually restricted to the south of the SCB, but great numbers were washed ashore at Monterey Bay during 1859 (Radovich, 1961). Since then, the species has been reported off California during 1941 (Hubbs 1948), 1969 (Hardwick and Spratt 1979), 1972 (Alvariño 1976), 1978 (McLain and Thomas 1983), 1982 and 1987 (pers. comm., Dr Paul Smith, Scripps of Oceanography, La Jolla, CA), and 2002 (pers. comm., Richard S. Schwartzlose, Scripps Institution of Oceanography, La Jolla, CA) which makes this species a particularly good index of northward faunal transport. Another subtropical species often reported northward of its usual range during warming periods is *Vellela* sp, the "by-the-wind-sailor."

Global warming induced variation

Unlike the above described variation scales, global warming is regarded as a consequence of anthropogenic-induced unidirectional change. Based on the Inter-governmental Panel on Climate Change (IPCC) scenarios, Everett *et al.* (1996) provided a general framework of possible consequences of global warming to marine ecosystems. They mention that highly productive systems are likely to be more pronounced near major ecosystem boundaries, that the rate of the warming trend might determine the abundance and distribution of new populations where a rapid change will likely favor smaller, low-priced, opportunistic species. However, they recognize the limitations of the available data to predict effects on global production, particularly because relevant fish population processes take place at regional or smaller scales for which general circulation models (GCMs) are insufficiently reliable.

One key question regarding marine ecosystem responses to global warming is whether biological production in upwelling systems (where most of the fish resources are captured) will increase. After analyzing upwelling time series for several upwelling systems of the world, Bakun (1990) speculated that productivity would be enhanced, as a greater difference between land and coastal ocean temperatures would induce increased wind speeds, hence increasing upwelling. On the other hand, after considering various sources of data for the wind-driven coastal upwelling system off California, Mann (1993) suggested that available data could be used to support the hypothesis that coastal upwelling increases during global cooling but decreases during global warming (Watson *et al.* 1997).

While coupled GCMs do not have the resolution to clearly understand or predict regional impacts of climate change, they have been used to create broad indications of global warming induced changes in the large-scale Pacific basin atmospheric and oceanic circulation and internal ocean structure, from which hypotheses of physical and biological changes in the California Current have been proposed. After cautioning about the high level of uncertainty that exists in modeling results, U.S. GLOBEC (1994) hypothesized several predictions of global warming effects on the California Current System.

For example, it was suggested that an atmospheric doubling of CO_2 will increase land-sea temperature differences (Trenberth 1993), intensifying equatorward winds in summer and leading to stronger coastal upwelling as suggested by Bakun (1990). However, other studies argue the effects will be more complex than the simple increase in wind strength everywhere (Hsieh and Boer 1992). Trenberth (1993) discussed the possible implications of the expected warming in the Eastern North Pacific of 1° to 1.5°C. He argues that northern regions will warm faster than equatorial ones resulting in a decreased strength of the mean atmospheric circulation, and increased monsoon circulation as the land heats more than the ocean as suggested by Bakun (1990); but, as proposed by Bernal (1993), this same CO_2 doubling might cause warming of mid-latitude oceans by 2°C at the surface, 1°C at 500 m and 0.5°C at 700 m, resulting in an increased stratification of the upper water column and probably making subsurface water pumping less effective.

U.S. GLOBEC (1992) also hypothesized some possible ecosystem effects such as: a) boundaries between physical and biological regions will move as a result of changes in wind forcing or advection; b) changes in atmospheric circulation and the predicted increase in land-sea temperature gradients could modify the strength, timing or even the occurrence of the spring transition, thus affecting reproduction success of many CCS inhabitants; c) stronger upwelling activity and increased stratification would result in greater mesoscale activity in summer (fronts and jets); d) mean southward transport in the core of the CCS is predicted to decrease; e) possible increased nutrient delivery to surface waters which should increase primary production and increase phytoplankton and higher trophic level biomasses. These predictions contrast U.S. GLOBEC (1992) studies with the works by Peterson *et al.* (1993) who argue global warming will cause the thermocline and nutricline to be depressed along the west coast of the Americas. In that case, reduced nutrient levels in the mixed layer would impact phytoplankton productivity and species composition and perhaps lengthen food chains, resulting in reduced fish biomass.

Climate change can be expected to result in distributional shifts in species, with the most obvious changes occurring near the northern or southern boundaries of species' ranges. Migration patterns will shift, causing changes in arrival times along the migration route. Growth rates are expected to vary (with the amplitude and direction species-dependent). Recruitment success could be affected by changes in time of spawning, fecundity rates, survival rate of larvae, and food availability (U.S. GLOBEC 1994).

Turning to observed rather than predicted changes, Tegner *et al.* (1996) found a reduction of 30-70 percent in the canopy coverage of kelp beds in southern California between 1911-1912 and the 1980s. However, the available data series are insufficient to asses the cause of long-term changes, given other sources of error.

Roemich and McGowan (1995) described the long term (1950-1994) decline of macrozoplankton volumes off California and suggested its possible connection to global warming. Barry *et al.* (1995) found changes in the invertebrate fauna of a California intertidal community between 1931-1933 and 1993-1994, including species' ranges shifting northward as consistent with predictions of change associated with global warming.

However, Royer (1989) had already suggested that the present warming is probably not a result of large scale global change, but part of the low frequency natural variation. Ware and Thompson (2000) analyzed reconstructed temperatures for the last 400 years and found that the average time between ENSO events has been decreasing since the middle of the 20th century and this trend could be associated with global warming. Historically, similar clusters of strong El Niño events have occurred as frequently, roughly every century.

All of the above examples of change, unfortunately, seem to extend within the limits of warming periods of described natural cycles and it remains uncertain whether they are a consequence of either natural fluctuations or human-induced changes. Within a longer perspective, sardines have been much more abundant in the southern California Bight before 1100 AD than at present, which could indicate warmer conditions (Lluch-Belda *et al.* 2001). Finney *et al.* (2002) found that Alaska salmon reconstructed abundance shows several synchronous patterns of variability with sardine and anchovy reconstructed abundance series from southern California.

CONCLUSIONS

The common mechanism of variation at all scales of interannual change seems to be related to changing patterns of ocean currents, warming periods being related to increased northward flow and relaxing of the California Current.

Although many other impacts should be expected, major changes in productivity at all trophic levels, organism advection, active invasion of areas, changes in the strength of cohorts and in the distribution of populations, growth of populations and species substitution are among the most documented effects.

Depending on the duration of the process, such changes may persist from a few years (1-3) to regime scale (~50 years), and the different scales of variation interact among them. These seem to be a consequence of natural cycles, thus the system should be expected to return to previous states. The interaction of these scales of change with anthropogenic-induced variation has yet to be unveiled.

ACKNOWLEDGEMENTS

The present paper originated from an invitation by Dr. Kenneth Sherman and has mostly been an integration of results that are part of a retrospective analysis of the California Current System for the Living Marine Resources panel (LMR) of Global Ocean Observing System (GOOS), based on an idea by Warren Wooster. This work had support from the Instituto Politécnico Nacional (CGEPI 20.05) and the Centro de Investigaciones Biológicas del Noroeste, S.C. (RP6). Lluch-Belda holds a fellowship from COFAA-IPN.

REFERENCES

Abbott, M.R. and B. Barksdale. 1991. Phytoplankton pigment patterns and wind forcing off Central California. J. Geophys. Res. 96:14649-14667.

Ahlstrom, E.H. 1954. Distribution and abundance of egg and larval populations of the Pacific sardine. Fish Bull. 56:83-140

Alvariño, A. 1976. Distribución batimétrica de *Pleuroncodes planipes* Stimpson (Crustáceo; Galateido). In: Simp. sobre biología y dinámica poblacional de camarones, Guaymas, Son. México Vol. 1. Instituto Nacional de la Pesca, México. 265-285

Bakun, A. 1990. Global climate change and intensification of coastal ocean upwelling. Science 247:198-201

Bakun, A. 1996. Patterns in the Ocean. Ocean Processes and Marine Population Dynamics. California Sea Grant / Centro de Investigaciones Biológicas del Noroeste. La Paz, B.C.S., Mexico. 323p

Balech, E. 1960. The changes in the phytoplankton population off the California Coast. In: Symposium on the Changing Pacific Ocean in 1957 and 1958. II, Section II: The biological evidence. Calif. Coop. Oceanic Fish. Invest. Rep. 7:127-132

Barber, R.T. and E.P. Chavez. 1983. Biological consequences of El Niño. Science 222:1203-1210

Barber, R.T. and E.P. Chavez. 1986. Ocean variability in relation to living resources during the 1982/83 El Niño. Nature 319:279-285.

Barnett, T.P., M. Latif, N. Graham, and M. Flugel. 1995. On the frequency-wave number structure of the tropical ocean-atmosphere system. Tellus 47: 998-1012

Barnston, A.G. and R.E. Livezey. 1987. Classification, seasonality and persistence of low-frequency atmospheric circulation patterns. Month. Weath. Rev. 115(6):1083-1126

Barry, J.P., C.H. Baxter, R.D. Sagarin and S.E. Gilman. 1995. Climate-related, long-term faunal changes in a California rocky intertidal community. Science 267:672-675

Baumgartner, T.R., A. Soutar and V. Ferreira-Bartrina. 1992. Reconstruction of the history of Pacific sardine and northern anchovy populations over the past two millennia from sediments of the Santa Barbara Basin, California. Calif. Coop. Oceanic Fish. Invest. Rep. 33:24-40

Beamish, R.J. 1995. Response of anadromous fish to climate change in the North Pacific. In: Peterson, D.L. and D.R. Johnson, eds. Human Ecology and Climate Change: People and Resources in the Far North. Taylor and Francis,Washington, DC. 123-136

Beamish, R.J. and D.R. Bouillon. 1993. Pacific salmon production trends in relation to climate. Can. J. Fish. Aquat. Sci. 50:1002-1016.

Bernal, P.A. 1979. Large-Scale biological events in the California current. Calif.

Coop. Oceanic Fish. Invest. Rep. 20:89-101
Bernal, P.A. 1993. Global climate change in the oceans: a review. In: H.A. Mooney, E.R. Fuentes and B. Kronberg, eds. Earth System Response to Global Change: Contrasts between North and South America, Academic Press, New York. 1-15
Bernal, P.A. and D.B. Chelton. 1984. Variabilidad biológica de baja frecuencia y gran escala en la Corriente de California, 1949-1978. In: Sharp, G.D. and J. Csirke, eds. Proceedings of the Expert Consultation to Examine Changes in Abundance and Species Composition of Neritic Fishery Resources, San José, Costa Rica, 18-29 April 1983. FAO Fisheries Report 2(291):713-730
Brinton, E. 1962. The distribution of Pacific euphausiids. Bull. Scripps Inst. Oceanogr. 8(2):51-270.
Brodeur, R.C., B.W. Frost, S.R. Hare, R.C. Francis and W.J. Ingraham. 1996. Interannual variations in zooplankton biomass in the Gulf of Alaska, and covariation with California Current zooplankton biomass. Calif. Coop. Oceanic Fish. Invest. Rep. 37:80-99.
California Department of Fish and Game. 2000. Review of some California fisheries for 1999. Calif. Coop. Oceanic Fish. Invest. Rep. 41:8-25.
Casas Valdés, M.M., E. Serviere Z., D. Lluch-Belda, R. Marcos and R. Aguilar-Ramírez. In press. Effects of climatic change on the harvest of the kelp *Macrocystis pyrifera* at the Mexican Pacific Coast. Bull. Mar. Sci.
Chelton, D.B. 1981. Interannual variability of the California Current - Physical factors. Calif. Coop. Oceanic Fish. Invest. Rep. 22:34-48
Chelton, D.B., P.A. Bernal and J.A. MacGowan. 1982. Large-scale interannual physical and biological interaction in the California Current. J. Mar. Res. 40(4):1095-1125
Chelton, D.B. and R.E. Davis. 1982. Monthly sea level variability along the western coast of North America. J. Phys. Oceanogr. 12:752-784
Clark, N.E., T.J. Blasing and H.C. Fritts. 1975. Influence of interannual climatic fluctuations on biological systems. Nature 256:302-305
Clarke, A.J. and S. Van Gorder. 1994. On ENSO coastal currents and sea levels. J. Phys. Oceanogr. 24:661-679
Clarke, A.J. and A. Lebedev. 1996. Long-term changes in the equatorial Pacific trade winds. J. Climate 9:1020-1029
Clarke, A.J. and A. Lebedev. 1999. Remotely driven decadal and longer changes in the coastal Pacific waters of the Americas. J. Phys. Oceanogr. 29(4):828-835.
Diaz, H.F. and G.N. Kiladis. 1992. Atmospheric teleconnections associated with the extreme phases of the Southern Oscillation. In: Diaz, H.F. and V. Markgraf, eds. El Niño. Historical and Paleoclimatical Aspects of the Southern Oscillation. Cambridge Univ. Press, Cambridge, U.K. 7-28. 476p
Ebbesmeyer, C.C., D.R. Cayan, D.R. McLain, F.H. Nichols, D.H. Peterson, and K. Redmond. 1991. 1976 step in the Pacific climate: Forty environmental

changes between 1968-1975 and 1977-1984. Proc. Seventh Annual Pacific Climate (PACLIM) Workshop. Tech. Rep. 26, Asilomar, CA. 115-126

Emery, W. and K. Hamilton. 1985. Atmospheric forcing of interannual variability in the northeast Pacific ocean: Connections with El Niño. J. Geophys. Res. 90(C1):857-868

Everett, J.T., A. Krovnin, D. Lluch-Belda, E. Okemwa, H.A. Regier and J.P. Troadec. 1996. Fisheries. In: M.C. Zinynowera, R.T. Watson and R.H. Moss, eds. Climate Change 1995. Impacts, Adaptations and Mitigation of Climate Change: Scientific-Technical Analyses. Cambridge Univ. Press, Cambridge, U.K. 511-537

Fiedler, P.C. 1984. Satellite observations of the 1982-1983 El Niño along the U.S. Pacific coast. Science 224:1251-1254

Fiedler, P.C. 1994. Seasonal and interannual variability of coastal zone color scanner phytoplankton pigments and wind in the eastern tropical Pacific. J. Geophys. Res. 99:18371-18384

Finney, B.P., I. Gregory-Eaves, V.M.S, Douglas and J.P. Smol. 2002. Fisheries productivity in the northeastern Pacific Ocean over the past 2,000 years. Nature 416:729-733

Ghil, M., and R. Vautard. 1991. Inter-decadal oscillations and the warming trend in global temperature time series. Nature 350:324-327

Graham, N.E. 1994. Decadal scale variability in the 1970s and 1980s: Observations and model results. Clim. Dyn 10:60-70

Graham, N.E. 1995. Simulation of recent global temperature trends. Science 267: 666-671

Hamilton, K. 1988. A detailed examination of the extratropical response to tropical El Niño/Southern Oscillation events. J. Climate 8:67-86

Hardwick, J.E. and J.D. Spratt. 1979. Indices of the availability of market squid, *Loligo opalescens*, to the Monterey Bay fishery. Calif. Coop. Oceanic Fish. Invest. Rep. 20:35-39

Hare, S.R. and N.J. Mantua. 2000. Empirical evidence for North Pacific regime shifts in 1977 and 1989. Prog. Oceanogr. 47:103-145

Hollowed, A.B. and W.S. Wooster. 1992. Variability of ocean winter conditions and strong year classes of Northeast Pacific groundfish. ICES Mar. Sci. Symp. 195:433-444

Horel, J.D., and J.M. Wallace. 1981. Planetary-scale atmospheric phenomena associated with the Southern Oscillation. Month.Weather Rev. 109: 813-829

Hsieh, W.W. and G. J. Boer. 1992. Global climate change and ocean upwelling. Fish. Oceanogr., 1:333-338

Hubbs, C.L. 1948. Changes in the fish fauna off western North America correlated with changes in ocean temperature. J. Mar. Res. 7:459-482

Hubbs, C.L. and L.P. Schultz. 1929. The northward occurrence of southern forms of marine life along the Pacific coast in 1926. Calif. Fish and Game

15(3):234-240

Ingraham, W.J., C.C. Ebbesmeyer and R.A. Hinrichsen. 1998. Imminent climate and circulation shift in Northeast Pacific Ocean could have major impact on marine resources. EOS 79(6):197

Ingraham, W. J. 2000. Getting to know OSCURS, REFM's Ocean Surface Current Simulator. http://www.refm.noaa.gov/docs/oscurs/get_to_know.htm, NOAA.

Jacobs, G.A., H.E. Hulburt, J.C. Kindle, E.J. Metzger, J.L. Mitchell, W.J. Teague, and A.J. Wallcraft. 1994. Decade-scale trans-Pacific propagation and warming effect of an El Niño anomaly. Nature 370:360-363

Johnson, J.H. 1962. Sea temperatures and the availability of albacore off the coasts of Oregon and Washington. Trans. Amer. Fish. Soc. 91(3):269-274

Kawasaki, T. 1983. Why do some pelagic fishes have wide fluctuations in their numbers? - biological basis of fluctuation from the view point of evolutionary ecology. In: Sharp, G.D. and J. Csirke, eds. Proceedings of the Expert Consultation to Examine Changes in Abundance and Species Composition of Neritic Fishery Resources, San José, Costa Rica, 18-29 April 1983. FAO Fisheries Report (291):1065-1080

Kerr, R.A. 1992. Unmasking a shifty climate system. Science 255:1508-1510

Lara-Lara, J.R., J.E. Valdez-Holguín and J. Jimenez-Pérez. 1984. Plankton studies in the Gulf of California during the 1982-1983 El Niño. Tropical Ocean-Atmosphere Newsletter 28:16-17

Latif, M. and T.P. Barnett. 1994. Causes of decadal climate variability over the North Pacific and North America. Science 266:634-637

Lavaniegos, B.E., J. Gómez-Gutiérrez, J.R. Lara-Lara and S. Hernández-Vázquez. 1998. Long-term changes in zooplankton volumes in the California Current System - the Baja California region. Mar. Ecol. Prog. Ser. 169:55-64

Lluch-Belda, D. , R. J. M. Crawford, T. Kawasaki, A. D. MacCall, R. H. Parrish, R. A. Schwartzlose and P. E. Smith. 1989. World-wide fluctuations of sardine and anchovy stocks: the regime problem. S. Afr. J. Mar. Sci. 8:195-205

Lluch-Belda, D. , S. Hernández, R. A. Schwartzlose. 1991. A hypothetical model for the fluctuation of the California sardine population (*Sardinops sagax caerulea*). In: Tanaka, T.S., Y. Kawasaki, Y. Toba and A. Tamiguchi, eds. Long-term variability of pelagic fish populations and their environment. New York, Pergamon Press. 293-300

Lluch-Belda, D. 2000. Centros de actividad biológica en la costa occidental de Baja California. In: Lluch-Belda, D., S.E. Lluch-Cota, J. Elorduy and G. Ponce, eds. *BACs: Centros de actividad biológica del Pacífico Mexicano.* La Paz, BCS, Centro Interdisciplinario de Ciencias Marinas del IPN, Centro de Investigaciones Biológicas del Noroeste, SC. y Consejo Nacional de Ciencia y Tecnología. 49-64

Lluch-Belda, D., R. Michael Laurs, D.B. Lluch-Cota and S.E. Lluch-Cota. 2001. Long term trends of interannual variability in the California Current

System. Calif. Coop. Oceanic Fish. Invest. Rep. 42:129-144

Lluch-Cota, D.B. 2002. Satellite measured interannual variability of coastal phytoplankton pigment in the tropical and subtropical eastern Pacific. Cont. Shelf Res. 22:803-820

Lluch-Cota, D.B., S. Hernández-Vázquez and S.E. Lluch-Cota. 1997. Empirical investigation on the relationship between climate and small pelagic global regimes and El Niño-Southern Oscillation (ENSO). FAO Fisheries Circular 934. 48p

Lluch-Cota, D.B. and G. Teniza-Guillén. 2000. BAC versus áreas adyacentes: una comparación de la variabilidad interanual de pigmentos fotosintéticos a partir del Coastal Zone Color Scanner (CZCS). In: D. Lluch-Belda, S.E. Lluch-Cota, J. Elorduy and G. Ponce, eds. BACs: Centros de actividad biológica del Pacífico Mexicano. Centro Interdisciplinario de Ciencias Marinas del IPN, Centro de Investigaciones Biológicas del Noroeste, SC. y Consejo Nacional de Ciencia y Tecnología, La Paz, B.C.S., Mexico. 199-218

Lluch-Cota, D.B., W.S. Wooster and S.R. Hare. 2001. Sea surface temperature variability in coastal areas of the Northeastern Pacific related to the El Niño-Southern Oscillation and the Pacific Decadal Oscillation. Geophys. Res. Lett. 28:2029-2032

Lluch-Cota, S.E., S. Alvarez-Borrego, E. Santamaría-del-Angel, F.E. Müllen-Karger and S. Hernández-Vázquez. 1997. The Gulf of Tehuantepec and adjacent areas: Spatial and temporal variation of satellite-derived photosynthetic pigments. Ciencias Marinas 23:329-340

Loder, J.W. and C. Garret. 1978. The 18.6-year cycle of sea surface temperature in shallow seas due to variations in tidal mixing. J. Geophys. Res. 83(C4): 1967-1970

Longhurst, A.R. 1966. The pelagic phase of *Pleuroncodes planipes* Stimpson (*Crustacea, Galatheidae*) in the California Current. Calif. Coop. Oceanic Fish. Invest. Rep. 11:142-154

Lynn, R.J. and J.J. Simpson. 1987. The California Current System: The seasonal variability of its physical characteristics. J. Geophys. Res. 92(C12):12,947-12,966

MacCall, A.D. 1996. Patterns of low-frequency variability in fish populations of the California Current. Calif. Coop. Oceanic Fish. Invest. Rep. 37:100-110

Mann, K.H. 1993. Physical oceanography, food chains and fish stocks, a review. ICES J. Mar. Sci. 50:105-119

McGowan, J. A., D. R. Cayan, and L. M. Dorman. 1998. Climate-ocean variability and ecosystem response in the Northeast Pacific. Science 281:210-217

Mantua, N.J., S.R. Hare, Y. Zhang, J.M. Wallace and R.C. Francis. 1997. A Pacific interdecadal climate oscillation with impacts on salmon production. Bull. Amer. Meteor. Soc. 78(6):1069-1079

McLain, D.R. and D.H. Thomas. 1983. Year-to-year fluctuations in the California Countercurrent and effects on marine organisms. Calif. Coop. Oceanic

Fish. Invest. Rep. 24:165-181
Miller, A.J. 1996. Recent advances in California Current modeling: decadal and interannual thermocline variations. Calif. Coop. Oceanic Fish. Invest. Rep. 37:69-79
Miller, A.J. and N. Schneider. 2000. Interdecadal climate regime dynamics in the North Pacific Ocean: theories, observations and ecosystem impacts. Prog. Oceanogr. 47:355-379
Minobe, S. 2000. Spatio-temporal structure of the pentadecadal variability over the North Pacific. Progr. Oceanogr. 47:381-408
Minobe, S. and N. Mantua. 1999. Interdecadal modulation of interannual atmospheric and oceanic variability over the North Pacific. Progr. Oceanogr. 43:163-192
Mo, K.C. and R.E. Livezey. 1986. Tropical-extratropical geopotential height teleconnections during the northern hemisphere winter. Month. Weather Rev. 114(12):2488-2515
Moser, H.G., P.E. Smith and L.E. Eber. 1987. Larval fish assemblages in the California Current region, 1954-1960. A period of dynamic environmental change. Calif. Coop. Oceanic Fish. Invest. Rep. 28:97-127
Mysak, L.A. 1986. El Niño, interannual variability and fisheries in the northeast Pacific Ocean. Can. J. Fish. Aquat. Sci. 43(2):464-497
Namias, J.R. and D.R. Cayan. 1984. El Niño: The implications for forecasting. Oceanus 27:41-47
Norton, J., D. McLain, R. Brainard and D. Husby. 1985. The 1982-1983 El Niño event off Baja California and Alta California and its ocean climate context. In: Wooster, W.S. and D.L. Fluharty, eds. El Niño North. Niño effects in the Eastern Subarctic Pacific Ocean. Washington Sea Grant Program. Seattle, WA. 44-72. 312p
Parés-Sierra, A. and J.J. O'Brien. 1989. The seasonal and interannual variability of the California Current System: A numerical model. J. Geophys. Res. 93(C3):3159-3180
Parrish, R.H., F.B. Schwing and R. Mendelssohn. 2000. Mid-latitude wind stress: the energy source for climatic shifts in the North Pacific Ocean. Fish. Oceanogr. 9(3):224-238
Peterson, C. H., R. T. Barber, and G. A. Skilleter. 1993. Global warming and coastal ecosystem response: How northern and southern hemispheres may differ in the eastern Pacific Ocean. In: H. A. Mooney, E. R. Fuentes, and B. Kronberg, eds. Earth System Response to Global Change: Contrasts Between North and South America. Academic Press, New York. 17-34
Polovina, J. J., G. T. Mitchum and G. T. Evans. 1995. Decadal and basin-scale variation in mixed layer depth and the impact on the biological production in the Central and North Pacific. Deep-sea Res. 42(10):1701-1716
Radovich, J. 1961. Relationship of Some Marine Organisms of the Northeast

Pacific to Water Temperatures. Fish Bull. (112):62

Reynolds, R.W. and T.M. Smith. 1994. Improved global sea surface temperature analyses using optimum interpolation. J. Climate 7:929-948

Rodríguez-Sánchez, R., D. Lluch-Belda, H. Villalobos-Ortiz and S. Ortega-García. 2001. Large-Scale Long-Term Variability of Small Pelagic Fish in the California Current System. In: Kruse, G.H., N. Bez, A. Booth, M.W. Dorn, S. Hills, R.N. Lipcius, D. Pelletier, C. Roy, S.J. Smith and D. Witherell, eds. Spatial Processes and Management of Marine Populations, Univ. of Alaska Sea Grant College Program, Anchorage, AK.447-462

Roemmich, D. and J. McGowan. 1995. Climatic warming and the decline of zooplankton in the California Current. Science 267:1324-1326

Roesler, C.S. and D.B. Chelton. 1987. Zooplankton variability in the California Current, 1951-1982. Calif. Coop. Oceanic Fish. Invest. Rep. 28:59-86

Royer, T.C. 1989. Upper ocean temperature variability in the northeast Pacific: Is it an indicator of global warming? J. Geophys. Res. 94(18):175-183

Royer, T.C. 1993. High-latitude oceanic variability associated with the 18.6-year nodal tide. J. Geophys. Res. 98(C3):4639-4644

Schlesinger, M.E. and N. Ramankutty. 1994. An oscillation in the global climate system of period 65-70 years. Nature 367: 723-726

Schwartzlose, R.A., J. Alheit, T. Baumgartner, R. Cloete, R.J.M. Crawford, W.J. Fletcher, Y. Green-Ruiz, E. Hagen, T. Kawasaki, D. Lluch-Belda, S.E. Lluch-Cota, A.D. MacCall, Y. Matsuura, M.O. Nevárez-Martínez, R.H. Parrish, C. Roy, R. Serra, K.V. Shust, N.M. Ward and J.Z. Zuzunaga. 1999. Worldwide large-scale fluctuations of sardine and anchovy populations. S. Afr. J. Mar. Sci. 21:289-347

Shkedy, Y., D. Fernández, C. Teague, J. Vesecky and J. Roughgarden. 1995. Detecting upwelling along the central coast of California during an El Niño year using HF-radar. Cont. Shelf Res. 15:803-814

Sinclair, M., M.J. Tremblay and P. Bernal. 1985. El Niño events and variability in a Pacific Mackerel (*Scomber japonicus*) Survival Index: support for Hjort's Second Hypothesis. Can. J. Fish. Aquat. Sci. 42:602-608

Soutar, A. and J.D. Isaacs. 1969. History of fish populations inferred from fish scales in anaerobic sediments off California. CalCOFI Calif. Coop. Oceanic Fish. Invest. Rep. 13:63-70

Strub, P.T., C. James, A.C. Thomas and M.R. Abbott. 1990. Seasonal and nonseasonal variability of satellite-derived surface pigment concentration in the California Current. J. Geophys. Res. 95(C7):11,501-11,530.

Tegner, M.J., P.K. Dayton, P.B. Edwards and K.L. Riser. 1996. Is there evidence for long-term climatic change in southern California kelp forests? CalCOFI Calif. Coop. Oceanic Fish. Invest. Rep. 37:111-126

Tibby, R. B. 1937. The relation between surface water temperature and the distribution of spawn of the California sardine *Sardinops caerulea*. California Fish&Game 23(2):132-137

Trenberth, K.E. 1990. Recent observed interdecadal climate changes in the Northern Hemisphere. Bull. Amer. Meteor. Soc 71:988-993

Trenberth, K. E. 1993. Northern hemisphere climate change: physical processes and observed changes. In: H. A. Mooney, E. R. Fuentes, and B. Kronberg, eds. Earth System Response to Global Change: Contrasts between North and South America, Academic Press, New York. 35-59

Unal, Y.S., and M. Ghil. 1995. Interannual and interdecadal oscillation patterns in sea level. Climate Dyn. 11:255-278

U.S. GLOBEC. 1992. Eastern Boundary Current Program: Report on Climate Change and the California Current Ecosystem. US GLOBEC Report 7, 99p

U.S. GLOBEC. 1994. A science plan for the California Current. U.S. Report No. 11. Univ. of California, Berkeley, CA.

Valdez-Holguín, J.E. and J.R. Lara-Lara. 1987. Primary productivity in the Gulf of California. Effects of El Niño 1982-1983 event. Ciencias Marinas 13:34-50

Wallace, J.M. and D.S. Gutzler. 1981. Teleconnections in the geopotential height field during the northern hemisphere winter. Mon. Wea. Rev. 109:784-812

Ware, D.M. 1995. A century and a half of change in the climate of the NE Pacific. Fish. Oceanogr. 4:267-277

Ware, D.W. and R.E. Thompson. 2000. Interannual to Multidecadal Timescale Climate Variations in the Northeast Pacific. J. Climate 13:3209-3220

Watson, R.T., M.C. Zinyowera and R.H. Moss. 1997. The regional impacts of climate change: An assessment of vulnerability: A Special Report of IPCC Working group II. Cambridge University Press, UK. 517p

White, W.B. and D.R. Cayan. 1998. Quasi-periodicity and global symmetries in interdecadal upper ocean temperature variability. J. Geophys. Res. 103: 21335-21354

Wooster, W.S. 1960. El Niño. In: Symposium on the Changing Pacific Ocean in 1957 and 1958. II, Section I: The physical evidence. Calif. Coop. Oceanic Fish. Invest. Rep. 7: 43-46

Wooster, W.S. and D.L. Fluharty. (eds). 1985. El Niño North: Niño Effects in the Eastern Subarctic Pacific Ocean. Washington Sea Grant Program. Seattle, WA. 312p

Wooster, W.S. and A.B. Hollowed. 1995. Decadal scale variations in the eastern subarctic Pacific. I. Winter ocean conditions. In: Beamish, R.J., ed. Climate change and northern fish populations. Can. Spec. Publ. Aquat. Sci. 121:81-85

Wyrtki, K. 1975. El Niño--the dynamic response of the equatorial Pacific. J.

Geophys. Res. 5:572-584

Xu, J-S. 1992. On the relationship between the stratospheric Quasi-Biennial Oscillation and the tropospheric Southern Oscillation. J. Atmos. Sci. 49: 725-734

Yasunari, T. 1985. Zonally-propagating modes of the global east-west circulation associated with the Southern Oscillation. J. Meteor. Soc. Japan 63:1013-1029

Zhang, Y., J.M. Wallace and D.S. Battisti. 1997. ENSO-like interdecadal variability: 1900-93. J. Climate 10:1004-1020

Zuria-Jordan, I.L., S. Alvarez-Borrego, E. Santamaría-del-Angel and F.E. Müllen-Karger. 1995. Satellite-derived estimates of phytoplankton biomass off southern Baja California. Ciencias Marinas 21:265-280

Large Marine Ecosystems of the World
G. Hempel and K. Sherman (Editors)

10

Sustainability of the Benguela: ex Africa semper aliquid novi

L. Vere Shannon and M. J. O'Toole

THE BENGUELA AS A LARGE MARINE ECOSYSTEM

The Benguela Current ecosystem is situated along the coast of south western Africa, stretching from east of the Cape of Good Hope, in the south, northwards into Angola waters and encompassing the full extent of Namibia's marine environment. It is one of the four major coastal upwelling ecosystems of the world which lie at the eastern boundaries of the oceans. Like the Humboldt, California and Canary systems, the Benguela is an important centre of marine biodiversity and marine food production. Its distinctive bathymetry, hydrography, chemistry and trophodynamics combine to make it one of the most productive ocean areas in the world, with a mean annual primary productivity of 1.25 kilograms of carbon per square metre per year—about six times higher than the North Sea ecosystem. This high level of primary productivity of the Benguela supports an important global reservoir of biodiversity and biomass of zooplankton, fish, sea birds and marine mammals, while near-shore and off-shore sediments hold rich deposits of precious minerals (particularly diamonds), as well as oil and gas reserves.

Based on the LME concept and definition proposed by Sherman (1994), subsequently elaborated on, some 64 Large Marine Ecosystems (LMEs) have been delineated. The Benguela, which appears as 29 on the list, fits the definition well, and the Benguela Current Large Marine Ecosystem (BCLME) is an obvious candidate for an LME programme. However, unlike most other LMEs identified for international action which are closed or semi-enclosed "concave" systems, e.g. Black Sea, Bay of Bengal, Baltic Sea, all of which have "hard" natural boundaries, the Benguela is a "convex" system situated at the crossroads between the Atlantic, Indian and Southern Oceans with open boundaries.

The Namib Desert which forms the landward boundary of the greater part of the BCLME is one of the oldest deserts in the world, predating the commencement of persistent upwelling in the Benguela (12 million years before present) by at least 40 million years. With the exception of the Congo River, the main impact of the discharges of rivers flowing into the Benguela tends to be episodic in nature and in terms of transboundary concerns these are limited to extreme flood events. The landward boundary of the greater part of the Benguela can thus be taken as the coast. Much of this coast is pristine and immensely beautiful.

What makes the Benguela upwelling system so unique in the global context is that it is bounded on both northern and southern ends by warm water systems, viz the equatorial eastern Atlantic and the Indian Ocean's Agulhas Current, and its retroflection area. The principal upwelling centre which is situated near Lüderitz in southern Namibia, is the most intense found in any upwelling regime and forms a natural internal divide within the Benguela, with the domains to the north and south of it functioning rather differently. Pronounced fronts exist at the boundaries of the upwelling system, but these display substantial spatial and temporal variability, at times pulsating in phase, and others not. Interaction between the BCLME and the adjacent ocean systems occurs over thousands of kilometers. For example, much of the BCLME, in particular off Namibia and Angola, is naturally hypoxic, even anoxic, at depth. This oxygen depleted water flows southwards at subsurface depths, and the hypoxia is compounded by depletion of oxygen from more localised biological decay processes. There are also teleconnections between the Benguela and processes in the North Atlantic and Indo-Pacific Oceans (e.g. El Niño). Moreover, the southern Benguela lies at a major choke point in the "Global Climate Conveyor Belt." Warm surface waters move from the Indo-Pacific into the Atlantic Ocean mainly in the form of rings shed from the retroflection of the Agulhas. The South Atlantic is the only ocean in which there is a net transport of heat towards the equator!). As a consequence, not only is the Benguela at a critical location in terms of the global climate system, but it is also potentially extremely vulnerable to any future climate change or increasing variability in climate.

So, from an LME perspective, where are the oceanic boundaries of the BCLME? The Benguela Current is generally defined as the integrated equatorward flow in the upper layers of the South-east Atlantic between the coast and the 0°meridian, so this would seem to be a pragmatic western boundary. It encompasses the entire upwelling region, upwelling fronts, and the EEZs of Angola, Namibia and South Africa. In the south, upwelling extends seasonally as far east as Port Elizabeth, and the area is dominated by the Agulhas Current and retroflection area. It would thus be appropriate to take the Agulhas as the Southern boundary and 27°E longitude (near Port Elizabeth) as the eastern limit.
The selection of a northern boundary is much more problematic. While the Angola-Benguela Front is an apparent natural boundary, the frontal zone is only

well defined in the surface layer. Moreover, there is evidence that there is a substantial flux of water (about 6 Sv) into the northern Benguela from the west and north (Lass *et al.* 2000), and it is likely that the Angola Dome, centred around 12°S, 5°E, is important in terms of regional (Benguela) ocean dynamics. Moreover, there is a well defined front further north at around 5°S, viz the Angola Front (Yamagata and Iizuka 1995) which is apparent at sub-surface depths and separates the Benguela part of the South Atlantic from the Gulf of Guinea and equatorial current systems. A northern boundary of the Benguela at 5°S would encompass the Angola Dome and the area in which the main oxygen minimum in the South Atlantic forms, the dynamics of which are inextricably linked with that off Namibia. The 5°S parallel is in general agreement with the transition between the BCLME and the Gulf of Guinea LME shown by Sherman (1994) and others. Further, in a geopolitical context, as 5°S is at the northern boundary of the Cabinda province of Angola, the full Angolan EEZ would fall within the scope of the BCLME, and this would mean that the issues and problems facing Angola's resource managers and which are not dissimilar to those further south, could be addressed within the context of the BCLME.

CHANGING STATE OF THE BENGUELA

The BCLME displays a high degree of variability over a broad spectrum of time and space scales. In the following paragraphs we shall only touch on aspects of small scale and mesoscale variability, highlighting some key characteristics of inter-annual variability, and then focus on decadal change. Papers listed under References address the subject more fully.

Short period and seasonal variability

In the extreme South, upwelling tends to be more seasonal than in the remainder of the Benguela, with the main upwelling season being out of phase with that in the North. This is a consequence of the seasonal shifts in the atmospheric pressure systems and increased influence of westerly winds during the austral winter in the South. Here, during summer, the free passage of easterly moving cyclones south of Africa result in upwelling events pulsed on time scales of about 3 to 10 days. In the central and northern Benguela, upwelling is most pronounced during winter and spring and winds have a distinct diurnal character (land-sea breeze effect). The ecosystem is well adapted to the pulsed nature of upwelling and intra-annual changes in physical forcing, and there is extensive literature on this. What is less well understood are the ecosystem effects of major

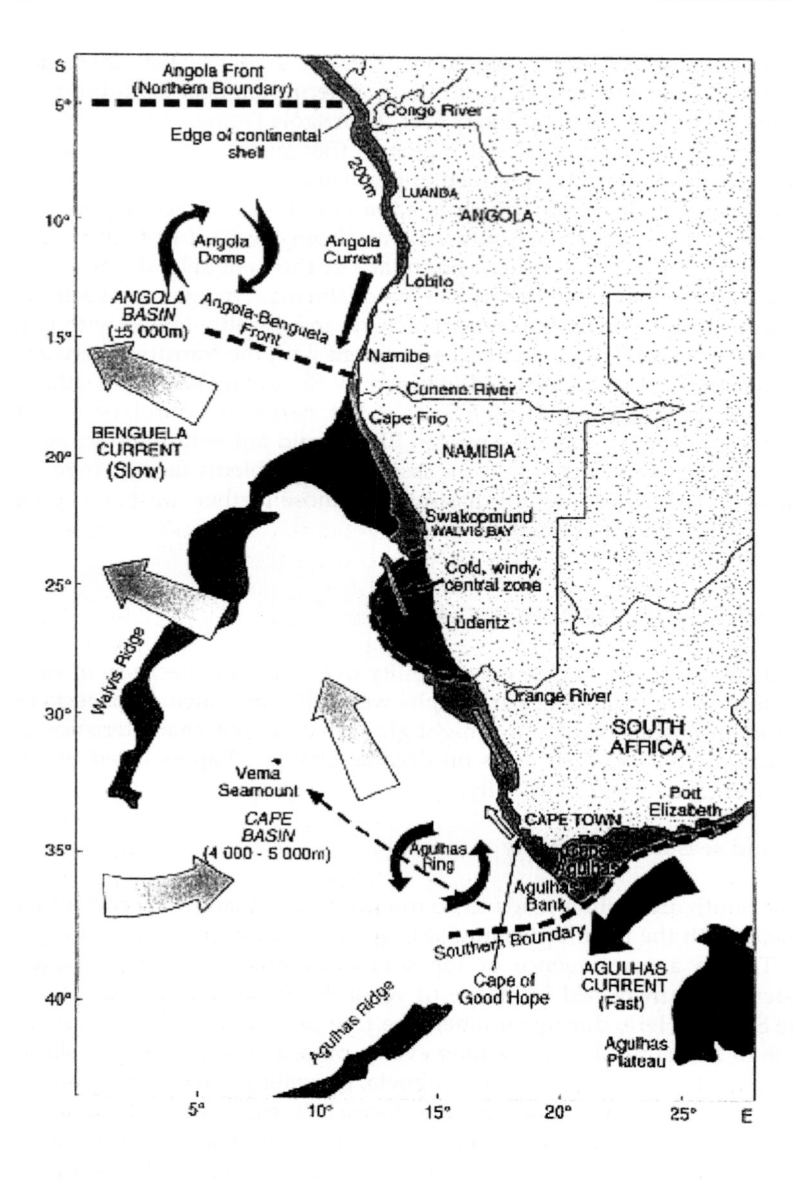

Figure 10-1. External and internal boundaries of the Benguela Current Large Marine Ecosystem, bathymetric features and surface (upper layer) currents.

storm events such as the 50 year storm which occurred in May 1984. Although short-lived, it is not inconceivable that events such as these might trigger a change of state of the ecosystem.

Inter-annual variability

In contrast with Pacific upwelling systems which have small seasonal and large inter-annual signals, the inter-annual variability in the Benguela is relatively small, and major events are less frequent. Perhaps more important, however, Benguela variability appears to be less predictable. Moreover, the occasional extreme sustained events in the Benguela do result in major perturbations of the ecosystem and mass mortalities of marine fauna have impacted on fisheries. The major variability and changes in the Benguela physical environment have manifested themselves in the following forms:

- Sustained intrusions of anomalously warm, nutrient poor equatorial/tropical water across the northern and southern boundaries of the ecosystem, viz Benguela Niños and Agulhas intrusions,
- Large scale changes in the windfield (intensity, direction and frequency) resulting in changes in the intensity and spatial distribution of upwelling, the position of the upwelling/oceanic fronts, warming or cooling of large areas of the system, altered stratification and changes in advection,
- Changes in the composition and advection of subsurface waters, particularly in changes in the concentrations of dissolved oxygen in the poleward undercurrent.

The term "Benguela Niño" was coined by Shannon *et al.* (1986) and refers to large scale episodic warm events that occur along the coast of southern Angola and Namibia every ten years on average, and which have a character not unlike the El Niño in the Pacific Ocean. Every few years the tropical eastern Atlantic becomes anomalously warm as a consequence of relaxation in the trade winds and the deepening of the thermocline and reduced loss of heat from the ocean to the atmosphere. Occasionally, every ten years on average, this warming is even more extreme, evidently as a consequence of a sudden relaxation of the winds off Brazil, and when this happens the warm water anomaly in the tropical Atlantic travels eastwards and southwards, trapped (guided) by the coast of Africa. The result is a large southward displacement of the Angola-Benguela front, and a flooding of the Namibian shelf by warm tropical water, sometimes very saline (1984), at other times low surface salinity (1995), depending on the orientation of the flow and the amount of fresh water from the Congo River present. Benguela Niños are accompanied by increased oxygenation of subsurface and bottom shelf waters either as a consequence of reduced deep flow of water southwards from Angola, or (more likely) reduced primary production and decay on the Namibian shelf. Benguela Niños occurred in 1934, 1949, 1963, 1984, 1995 and probably around 1910, in the mid-1920s and in 1972-1974. The most recent event and its biological impact has been well documented by Gammelsrød *et al.* (1998). Benguela Niños are not necessarily in phase with Pacific El Niños, but there is increasing evidence that they are a regional response to changes in the global atmosphere-ocean system.

Just as Benguela Niños are characterized by extreme disturbance of the Angola-Benguela front, extreme disturbances in the retroflection (turning back) of the Agulhas Current at the southern boundary of the Benguela can be manifest as a major incursion of Agulhas water moving into the Benguela system around the Cape of Good Hope. These intrusions may be in the form of shallow surface filaments of warm water (the more usual case) but, on occasions, warm rings shed from the Agulhas Current can take a more northerly path than usual and impact on subsurface and deeper currents along the edge of the continental shelf, as was the case in 1989. A well-documented Agulhas intrusion occurred in 1986 (Shannon *et al.* 1990). Previous large-scale intrusions may have occurred in 1957 and 1964. The most recent event took place during the post austral summer 1997-1998.

Whereas Agulhas intrusions result in the input of anomalously warm water into the Benguela, incursions of cold sub-Antarctic water do occur, and occasionally the effect of these can be felt as far north as 33°S (north of Cape Town), as was the case early in 1987. It seems that these rather unusual events are associated with the shedding of rings at the Agulhas retroflection, and they have an appearance of a compensatory northward flow from the Subtropical Convergence following the formation of a ring. Their biological consequences in the Benguela are unknown.

During some years, the most recent being 1993-1994, the oxygen depletion of shelf waters in the Benguela is unusually severe, and this results in widespread anoxia and hypoxia in the system. Although most pronounced in the northern Benguela, episodic depletion of oxygen does occur in the southern Benguela (e.g. in autumn of 1994). These large scale hypoxic and anoxic conditions appear to coincide with quiescent conditions which follow periods of sustained and enhanced upwelling, as a consequence of increased primary production and subsequent decay of phytoplankton blooms (e.g. red tides). It also appears likely that changes in the composition and flow of subsurface waters, in particular the concentrations of dissolved oxygen in the southward moving undercurrent along the west coast, may be contributing factors, and this suggests that processes taking place off

Angola may exert an important influence on the entire upwelling domain via advection. Large scale hypoxia and anoxia result in massive mortalities of marine organisms and changes in distribution and abundance of fish such as hakes (e.g. Hamakuaya *et al.* 1998).]

Decadal changes and regime shifts

There is a growing body of evidence which suggests that marine ecosystems undergo decadal-scale fluctuations driven by variability in climate. This is most apparent at the higher trophic levels (fish). For example, in the Humboldt Current system decadal period switches in dominance between sardine and anchoveta have occurred this century, and it is also apparent from the sediment record (from analysis of fish scale deposits) that species alternations have occurred during past centuries, i.e. prior to the advent of fishing and other anthropogenic impacts. Moreover, it has been demonstrated that populations of sardine in different parts of the Pacific (off Chile/Peru, California/Mexico and Japan) have undergone synchronous fluctuations this century. In the Benguela ecosystem there have been corresponding fluctuations in sardine abundance, although these appear to be out of phase with those in the Pacific. Also, there is evidence in the southern Benguela, at least, of species alternations, regime shifts, between anchovy and sardine, and in the northern Benguela the sediment record suggests that both species undergo decadal-scale fluctuations, at times alternating species dominating, at others the two species fluctuating synchronously (e.g. Shackleton 1986). What is also apparent from the sediment record is that there can be long periods (several decades) where both species are absent or only present at a low biomass.

There are few long-term data series of environmental parameters for the Benguela region. The available indices do (see Figure 10-2), nevertheless, show that changes and fluctuations on decadal or longer time scales have occurred in the Benguela this century and, superimposed on this variability, a progressive warming of surface waters of about 0.7° C from 1920 is apparent throughout the Benguela and South-east Atlantic.

Figure 10-2 shows an increasing trend in equatorward windstress in the off-shore northern Benguela from the late 1940s until the 1990s. This trend, which is also confirmed by a much sanitized COADS (Comprehensive Ocean Atmosphere Data Set) data set, appears to be real. The analysis of Taunton-Clark and Shannon (1988) showed that the 1920s and 1930s were cool years in the region and that a change to warmer conditions took place during the 1940s, followed by a gradual strengthening of the south-easterly trade winds over the next few decades. A system-wide change occurred in the late 1960s, to be followed by an extended warm period. The warming trend accelerated during the 1980s and this decade was the warmest this century in the South-east Atlantic. Longshore wind stress in the Benguela increased sharply after 1974 and the shelf waters along the west coast of South Africa and Namibia were abnormally cold in the early 1980s. In the extreme southern Benguela (at least), the wind record displays a sharp change in 1982/1983, with significantly lighter winds blowing for the remainder of the

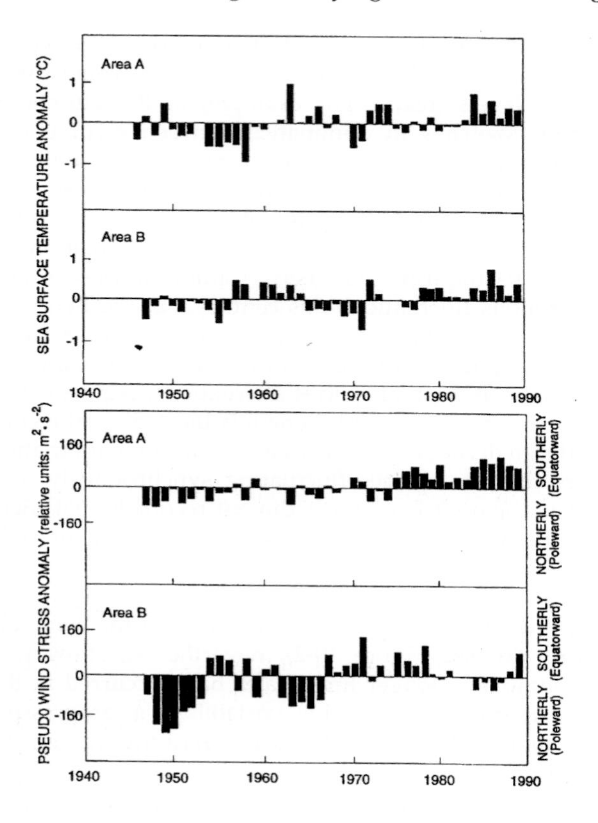

Figure 10-2. Sea surface temperature (°C) and pseudo windstress ($m^2 \cdot s^{-2}$) anomalies in the offshore northern Benguela (area A: 10° - 15°S, 0°- 5°E) and southern Benguela (area B: 30° - 35°S, 10° - 15°E) modified from Shannon *et al.* (1992).

decade there (Shannon *et al.* 1992). It is perhaps significant that there is substantial decadal-scale variability evident in the post 1960 winds off Brazil

(Carton and Huang 1994). The Gulf of Guinea temperature records also suggest an apparent periodicity between large sustained warm events of about 10 years.

While decadal-scale changes are evident from some of the physical parameters, these are not so apparent in the chemistry and plankton because of the fragmented nature of the available records. The available information does, however, point to the likelihood that comparable changes in the plankton have

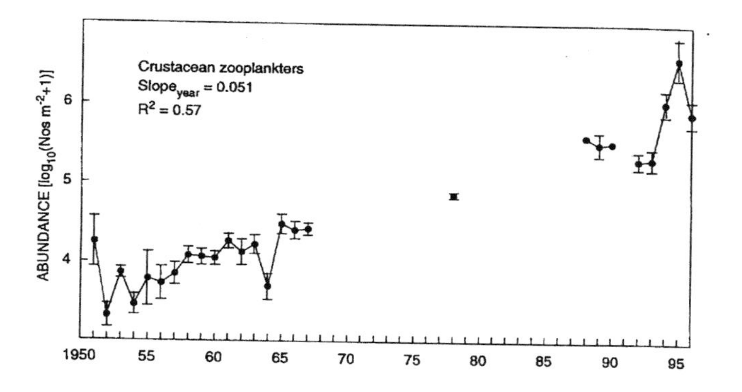

Figure 10-3. Change in crustacean zooplankton abundance, St. Helena Bay area (southern Benguela), March-June, 1951-1995 from Verheye *et al.* 1998.

taken place, and by extrapolation from the biology and physics, that changes in nutrient supply have also occurred. Brown *et al.* (1991) analyzed chlorophyll *a* (a proxy for phytoplankton biomass) data available for the southern Benguela between 1971 and 1989. Although there appears to be a decreasing trend during the two decades from about 3.5mg/m^3 to 2.0mg/m^3, the trend is not statistically significant because the data are highly patchy. In what is perhaps the most important paper on zooplankton in the Benguela during recent years, Verheye and Richardson (1998) found that the abundance of animals in all main taxonomic groups in the St. Helena Bay area increased by at least tenfold between 1951 and 1996. (These measurements applied to the main pelagic fish recruitment season, viz March-June). This is illustrated in Figure 10-3. Total zooplankton abundance expressed in numbers of animals increased by more than one hundredfold. The increase was accompanied by a significant shift in the community structure of near shore zooplankton, with a trend towards smaller size organisms. Whether the observed change is a consequence of reduced predation pressure by anchovy and sardine ("top down control") or climatologically induced, viz upwelling,

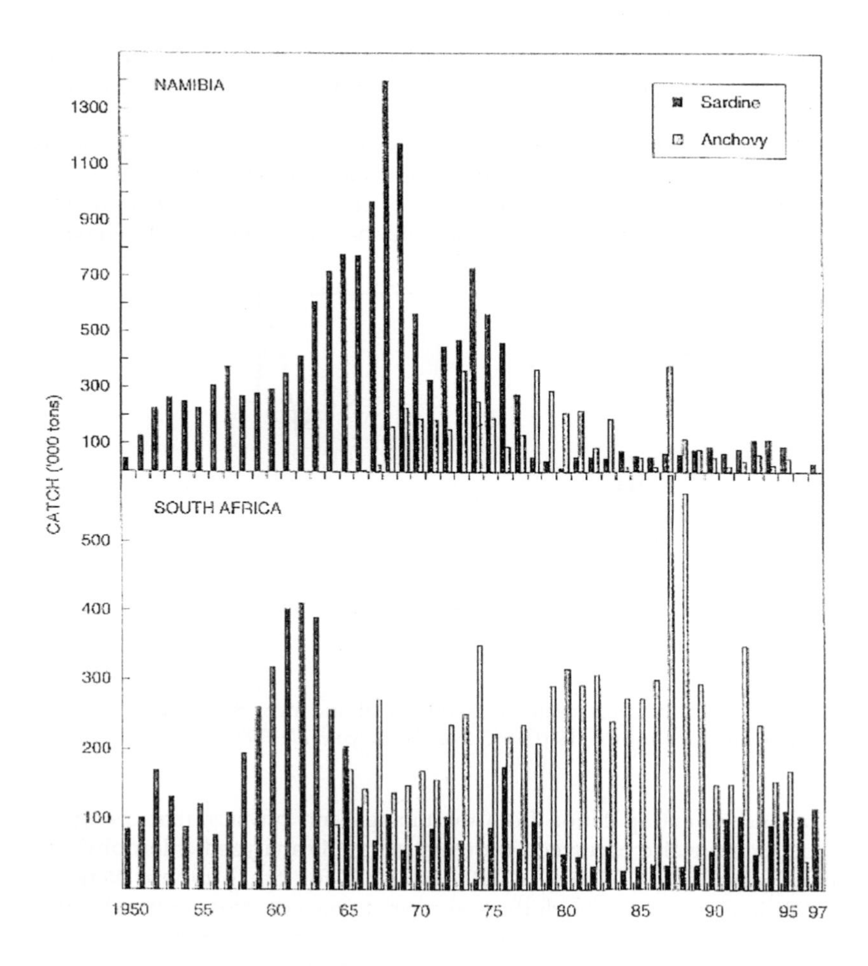

Figure 10-4. Sardine and anchovy catches off Namibia and South Africa, 1950-1997, modified from Crawford (1999)

primary production and entrainment ("bottom up control"), is not clear at this stage. Other changes in zooplankton include the decrease in abundance on the west coast of the copepod *Calanus agulhensis* between 1988 and 1991 and the relative scarcity of the euphausiid *Nyctiphanes capensis* now in comparison with the 1950s (*Euphausia lucens* is currently dominant).

Major changes in the principal harvested species of finfish and crustaceans have occurred this century. While there is little doubt that the declines in stocks of hake, sardine and rock lobster during the 1960s and 1970s were consequences of over-fishing, changes in the environment have been contributory factors. In the following paragraphs we summarize the main changes in key harvested species and their predators which have taken place, and comment on possible environmental links.

Sardine and anchovy

Sardine *Sardinops sagax* and anchovy *Engraulis japonicus* mainly occur off Namibia and South Africa, and to a lesser extent off southern Angola. From the sediment record these species have historically (i.e. prior to fishing) displayed large fluctuations. In the more recent past the biomass of the sardine resource declined sharply off Namibia following the 1963 Benguela Niño during a period when the sardine was available close to Walvis Bay and fishing mortality was high (Stander and De Decker 1969). A second collapse occurred there after 1974, again following a protracted, but less intense, Benguela Niño. The resource showed a slow recovery subsequently, but this was retarded (Figure 10-4) by the most recent 1995 Benguela Niño. Anchovy was abundant off Namibia in the 1970s but has been less significant in the catches there during the past decade. In the southern Benguela, the sardine stock declined in the late 1960s and again after 1976, only recovering as a result of successive good year classes in the late 1980s-1990s (Figure 10-4). In the South, the anchovy which was the dominant species during the 1970s and 1980s appears to have been partly replaced by the increasing biomass of sardine. This evident species switching is reflected quite well in the catch record for South Africa, but less so in Namibia (Figure 10-4).

Apart from the system wide changes in the abundance of species such as anchovy and sardine, the Benguela exhibits equatorward and poleward shifts associated with meridional shifts in the major wind belts which appear to be in synchrony with shifts in the Canary Current System. In addition, species switching and regime shifts occur in the Benguela which appear to be out of phase with those in Pacific stocks, displaying a characteristic periodicity of about 50 years. In the case of the sardine *Sardinops sagax,* the species is most abundant in the Pacific when it is least abundant in the Benguela and vice versa.

Sardinellas

The most important commercial pelagic species off Angola are the sardinellas, *Sardinella aurita* in the south and S. *maderensis* north of 10ºS. The main spawning area is in the area between Pointe Noire and the Congo River, with peak spawning in March-April. Longshore migration of both species occurs seasonally, with an equatorward shift during the first part of the year (to spawning grounds) with a return migration of adults occurring later in the year. There is some evidence that a southward shift in the distribution of sardinellas occurred in the mid-1960s which was followed by an equatorward displacement during the early 1980s, congruent with the changes in the distributions of sardines and hakes in the Benguela proper. These displacements of stocks which occur across national boundaries have important implications for resource management.

Hakes

Three species of hakes occur in the Benguela viz *Merluccius capensis, M. paradoxus* and *M. polli.* Relatively little is known about their behavior and responses to environmental variability and change. Adult hakes are good swimmers, undergo diurnal vertical migrations and can tolerate a range of temperatures. Being relatively opportunistic feeders, long lived, and fished over a variety of age classes, hakes would be robust to all but major environmental perturbations. There is some published evidence (Shannon *et al.* 1988) which suggests that hakes do better when SSTs (sea surface temperatures) are low, or at least the recruits seem to be more available and at higher densities during cool periods, e.g. in 1992 in the case of *M. capensis* off Namibia and in 1987 in the southern Benguela (*M. paradoxus*). There is, from catch results, an indication that longshore shifts in hake stocks in the Benguela might occur. Although hake can evidently tolerate low oxygen levels, down to 0.25 $ml \cdot l^{-1}$, they would avoid areas with extremely low concentrations of dissolved oxygen. Thus it is likely that the system wide low oxygen event in 1993-4 which was followed by the 1995 Benguela Niño did retard the expected recovery of the hake resources off Namibia (e.g. Woodhead *et al.* 1998). Nevertheless, on a time scale of decades, the main impact on hakes appears to be due to fishing mortality (Figure 10-5) rather than to changes in the environment.

Rock Lobster

The Benguela spiny lobster, *Jasus ialandii,* has been fished for centuries, and there is evidence that the overall biomass decline started in the early 1900s and then accelerated after 1950. Toward the end of the 1980s there was a sharp decline in

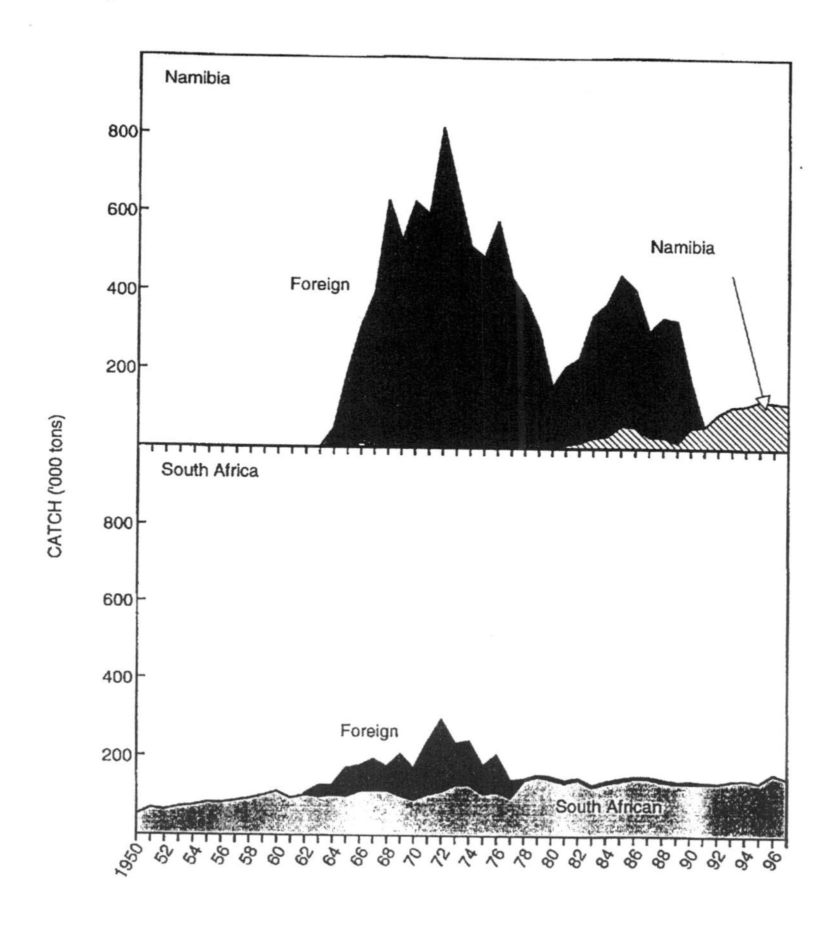

Figure 10-5. Catches of hakes off Namibia and South Africa from 1950 (Data kindly provided by Mr. Michael Stuttaford, author of Fishing Industry Handbook, South Africa, Namibia and Moçambique, published by Marine Information CC and updated). Note the extent of foreign exploitation of Namibia's resources prior to independence in 1990.

production of rock lobster throughout the Benguela system. In the southern Benguela, the decreased production resulted from reduced growth rates. Off Namibia, it has been attributed to over-exploitation and changes in availability related to fluctuations in the levels of oxygen depletion of bottom waters. There is relatively little meridional migration of rock lobster, and it seems improbable that fishing would have impacted all fishing zones of the species at the same time, unless what happened was a consequence of fishing out of larger animals and reduced predation on smaller lobsters with resultant increased competition for food. It seems likely that the resource responded to some large-scale change in the environment. In the south, the reduced rates of growth may have been caused by a reduced biomass of ribbed mussels. Alternatively the changes may have been linked with changes in primary production and a regime shift in the benthic food web. Whatever the cause, the west coast rock lobster growth rate appears to be responding in a manner analogous to that of a depressed population trapped in a predator pit.

Impact on predators

Crawford *et al.* (1992), Crawford (1999), and others, have highlighted the relationship between changes in abundance of certain resources in the Benguela ecosystem and changes in the diets of predators such as seabirds and seals. In the southern Benguela numbers of Cape cormorants that breed in any years are closely related to the biomass of anchovy. Breeding of swift terns and Cape gannets also reflects changes in anchovy biomass. Similarly, the distribution and breeding of the African penguin is related to the abundances of anchovy and sardine. Following the collapse of the sardine stocks in the 1960s, penguin colonies in southern Namibia and near Cape Town crashed to low levels and the centre of their distribution shifted southwards and eastwards with increasing reliance on anchovy. The trend was reversed in the mid-1980s when the sardine stocks began to recover. Not only do the changes in diet of top predators provide information on possible regime shifts, but trends in abundances of distribution of the predators provide proxy information on the performance of some harvested resources. For example, with the demise in the sardine in the northern Benguela in the late 1960s and 1970s, there was an evident increasing reliance on pelagic gobies. In another example the severity of the 1993/1994 black tide/low oxygen event and its system-wide impact was highlighted by the response of the seal population north in the northern and central Benguela, where the high mortality rates of pups (and adults) from starvation caused the seal stock to decline.

FRAGMENTED MANAGEMENT: A LEGACY OF THE PAST

Following the establishment of European settlements at strategic coastal locations where victuals and water could be procured to supply fleets trading with the East Indies, the potential wealth of the African continent became apparent. This resulted in the great rush for territories and the colonization of the continent, mostly during the nineteenth century. Boundaries between colonies were hastily established, often arbitrary and generally with little regard for indigenous inhabitants and natural habitats. Colonial land boundaries in the Benguela region were established at rivers (Cunene, Orange). Not only were the languages and cultures of the foreign occupiers different (Portuguese, German, English, Dutch) but so were the management systems and laws which evolved in the three now independent and democratic countries of the region—Angola, Namibia and South Africa. Moreover, not only were the governance frameworks very different, but a further consequence of European influence was the relative absence of inter-agency (or inter-ministerial) frameworks for management of the marine environment and its resources and scant regard for sustainability. To this day, mining concessions, oil/gas exploration, fishing rights and coastal development have taken place with inadequate proper integration and scant regard for other users. For example, exploratory wells have been sunk in established fishing grounds and the well-heads (which stand above the sea bed) subsequently abandoned. Likewise the impact of habitat alterations due to mining activities and ecosystem alteration (including biodiversity impacts) due to fishing have not been properly assessed.

Prior to the coming into being of the United Nations Law of the Sea Convention and declaration and respecting of sovereign rights within individual countries' Exclusive Economic (or Fishing) Zones, there was an explosion of foreign fleets fishing off Angola, Namibia and South Africa during the 1960s and 1970s, an effective imperialism and colonization by mainly First World countries of the Benguela and the rape of its resources. (This is highlighted quite dramatically in Figure 10-5). This period also coincided with liberation struggles in all three countries, and associated civil unrest. In the case of Namibia, over whom the mandate by South Africa was not internationally recognized, there was an added problem in that, prior to independence in 1990, an EEZ could not be proclaimed. Although an attempt was made to control the foreign exploitation of Namibia's fish resources through the establishment of the International Commission for the South-east Atlantic Fisheries (ICSEAF), this proved to be relatively ineffectual at husbanding the fish stocks. Until fairly recently, environmental issues and sustainable management were low priorities on the political agenda in South Africa. Another consequence of the civil wars has been the population migration to the coast and localized pressure on marine and coastal resources (e.g. destruction of coastal forests and mangroves) and severe pollution of some embayments.

While mineral exploration and extraction and developments in the coastal zones obviously occur within the geographic boundaries of the three countries, i.e. within the EEZs, and can to a large degree be independently managed by each of the countries, mobile living marine resources do not respect the arbitrary geographic borders. This has obvious implications for the sustainable use of these resources, particularly so in the case of straddling and shared fish stocks.

Thus, the legacy of the colonial and political past is that the management of resources in the greater Benguela area has not been integrated within countries or within the region. The real challenge of an LME project in the Benguela is to develop a viable joint and integrative mechanism for the sustainable environmental management of the region as a whole, i.e. at the ecosystem level.

REGIONAL CO-OPERATION: THE KEY TO SUCCESS

Nearly all of the problems in the BCLME which require scientific investigation and management action are common to all three countries, Angola, Namibia and South Africa. For example most of the regions' important harvested resources are shared between countries or move across national EEZ boundaries at times. Environmental variability and change impact on the ecosystem as a whole and there is poor predictability of its consequences. Mining impacts and pollution, while seemingly localized, are really generic issues. So are harmful algal blooms and the loss of biodiversity. Perhaps the greatest problem, however, is the lack of appropriate capacity in the region (both human and infrastructure) and the enormous capacity gradient from south to north. Putting this together with the existing fragmented management suggests that, if ever there were a case for collaboration between countries and concerted action to address problems collectively, then this was so in the Benguela region, with the enablement of the international community.

Namibia made the first move in 1995 when its Ministry of Fisheries and Marine Resources hosted an International Workshop/Seminar on "Fisheries Resource Dynamics in the Benguela Current Ecosystem" in partnership with the German Organisation for Technical Co-operation (GTZ), the Norwegian Agency for Development Co-operation (NORAD) and the Intergovernmental Oceanographic Commission (IOC) of UNESCO. This meeting proved to be a milestone in regional co-operation, for out of it evolved two major Benguela initiatives. The first, BENEFIT (BENguela-Environment-Fisheries-Interaction-Training) was launched in April 1997 jointly by Angola, Namibia and South Africa with foreign partners, "To develop the enhanced science capacity required for the sustainable utilization of living resources of the Benguela ecosystem by (a) improving knowledge and understanding of the dynamics of important commercial stocks, their environment and linkages between environmental processes and stock dynamics, and (b) building appropriate human and material capacity for marine science and technology in the countries bordering the Benguela ecosystem." The

BENEFIT Programme which is driven by the region for the region has been a catalyst for stimulating further collaboration within the BCLME. Several joint ocean surveys have been undertaken generating a number of publications and reports and improved understanding of, for example, the Angola-Benguela Front. Training courses have been held, and fifteen joint projects have been funded. In the first half of 1999 alone, some 50 persons from Southern African Development Community (SADC) countries have been trained at sea on the BENEFIT cruises. Strong links have been built between BENEFIT and other parallel but distinctly different programmes, viz South Africa's established and internationally acclaimed Benguela Ecology Programme (BEP), ENVIFISH (a three year European Union funded project between seven EU states and Angola, Namibia and South Africa which focuses primarily on the application of satellite data in environment-fisheries research and management, and which commenced in October 1998) and VIBES (a bilateral French-South African initiative which focuses on the variability of pelagic fish resources in the Benguela and the environment and spatial aspects of the system, which also commenced in 1998).

The second regional initiative for which the seed was sown at the 1995 Workshop/Seminar was the development of a programme to enhance the sustainable integrated management of the Benguela. Inspired by the success of BENEFIT, the tangible fruits of regional collaboration, and progress being made on sustainable management of other LMEs, Angola, Namibia and South Africa in partnership requested support from the Global Environment Facility (GEF), a fund established in 1991 under the management of The World Bank. Following the award of a grant from the GEF in 1998, a comprehensive project proposal has been developed in the region, which hopefully will attract incremental international funding and assistance in developing a proper framework for the sustainable integrated management of the BCLME. This will be very different from BENEFIT (which focuses on science capacity for fisheries) in that it aims to develop enabling management mechanisms to address a broad spectrum of environmental issues, including diamond mining, offshore oil and gas exploration and extraction, coastal development and modification, environmental variability and ecosystem change, habitat loss and degradation, pollution, loss of biodiversity etc. in addition to fisheries. Socio-economic factors and the need for sustainability will play an overarching role. The full implementation phase of the BCLME was approved by the GEF Council in late 2001, and the programme commenced in March 2002.

What has emerged so clearly from BENEFIT and related activities and the embryonic BCLME Management Programme is that there is a strong desire in Angola, Namibia and South Africa to work together to solve common problems in the Benguela region, to share expertise, to build capacity and to develop a collective approach to ensure the sustainability of the Benguela ecosystem.

TRANSBOUNDARY CONSIDERATIONS

In developing an enabling management mechanism for an LME, priority has to be given to addressing the principal common or transboundary issues (transboundary here refers to cross internal geopolitical, not the external boundaries of the ecosystem). These include inter alia:

- Regional/national issues with transboundary causes/sources
- Transboundary issues with national causes/sources
- National issues that are common to at least two of the countries and that require a common strategy and collective action to address
- Issues that have transboundary elements or implications (e.g. fishery practices on biodiversity/ecosystem resilience).

The first step in the process is therefore to identify the key transboundary issues and the associated problems. The second is to identify the root causes of these problems, and the third step is to specify the affordable and implementable remedial action which is needed. This requires broad inter-country consultation involving all stakeholders. Within the context of the developing BCLME Management Programme (GEF), there has been wide stakeholder participation in the process, and regional planning workshops involving a large number of local and international experts were held in 1998 and 1999. From this broad consultative process a consensus view has emerged that there are three main transboundary issues in the Benguela which need to be addressed to ensure sustainable integrated management viz (a) utilization of resources, (b) environmental variability and (c) ecosystem health and pollution. Within the context of these, many problems requiring solution were identified, of which seven were seen as major problems common to all three countries. These seven perceived problems and their transboundary characteristics are as follows:

Problem (i): Decline in BCLME commercial fish stocks and non-optimal harvesting of living resources
Transboundary Characteristics: Country boundaries do not coincide with ecosystem boundaries; most of the regions' important harvested resources are shared between countries, or move across national boundaries at times. Over-harvesting of a species in one country can therefore lead to depletion of that species in another, and in changes to the ecosystem as a whole. Moreover, many resource management difficulties are common to all the countries.

Problem (ii): Uncertainty regarding ecosystem status and yields in a highly variable environment
Transboundary Characteristics: The Benguela environment is highly variable and the ecosystem is naturally adapted to this. However, sustained environmental events such as Benguela Niños, Agulhas intrusions and changes in winds, impact on the system as a whole, compounding the negative effects of fishing, while the poor predictive ability limits the capacity to manage effectively system wide. In addition, the BCLME is believed to play a significant role in global ocean and climate processes and may be an important site for the early detection of global climate change.

Problem (iii): Deterioration in water quality – chronic and catastrophic
Transboundary Characteristics: Although most impacts of chronic deterioration in water quality are localized (national issues), they are common to all of the countries and require collective action to address. Moreover, chronic pollution can favour the development of less desirable species, and result in species migration. Catastrophic events (major oil spills, maritime accidents) can impact across country boundaries, requiring co-operative management and sharing of clean-up equipment and manpower.

Problem (iv): Habitat destruction and alteration, including inter alia modification of seabed and coastal zone and degradation of coastscapes
Transboundary Characteristics: Although most impacts may appear localized, habitat alteration or loss due to fishing and mining can cause migration of fauna and system-wide ecosystem change. Uncertainties exist about the regional cumulative impact on benthos resulting from mining and associated sediment re-mobilisation. Inadequately planned coastal developments result in degradation of coastscapes and reduce the regional value of tourism.

Problem (v): Loss of biotic integrity (changes in community composition, species and diversity, introduction of alien species, etc.) and threat to biodiversity/endangered and vulnerable species
Transboundary Characteristics: Past over-exploitation of targeted fish species has altered the ecosystem, impacting at all levels, including on top predators and reducing the gene pool. Some species, e.g. African penguin, are threatened or endangered. Exotic species have been introduced into the Benguela. (This is recognized as a global transboundary problem).

Problem (vi): Inadequate capacity to monitor/assess ecosystem (resources and environment, and variability thereof)

Transboundary Characteristics: There is inadequate capacity, expertise and ability, in the region to monitor and assess adequately the shared living resources and to monitor environmental variability. Moreover, there is unequal distribution of this capacity between the three countries.

Problem (vii): Harmful algal blooms (HABs)
Transboundary Characteristics: HABs occur in all three countries, who face similar problems in terms of impacts and management, and which require collective regional action to address.

A host of root causes of the common problems in the BCLME listed above have been identified, and these can be grouped into seven broad generic categories, viz:

- Complexity of ecosystem and high degree of variability (resources and environment)
- Inadequate capacity development (human and infrastructure) and training
- Poor legal framework at the regional and national levels
- Inadequate implementation of available regulatory instruments
- Inadequate planning at all levels
- Insufficient public involvement
- Inadequate financial mechanisms and support

The above problems and generic root causes are illustrated schematically in Figure 10-6.

The next step in the process is the specification of agreed collective actions which must be taken within the BCLME. These must be affordable, implementable and sustainable.

TOWARDS A SUSTAINABLE FUTURE

Correcting decades of over-exploitation of resources in the Benguela ecosystem and fragmented management actions (the consequence of the colonial/political past and greed) will require a substantial coordinated effort in the years ahead, to be followed by sustained action on a permanent basis. A task of this magnitude will require careful planning, not only by the government agencies in the three countries bordering on the Benguela but also by all the other stakeholders, including the international community. There already exists the willingness on

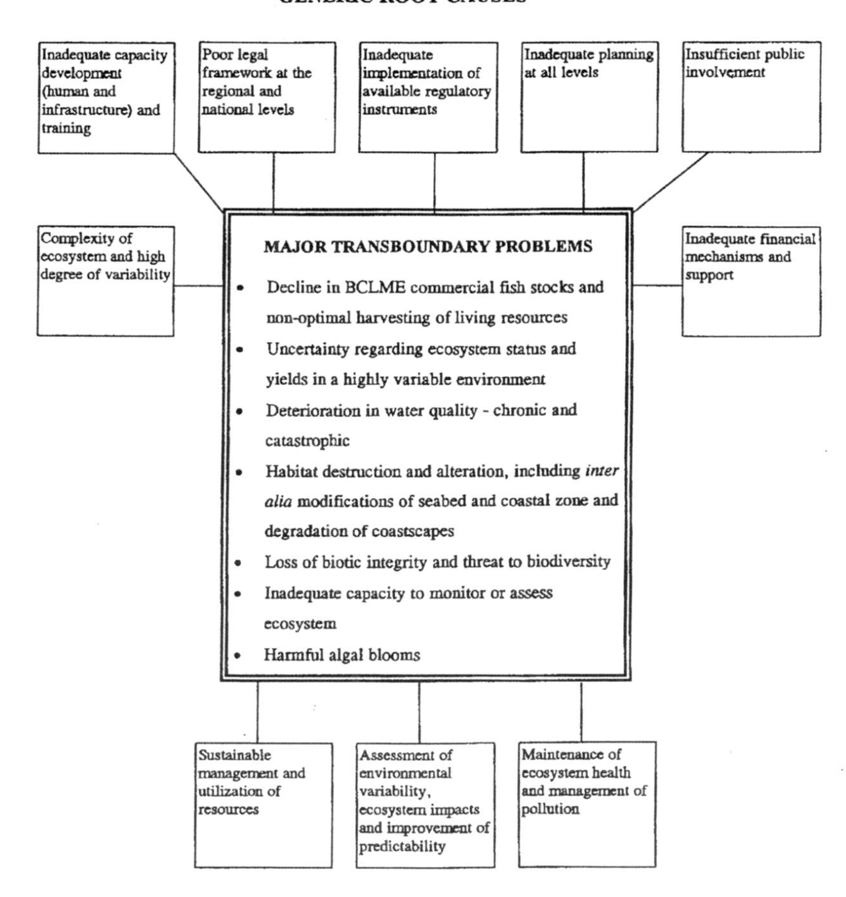

GENERIC ROOT CAUSES
Inadequate capacity development (human and infrastructure) and training
Poor legal framework at the regional and national levels
Inadequate implementation of available regulatory instruments
Inadequate planning at all levels
Insufficient public involvement
Complexity of ecosystem and high degree of variability
Inadequate financial mechanisms and support
MAJOR TRANSBOUNDARY PROBLEMS
• Decline in BCLME commercial fish stocks and non-optimal harvesting of living resources
• Uncertainty regarding ecosystem status and yields in a highly variable environment
• Deterioration in water quality - chronic and catastrophic
• Habitat destruction and alteration, including *inter alia* modifications of seabed and coastal zone and degradation of coastscapes
• Loss of biotic integrity and threat to biodiversity
• Inadequate capacity to monitor or assess ecosystem
• Harmful algal blooms
Sustainable management and utilization of resources
Assessment of environmental variability, ecosystem impacts and improvement of predictability
Maintenance of ecosystem health and management of pollution
AREAS WHERE ACTION IS REQUIRED

Figure 10-6. Major transboundary problems, generic root causes and areas requiring action.

the part of the key players to collaborate to achieve this objective, but the real challenge will be to develop systems that take cognizance of the naturally highly variable and potentially fragile nature of the BCLME and its coastal environments within the context of a changing society and world.

In order to address the main transboundary problems and their root causes (see previous section), the stakeholders have identified three broad areas where immediate action is necessary, viz improving management and utilization of resources, assessing impacts and improving predictability of environmental variability, and maintaining ecosystem health and reducing pollution (see Figure 10-6). In all of these areas capacity development and training is seen as a high priority activity. Policy development and harmonization and regional networking and collaboration in surveys, monitoring, assessment etc. are likewise seen as very important. However, the success of any management action will clearly be dependent on the proper understanding of the underlying ecosystem processes and the linkages between the Benguela and the larger ocean-atmosphere environment. This will require a concentrated research effort in partnership with the international science community, building on existing local and regional activities such as the BEP and the fisheries-based BENEFIT programmes, and international programmes such as Climate Variability and Prediction Program (CLIVAR), Global Ocean Ecosystem Dynamics (GLOBEC) and Global Ocean Observing system (GOOS). The question is, can this be best done through an LME-type approach? Furthermore, is the present LME definition adequate or even appropriate for an ecosystem such as the Benguela?

From an LME research and management perspective, system boundaries cannot ignore regional political and economic realities, and the interdependence of countries. In the case of the Benguela, it would make little sense to regard the relatively shallow Angola-Benguela Front as the northern extent of the ecosystem for reasons previously given. Moreover, it is our contention that the present LME definition is applicable to closed or semi-closed systems with "concave" fixed boundaries, but perhaps less so to open "convex" systems such as the Benguela. At first appearance the Benguela upwelling area does have distinct boundaries in the form of shallow fronts which demarcate various epi-pelagic domains. However, the sources of much of the significant variability and change in the Benguela ecosystem lie outside the system. Regional human impact on the ecosystem (fishing, mining, coastal development, etc.) thus takes place within the context of substantial externally forced variability—mainly natural, but there is also increasing evidence of anthropogenic climate change superimposed thereon. Unlike concave fixed boundary ecosystems such as the Black Sea, where corrective management can largely ignore external forcing, in an open ecosystem such as the Benguela, sustainable integrated management has to take the external forcing into account, and that means looking beyond the narrow confines of the various fronts. The approach to a highly variable LME like the Benguela will necessarily be very different to that for closed or semi-closed systems. What must therefore clearly be recognized is that the Benguela does not function in isolation, but rather as part of the global ocean system.

Sustainable integrated management of the Benguela requires a collective and proactive approach by Angola, Namibia and South Africa, not a reactive "knee-jerk" response to problems. Apart from the joint actions which the three countries are committed to, visionary thinking and innovative management on the part of the governments will be required. Within the next decade it is likely that the first signs of global environmental change will become apparent, and governments which choose to ignore this probability do so at their peril. Management cannot proceed in the absence of good advice based on good science, and accordingly the regional research structures will need to be strengthened, not undermined. Access to international expertise and collaboration with other players in the Atlantic is essential, particularly in terms of modeling and improving predictability. Moreover, links will need to be established between the BCLME and the Gulf of Guinea LME and activities in the LMEs of our neighbors the other side of the Atlantic (Patagonian Shelf, Brazil Current, North Brazil Shelf) and the proposed Agulhas Current LME. In this respect the establishment of operational networking between the various LMEs would perhaps go a long way towards giving effect to the UNCED declaration. If this is done, then certainly the LME approach will stand a good chance of success.

Returning to the title of the paper, "The Benguela: Ex Africa semper aliquid novi," so what then is new? Simply put, it is the determination of three countries which have been subjected to centuries of oppression and exploitation to take joint action to correct the wrongs of the past and demonstrate to the rest of the world how a fragile and variable marine ecosystem can be managed sustainably.

ACKNOWLEDGEMENTS

The paper, "The Benguela: Ex Africa semper aliquid novi," is largely based on material contained in the BENEFIT Science Plan (Anon 1997) and in documents prepared for GEF/UNDP as inputs for approval by the GEF of the BCLME Programme. Financial support from the BCLME PDF/B grant via UNDP is gratefully acknowledged.

REFERENCES

Andrews, W.R.H. and L. Hutchings. 1980. Upwelling in the southern Benguela Current. Prog. Oceanogr. 9:81 p

Anon. 1997. BENEFIT Science Plan, BENEFIT Secretariat, Windhoek, Namibia 90p

Bailey, G.W. and J. Rogers. 1997. Chemical oceanography and marine geoscience off southern Africa: Past discoveries in the post-Gilchrist era, and future prospects. Trans. Roy. Soc. S. Afr. 52(1): 51-79

Brown, P.C., S.J Painting and K.L. Cochrane. 1991. Estimates of phytoplankton and bacterial biomass and production in the northern and southern Benguela ecosystems. S. Afr. J. mar. Sci. 11: 537-564

Carton, J.A. and B. Huang. 1994. Warm events in the tropical Atlantic. J. Phys. Oceanogr. 24: 888-903

Chapman, P. and L.V. Shannon. 1985. The Benguela ecosystem. 2. Chemistry and related processes. In: Barnes, M., ed. Oceanography and Marine Biology: An annual review. Aberdeen Univ. Press. 23: 183-251

Crawford, R.J.M. 1999. Seabird responses to long-term changes in prey resources off southern Africa. In: Proceedings of 22nd International Ornithological Congress, Durban. Adams, N. and R. Slotow, eds. Birdlife South Africa; Johannesburg. 688-705

Crawford, R.J.M., Underhill, L.G., Raubenheimer, C.M., Dyer, B.M. and J. Martin. 1992. Top predators in the Benguela ecosystem—implications of their trophic position. In: Payne, A.I.L., Brink, K.J., Mann, K.J. and R. Hilborn, eds. Benguela Trophic Functioning. S. Afr. J. mar. Sci. 12: 95-99

Croll, P. 1998. Benguela Current Large Marine Ecosystem (BCLME) First Regional Workshop UNDP, 22-24 July 1998, Cape Town, South Africa (Moderator's report on Workshop). 73 p

De Decker, A.H.B. 1970. Notes on an oxygen-depleted subsurface current off the west coast of South Africa. Investl. Rep. Div. Sea Fish. S. Afr. 84:24p

Dubnov, V.A. 1972. Structure and characteristics of the oxygen minimum layer in the Southeastern Atlantic. Oceanology 12(2): 193-201

Gammelsrød, T., O'Toole, M.J., Bartholomae, C.J., Boyer, D. and V.L. Filipe. 1998. Intrusion of warm surface water layers at the Southwest African coast in February-March 1995: The Benguela Niño '95: S. Afr. J. mar. Sci. 19. 99 41-56

Gordon, A.L. and K.T. Bosley. 1991. Cyclonic gyre in the tropical South Atlantic. Deep-Sea Res. 38 (Suppl. 1A) S323-S343

Hampton, I., Boyer, D.C., Penney, A.J., Pereira, A.F. and M. Sardinha. 1999. Integrated overview of fisheries of the Benguela Current region. BCLME Thematic Report 1, UNDP, Windhoek, 89 p

Hamukuaya, H., O'Toole, M.J., and P.M.J. Woodhead. 1998. Observations of severe hypoxia and offshore displacement of Cape hake over the Namibian shelf in 1994. S. Afr. J. mar. Sci.19

Hart, T.J. and R.I. Currie. 1960. The Benguela Current. Discovery Rep. 31: 123-297

Hutchings, L. and J.G. Field. 1997. Biological oceanography in South Africa, 1896-1996: Observations, mechanisms, monitoring and modeling. Trans. Roy. Soc. S. Afr. 52(1): 81-120

Hutchings, L., Pitcher, G.C., Probyn, T.A. and G.W. Bailey. 1995. The chemical and biological consequences of coastal upwelling. In: Summerchanges, C.P., Emeis, K-C., Angel, M.V., Smith, R.L. and B. Zeitzschel,eds. Upwelling in the Ocean: Modern Processes and Ancient Records. John Wiley & Sons Ltd: 65-81

Lass, H.U., M. Schmidt, V. Mohrholz and G. Nausch. 2000. Hydrographic and current measurements in the area of the Angola-Benguela front. J. Phys. Oceanogr., 30(2000), S. 2589-2609

Monteiro, P.M.S. 1996. The oceanography, the biogeochemistry and the fluxes of carbon dioxide in the Benguela upwelling system. Ph.D thesis, Univ. Cape Town, S. Afr.: 354p

Moroshkin, K.V., Bubnov, V.A. and R.P. Bulatov. 1970. Water circulation in the eastern South Atlantic Ocean. Oceanology 10(1): 27-34

Nelson, G. 1980. Poleward motion in the Benguela area. In: Neshyba, S.J., Moers, C.N.K., Smith, R.L. and Barber, R.T., eds. Poleward Flows along Eastern Ocean Boundaries. Springer, New York: 110-130 (Coastal and Estuarine Studies 34)

Nelson, G. and L. Hutchings. 1983. The Benguela upwelling area. Prog. Oceanogr. 12(3): 333-356.

Payne, A.I.L., Gulland, J.A. and K.J. Brink (Eds.). 1987. The Benguela and Comparable Ecosystems. S. Afr. J. mar Sci, 5: 957 p

Parrish, R.H., Bakun, A., Husby, D.M. and C.S. Nelson. 1983. Comparative climatology of selected environmental processes in relation to eastern boundary current pelagic fish reproduction. In: Sharp, G.D. and J. Csirke eds. Proceedings of the Expert Consultation to Examine changes in Abundance and Species Composition of Neritic Fish Resources, San Jose, Costa Rica, April 1983. F.A.O. Fish. Rep. 291(3): 731-777

Pillar, S.C., Moloney, C.L., Payne, A.I.L. and F.A. Shillington, eds. 1998. Benguela Dynamics: Impacts of Variability on Shelf-Sea Environments and their Living Resources. S. Afr. J. mar. Sci. 12:1108p

Shackleton, L.Y. 1986. An assessment of the reliability of fossil pilchard and anchovy scales as fish population indicators off Namibia. MSc. Thesis, Univ. Cape Town, S. Afr. 141p

Shannon, L.V. 1985. The Benguela ecosystem. 1. Evolution of the Benguela, physical features and processes. In: Oceanography and Marine Biology. An Annual Review 23

Shannon, L.V., Agenbag, J.J. and M.E.L. Buys. 1987. Large and mesoscale features of the Angola-Benguela front. In: Payne, A.I.L., Gulland, J.A. and K.H. Brink, eds. The Benguela and Comparable Ecosystems, S. Afr. J. mar. Sci. 5: 11-34

Shannon, L.V., Agenbag, JJ., Walker, N.D. and J.R.E. Lutjeharms. 1990. A major perturbation in the Agulhas retroflection area in 1986. Deep Sea Res. 37(3): 493-512

Shannon, L.V. and thirteen co-authors. 1992. The 1980s—A decade of change in the Benguela ecosystem. In: Payne, A.I.L., Brink, K..H., and R. Hilborn, eds. Benguela Trophic Functioning, . S. Afr. J. mar. Sci 12:271-296

Shannon, L.V., Boyd, A.J., Brundrit, G.B. and J. Taunton-Clark. 1986. On the existence of an El Niño-type of phenomenon in the Benguela system. J. Mar. Res. 44(3): 495-520

Shannon, L.V., Crawford, J.R.M., Brundrit, G.B. and L.G. Underhill. 1988. Responses of fish populations in the Benguela ecosystem to environmental change. J. Cons. Perm. Int. Explor. Mer. 45(1): 5-12

Shannon, L.V. and G. Nelson. 1996. The Benguela: Large scale features and processes and system variability. In: Wefer, G. and G. Siedler, eds. South Atlantic Circulation: Past and Present. Springer Verlag. 163-210

Shannon, L.V. and J.J. O'Toole. 1998. Proceedings of the International Symposium on Environmental Variability in the South-east Atlantic, Swakopmund., Namibia, March 1998

Shannon, L.V. and J.J. O'Toole. 1999. Integrated overview of the oceanography and environmental variability of the Benguela Current region. BCLME Thematic Report 2, UNDP, Windhoek, 57p

Shannon, L.V. and S.C. Pillar 1986. The Benguela ecosystem. 3. Plankton. In: Barnes, M., ed. Oceanography and Marine Biology. An Annual Review. 24: 65-170. Aberdeen; University Press

Sherman, K. 1994. Sustainability, biomass, yields, and health of coastal ecosystems: an ecological perspective. Mar. Ecol. Prog. Ser. 112: 277-301

Shillington, R.P. 1998. The Benguela upwelling system off Southwestern Africa, Coastal segment (16, E). In: Robinson, A.R. and K.H. Brink, eds. The Sea Vol.II, 583-604

Stander, G.H. 1964. The Benguela Current off South West Africa. Investl. Rep. Mar. Res. Lab SW Afr. 12: 43pp. plus 77 pages figures

Stander, G.H. and A.H.B. De Decker. 1969. Some physical and biological aspects of an oceanographic anomaly off South West Africa in 1963. Investl. Rep. Div. Sea Fish. S. Afr. 81, 46 p

Stuttaford, M. (Ed). 1996. Fishing Industry Handbook South Africa, Namibia & Moçambique. Marine Information C.C., Stellenbosch. 434p

Taunton-Clark, J. and L.V. Shannon. 1988. Annual and interannual variability in the South-east Atlantic during the 20th century. S. Afr. J. mar. Sci. 6:97-106

Verheye, H.M., Hutchings, L., Richardson, A.J., Marska, G. and D. Gianakouras. 1998. Long-term trends in the abundance and community structure of zooplankton during anchovy and sardine recruitment on the west coast of South Africa (1951 to present). S. Afr. J. mar. Sci. 19:317-332

Verheye, H.M. and A.J. Richardson. 1998. Long-term increases in crustacean zooplankton abundance in the southern Benguela upwelling region (1951-1996): bottom-up or top-down control? ICES Journal of Marine Science. 55: 803-807

Voituriez, B. and A. Herbland. 1982. Comparison des systemes producitfs de l'Atlantique Tropical Est.: domes thermiques, upwellings côtiers et upwelling equatorial. Rapp P.V. Reun Con. Int. Explor. Mer. 180: 114-130

Wattenberg, H. 1938. Die Verteilung des Sauerstoffs im Atlantischen Ozean. Wiss. Ergeb. Dt. Atlant. Exped. "Meteor" 1925-1927, 9: 132 p

Woodhead, P.M.J., Hamukuaya, H., O'Toole, M.J., Strømme, T. and S.S. Kristmannsson. Submitted. Recruit mortalities of Cape hake, following exclusion from shelf habitat by persistent hypoxia in the northern Benguela Current. 1998 ICES Recruit Dynamics Symposium Proceedings. ICES J. mar.sci.

Yamagata, T. and S. Iizuka. 1995. Simulation of tropical thermal domes in the Atlantic : a seasonal cycle. J. Phys. Oceanogr. 25(9): 2129-2140

Large Marine Ecosystems of the World
G. Hempel and K. Sherman (Editors)

11

Decadal Environmental and Ecological Changes in the Canary Current Large Marine Ecosystem and Adjacent Waters: Patterns of connections and teleconnection

C. Roy and P. Cury

ABSTRACT

In order to explore decadal changes in the Canary Current Large Marine Ecosystem (LME), wind and SST data collected by merchant ships are extracted from the Comprehensive Ocean Atmosphere Data Set (COADS) database from 1950 to 1995 within sixteen boxes along the Canary Current LME and adjacent waters from 10°N up to 43°N. Potential biases are reviewed and discussed when considering long-term changes and large spatial resolution. The heterogeneity of the seasonal patterns in the Canary current is described. Decadal environmental changes are explored using Sea Surface Temperature (SST) data. In the early and mid-1970s, an intensive cooling had a large latitudinal extension that affected the Canary Current LME and adjacent waters from Spain to Senegal. This cooling period was followed by a warming period, which started in the 1980s. Fisheries, fish stock abundance as well as fish population distributions have drastically changed during the last five decades. The sardine populations off Morocco widely fluctuated in a disconnected manner and a substantial increase in biomass was observed in fishing zone C (located between 20°N and 27°N) during the mid-1970s and mid-1980s. Several outbursts of fish populations (*Micromesistius poutassou, Macrorhamphosus scolopax* and *M. gracilis, Octopus vulgaris, Balistes carolinensis*) were observed during the last three decades in the Canary Current. In the case of the latter species it was connected with the outburst populations which took place in the Gulf of Guinea LME. Upwelling intensity and sea surface temperature are strongly linked and are thought to affect both spatial distribution and abundance of fish in the Canary Current LME. Ecosystems appear to be connected at very large scales through climatic teleconnection between the Northern and Central Atlantic but also between the Pacific and the Atlantic as the North Atlantic Oscillation (NAO) index and the El Niño Southern Oscillation (ENSO) events seem to have a major impact on the dynamics of the upwelling in the Canary Current.

THE CANARY CURRENT UPWELLING LME

The Canary Current LME is part of the eastern boundary of the central north Atlantic. This LME constitutes one of the four major upwelling systems of the world oceans. The main oceanographic features of this LME have been described in several papers dating from the 1970s. Wooster *et al.* (1976), using a compilation of merchant ship SST and wind data from 43°N to 7.5°N, summarized the seasonal patterns of the Canary Current coastal upwelling. In the mid 1970s, intensive process oriented studies were carried out along the coast to study the Canary Current upwelling with the main focus being off the Sahara coast between 20°N and 26°N (see Hempel 1982 for a collection of papers presenting detailed results of the CINECA [Cooperative Investigations of the Northern part of the Eastern Central Atlantic] program). The mean patterns of the circulation in the region were summarized by Mittelstaedt (1983) and Bas (1993). An updated review of the characteristics of upwelling in general, and of the Canary Current in particular, can be found respectively in Robinson and Brink (1998) and Barton (1998).

After the intensive process oriented studies carried out during the 1970s, most of the research and data collection effort was directed toward stock assessment and fisheries related studies. Except from some intensive surveys by the former GDR and USSR research groups off Mauritania, few oceanographic cruises have been carried out by the national fisheries research centres. To our knowledge, there is no long-term time series of subsurface data available. A network of coastal stations, where daily SST and (sometimes) nutrient data are collected, is maintained by several fisheries research centres along the coast (Cury and Roy 1991, Durand *et al.* 1998) but the accessibility of these data remains limited. The interannual and long-term variability of this LME over the last 40 years has not yet been summarized but several studies presented detailed information on the variability in a given area. Arfi (1985), using wind measurement at a coastal station, studied the variability of the wind-driven upwelling off Cape Blanc (Mauritania 21°N) from 1955 to 1982. The link between wind forcing and temperature fluctuations at 21°N was further explored by Ould-Dedah *et al.* (1999). An analysis of the interannual variability of the Senegalese upwelling (14°N) from 1963 to 1986 was presented by Roy (1989) using wind data from a coastal station. At a regional scale, Roy (1991) summarized the interannual variability of the wind induced upwelling process along the coast from 12°N to 30°N using merchant ship data. Using SST derived from AVHRR satellite data, Nykjaer and Van Camp (1994) presented a regional overview of the seasonal and interannual variability, over a limited period of time, from 1981 to 1991.

DATA USED

The present analysis gives an overview of the pattern of variability of the physical characteristics of the Canary Current LME and adjacent waters. Using an updated version of the CODE software (Mendelssohn and Roy 1996), wind and SST data collected by merchant ships are extracted from the COADS release 1-ab database (Woodruff *et al.* 1987) for sixteen rectangular boxes following the Canary Current upwelling coast from 10°N up to 43°N (Figure 11-1). The shipping lanes from Europe to the Indian Ocean follow the shape of the West African coast south of 28°N; in these regions data density is high along the coast. Between 28°N to 32°N, the shipping lanes move offshore and data density sharply decreases. Further north, along the coast of Morocco, data density increases again and reaches a maximum off Spain and Portugal. The options regarding the source of the data selected by CODE were set in order to exclude all data from the former USSR vessels, a fishing fleet that was quite active in the 1980s and early 1990s off Mauritania. The activity of this fishing fleet was highly seasonal and concentrated over the shelf, while the shipping lanes are located further offshore. Blending these data with the merchant ship data can introduce a bias of the seasonal cycle because of the wind and SST cross-shore gradient. Both estimated wind (Beaufort scale) and measured wind (using an anemometer) data were selected. A correction factor was applied to the estimated wind data in order to avoid the introduction of a spurious trend due to the decreasing percentage of estimated wind data during the last twenty years. For each of the sixteen boxes,

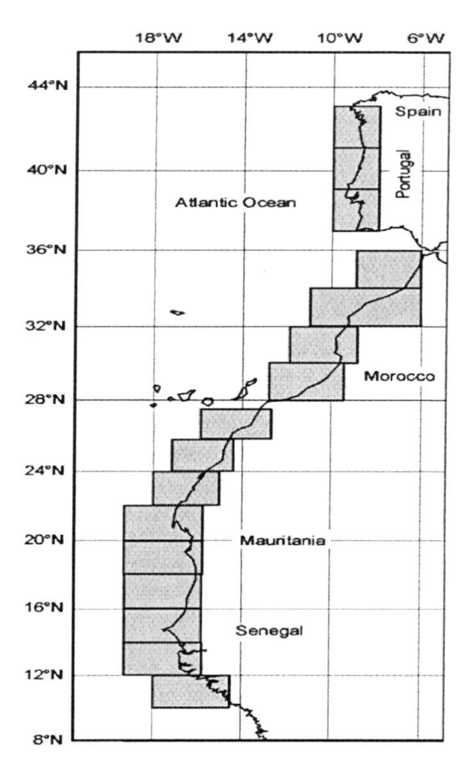

Figure 11-1. The Canary Current LME and adjacent waters, from 46°N to 8°N. The 16 coastal boxes where the COADS data have been extracted are shaded.

monthly time series from 1950 to 1995, of SST, scalar wind speed and of a monthly Coastal Upwelling Index (CUI) are constructed. CUI is the offshore component of the wind induced Ekman transport and it is used as an index of the strength of the upwelling process (Bakun 1973). The mean monthly seasonal cycle of SST, scalar wind speed and CUI are calculated and are later used to compute monthly anomalies for each of the three parameters. Offshore SST (1400km from the coast) was also extracted from da Sylva *et al.* (1994) in order to compute time series of the offshore-coastal SST deficit and the corresponding mean seasonal cycle.

SEASONAL PATTERNS

The mean seasonal variability for the offshore-coastal SST deficit, SST and CUI are briefly analysed in this section. SST deficits greater than 2°C can be used to trace the seasonal and latitudinal variability of the upwelling (Figure 11-2). The seasonal pattern of the SST deficit is similar to the general pattern first described by Wooster *et al.* (1976).

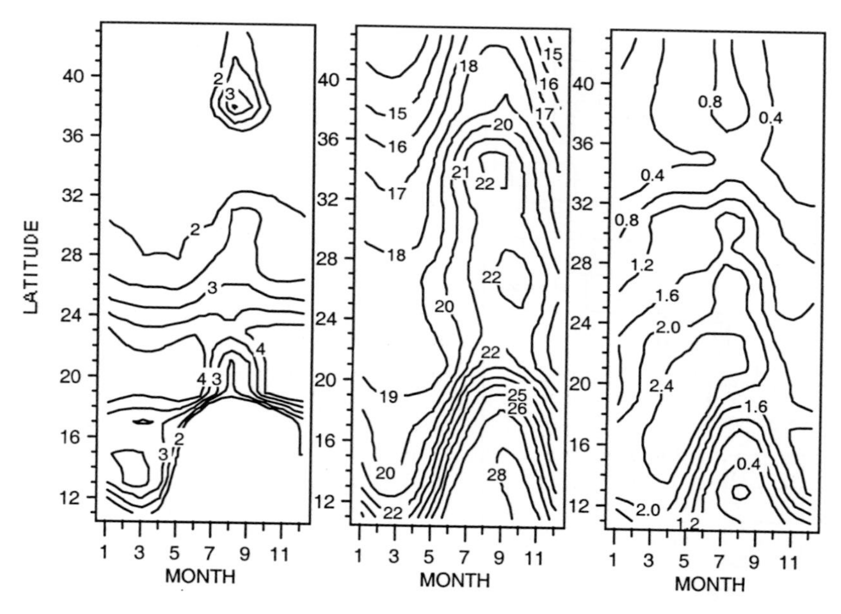

Figure 11-2. Seasonal pattern of coastal SST deficit (°C, left panel), SST (°C, middle panel) and CUI ($m^3 \cdot s^{-1} \cdot m^{-1}$, right panel) along the coast of the Canary Current LME.

Upwelling is permanent between 19°N and 28°N. South of 21°N, upwelling occurs only in winter; the alternation of a winter upwelling season with a monsoon-type warm season strongly enhances the SST seasonal cycle. The positive SST deficit that occurs in summer between 28°N and 32°N, is the result of the northern Moroccan upwelling. Further north, between 32°N and 37°N, the orientation of the coast and the presence of the Gibraltar Straits are not upwelling favourable. North of 37°N, upwelling is active in summer off Portugal and Spain. In this region, upwelling disappears in winter, except during brief periods of favourable winds (Barton 1998). The seasonal pattern of SST (Figure 11-2) is similar to the general pattern first described by Wooster *et al.* (1976).

The prominent feature is the strong seasonal variability south of 21°N. In that region, the amplitude of the seasonal cycle reaches 8°C between 14°N and 16°N. Data from coastal stations show that the amplitude of the seasonal cycle reaches up to 16°C (from 14°C in winter to 30°C in summer) in this area. Between 21°N and 29°N, the low amplitude of the SST seasonal cycle reflects the year-round persistence of the upwelling process. The CUI seasonal and latitudinal patterns over the region match closely the pattern described by Wooster *et al.* (1976) but with values up to 40 percent higher (Figure 11-2). One possible cause of the difference between our estimation and the previous one is the correction factor that we applied to the COADS wind data in order to take into account the errors introduced by the use of the WMO1100 scale. This scale was used to convert wind data from Beaufort units to $m \cdot s^{-1}$ and it tends to underestimate the winds for Beaufort numbers less than about 6 and overestimates for Beaufort numbers greater than 6. Off West Africa, a comparison between the corrected and uncorrected wind data shows that the mean value of the corrected wind can be from 10 to 15 percent higher than the uncorrected ones, depending on the percentage of wind data expressed in Beaufort units. This will lead to an increase of respectively 20 percent to 30 percent of the corresponding wind stress and CUI.

When comparing the spatio-temporal pattern of CUI with the pattern of coastal SST deficit, the global picture remains similar with a winter upwelling south of 19°N, a permanent upwelling between from 19°N to 28°N and a summer upwelling off the Portuguese and Spanish west coasts. CUI provides some indication of the seasonal pattern of the upwelling in the central region where upwelling is a permanent feature. In that region, the wind-induced upwelling is maximum in summer (July-August) and minimum in early winter (November-December).

DECADAL CHANGES IN THE ENVIRONMENT

SST

In the following section, SST is used to track changes in the intensity of the Canary Current LME upwelling. An intensification (relaxation) of the upwelling process enhances (reduces) the upward flux of cold water along the coast; the offshore extension of the cold upwelled water is also enhanced (reduced) during an intensification (relaxation) of the upwelling process. As a result, negative (positive) SST anomalies are expected during an intensified (relaxed) phase of the upwelling.

For each coastal box, monthly SST anomalies are derived from the monthly time series of SST. To get a synthetic view of the variability, monthly anomalies are then averaged by quarter of the year. The data are presented on a time/latitude diagram and smoothed to get the long-term pattern of the variability (Figure 11-3). In the early and mid-1970s, an intensive cooling (negative SST anomalies) characterizes the variability of the SST anomalies during the first quarter (Figure 11-3). It has a large latitudinal extension, affecting the Canary Current LME and adjacent waters from Spain to Senegal. South of 16°N, the low frequency variability is characterized by a succession of warm and cold periods. Between 16°N and 23°N, the variability appears to increase after the mid-1970s with a succession of warm and cold periods, similar to what is found further south. North of 23°N the variability decreases, the cooling of the 1970s being the major climatic event.

The SST anomalies' variability during the second quarter shares some common characteristics with the variability during the first quarter. There is an extensive cooling affecting the whole region in the 1970s but, while being less pronounced, this cooling appears to start in the mid 1960s and extends to the late 1970s. South of 16°N, the variability seems also to be characterized by a succession of warm and cold periods, but with a slightly different pattern than during the first quarter. Further north, a pronounced warm event developed in the early 1960s between 25°N and 43°N. Following the cooling of the 1970s, the variability north of 25°N remains weak with a slight warming.

During the third quarter, SST anomalies are characterized by a pronounced cooling extending from the mid-1960s to the mid-1980s. It affects the region north of 18°N up to 35°N. Another emergent pattern with a large latitudinal extension is the strong warming that developed north of 20°N during the late 1980s and early 1990s. From 1950 to 1965, the latitudinal variability is contrasted with warm anomalies in the central region (maximum intensity at 20°N) and cold anomalies at both high and low latitudes. During the fourth quarter, the global 1970s cooling pattern over the region is again a notable feature of the SST anomalies. It reaches its maximum intensity between 15°N and 30°N and extends from 1972 to 1975. A warming affecting all the Canary Current LME since the 1980s follows this cooling.

As a summary, time series in selected regions are presented to highlight the dominant patterns of variability of SST anomalies by quarter from 1950 to 1995 (Figure 11-4.1 to 11-4.4). The cooling in the 1970s is a major feature that has affected the whole region during and outside the upwelling seasons. The southern part of the regions presents a quite dynamic pattern of variability that is unique to the area. This is confirmed by a comparative analysis of the variability of SST anomalies times series using the standard deviation as an index of the variability (Figure 11-5). It shows that the SST variability reaches a maximum during the first and second quarter in the southern part of the region. During the third and fourth quarters, the interannual variability is maximum around 20°N. This suggests that the variability of the upwelling strongly enhances the interannual variability of SST anomalies in the southern part of the Canary Current LME.

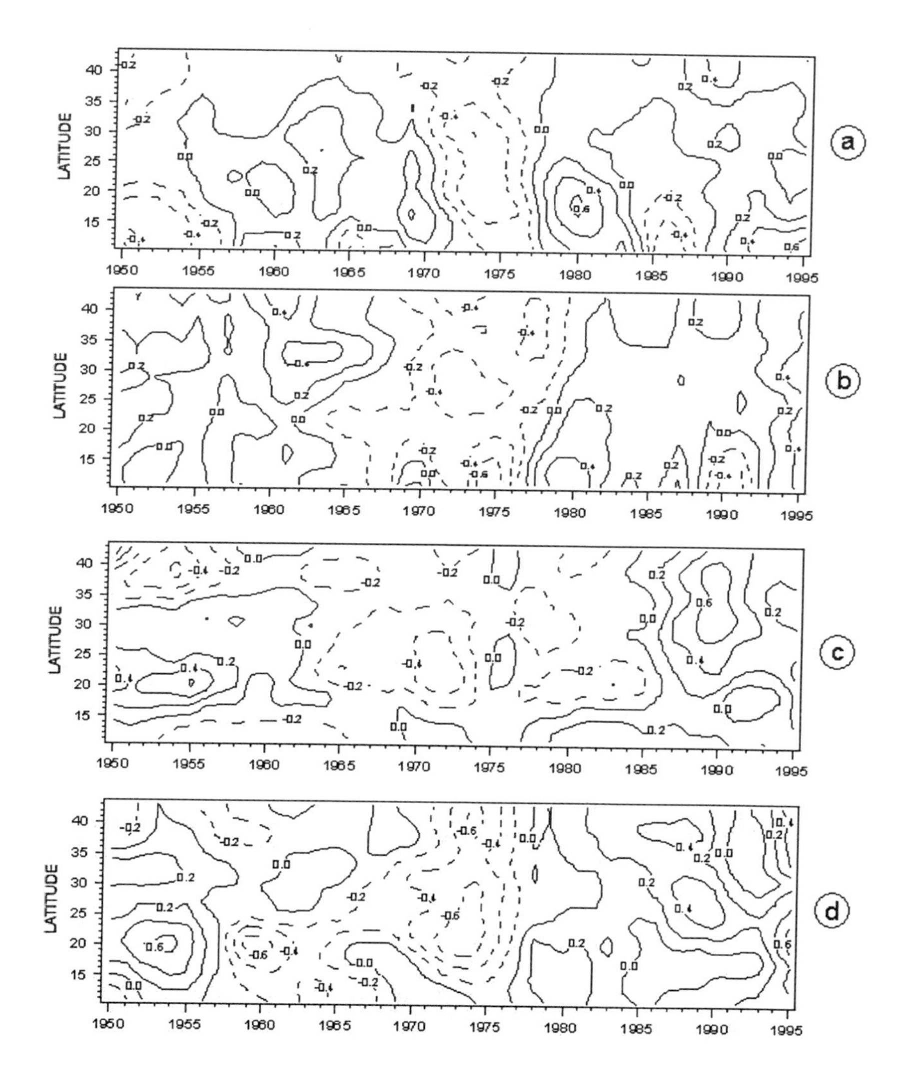

Figure 11-3. Time/latitude (°N) plot of SST anomalies (°C) in the Canary Current LME and adjacent waters from 43°N to 11°N and from 1950 to 1995 for the first (a), second (b), third (c) and fourth quarter (d).

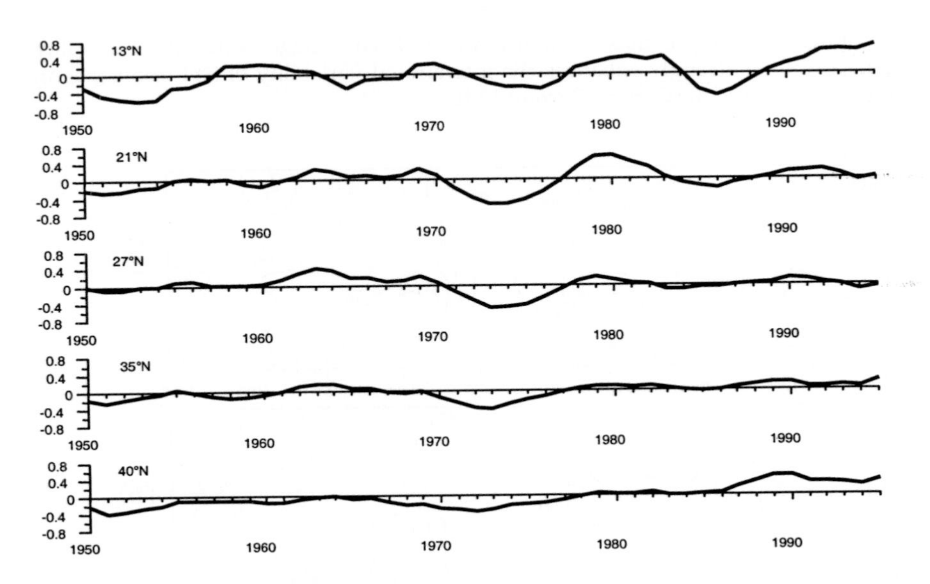

Figure 11-4.1. Trend of the SST anomalies (°C) at selected latitudes in the Canary Current LME and adjacent waters during the first quarter, from 1950 to 1995.

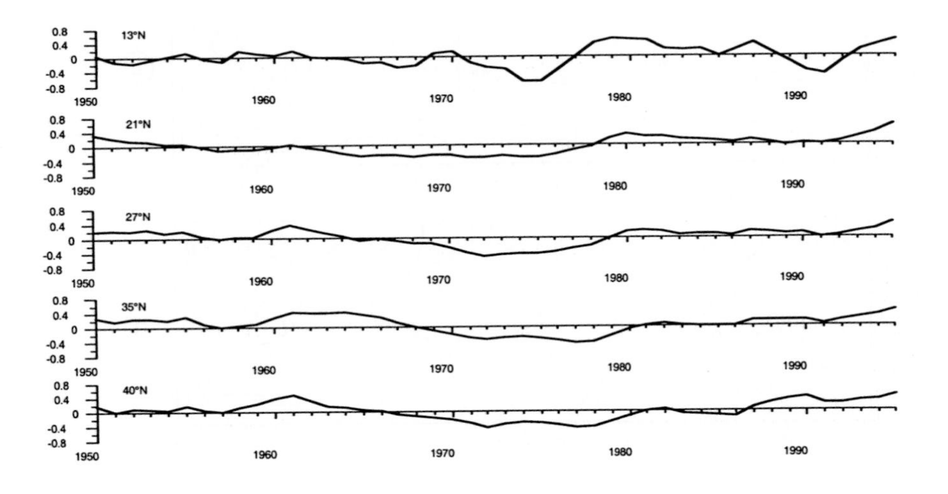

Figure 11-4.2. Trend of the SST anomalies (°C) at selected latitudes in the Canary Current LME and adjacent waters during the second quarter, from 1950 to 1995.

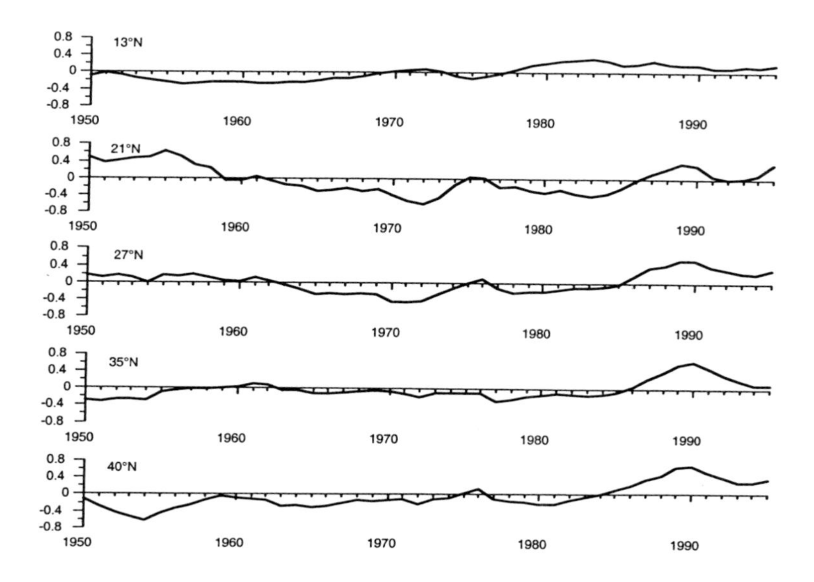

Figure 11-4.3. Trend of the SST anomalies (°C) at selected latitudes in the Canary Current LME and adjacent waters during the third quarter, from 1950 to 1995.

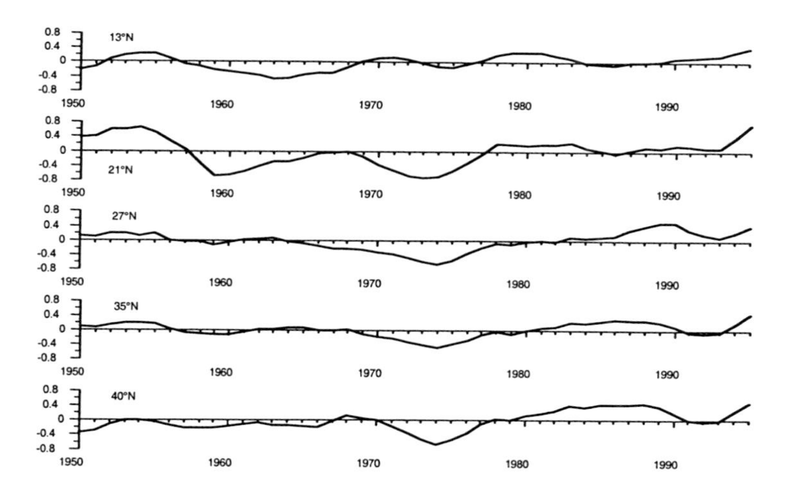

Figure 11-4.4. Trends of the SST (°C) anomalies at selected latitudes in the Canary Current LME and adjacent waters during the fourth quarter, from 1950 to 1995.

Wind

Wind is the driving force of the upwelling process and interannual fluctuations or decadal trends of the alongshore wind component modulate the upwelling and can have drastic consequences on marine living resources. In the Canary Current LME, there are just a few stations where long-term time series of wind data are available and these data are not easily obtained. We had to rely on wind data extracted from the COADS database to set up an homogeneous set of time series over the region. Within the Canary Current, the wind speed monthly times series built using the COADS database are characterized by a pronounced upward trend over the last 40 years (Figure 11-6). Before going further into the analysis, an evaluation of the intensity of the trend and a comparison over the region is performed. For each time series, the long-term trend is extracted by applying a low pass filter to the monthly time series. The trends appear to be similar within the region; they all are almost linear and show an increase of wind speed of about 20 percent over the last 40 years (Figure 11-7). By normalizing the wind data within each coastal box, differences between the resulting trends become almost barely discernible (Figure 11-7). Similar results were obtained by comparing low pass filtered quarterly wind data (Figure 11-7); the slope of the quarterly trends is independent of the season being considered. Wind during winter increased as much as wind increased during summer. This analysis shows that the slope of the trend in the COADS wind data is surprisingly independent of both location and season. Coastal stations wind data show that there is an important inter-annual variability but none of the data that have been analysed shows a trend compatible with the COADS wind trend (Arfi 1985, Roy 1989). Using the COADS data, Roy and Mendelssohn (1998) showed that the correlation between the alongshore wind component and SST during upwelling seasons significantly increases when detrended wind data are used. These arguments raise some concern about the reality of the trend in the COADS wind data.

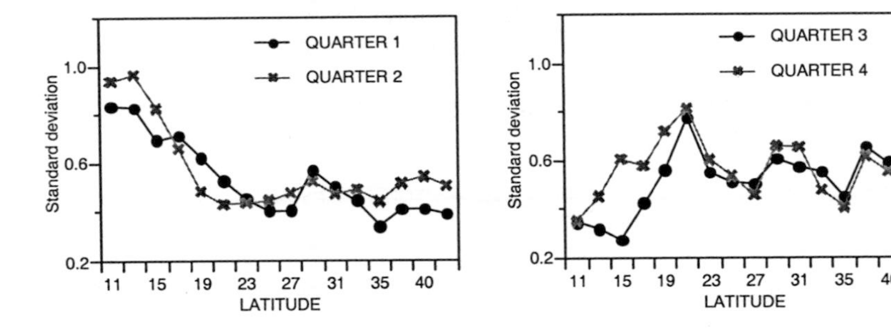

Figure 11-5. Standard deviation of the quarterly time series (1950 to 1995) of SST anomalies in the Canary Current LME and adjacent waters from 11°N to 43°N.

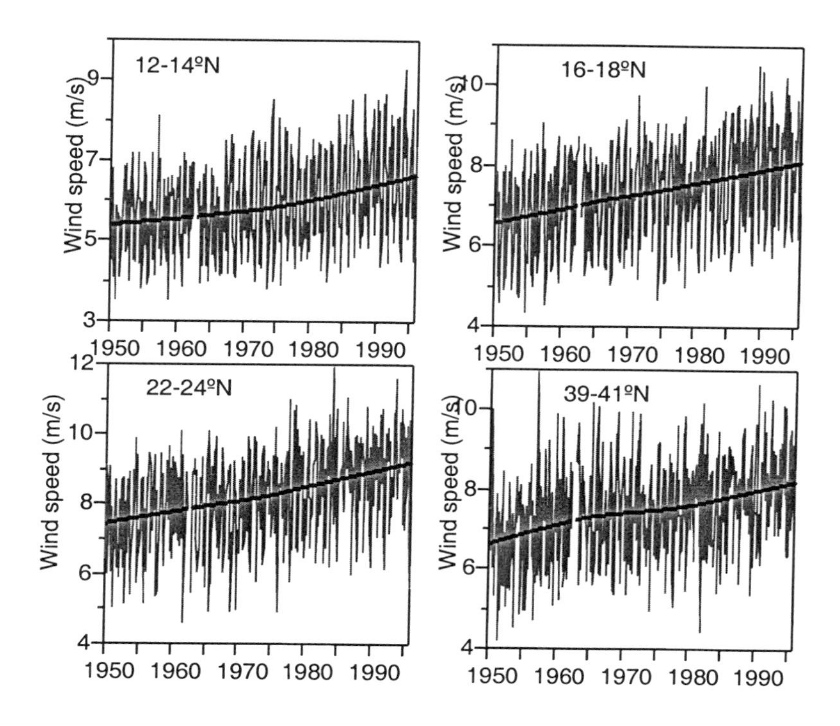

Figure 11-6. Trend in the scalar wind speed (m·s^{-1}) derived from the COADS database at several locations in the Canary Current LME and adjacent waters.

Furthermore, a 20 percent increase of wind speed results in a 40 percent increase of wind stress and of upwelling intensity according to the Ekman transport equation. Such a dramatic enhancement of the upwelling over the last 40 years would certainly have had an enormous impact on the ecosystem. Moreover such drastic environmental changes have never been observed in any upwelling systems and are not reflected in the other environmental time series. The reality of the COADS wind trends has been the subject of considerable debate (Cardone *et al.* 1990). Ward and Hoskins (1996), using pressure derived wind, showed that there is no statistically significant strengthening of the atmospheric circulation at a global scale; they concluded that a correction needs to be applied to the reported wind data in order to minimize the trend. On the other hand, Bakun (1990) proposed a mechanism whereby global greenhouse warming could intensify the alongshore surface wind.

Using analysed wind fields (a blend of pressure derived wind and ship reported wind), he gave evidence of a positive trend in the alongshore wind stress in several regions. No definitive agreement on the reality of the trends in the COADS wind data has been reached yet. Ward and Hoskins (1996) concluded that although there is no evidence for a global intensification of the circulation strength, regional patterns of upward and downward trends may also exist. Following these arguments, we considered that the upward trend in the COADS wind data in the Canary Current is an artifact and that wind data have to be detrended before being used. We are aware that removing the trend is based on a strong assumption, since possible natural causes such as Bakun's (1990) greenhouse effect on the coastal upwelling region are then not considered. A detailed understanding of the origin of the bias in the COADS data is a necessary step before being able to separate the natural long term variability from the data related bias.

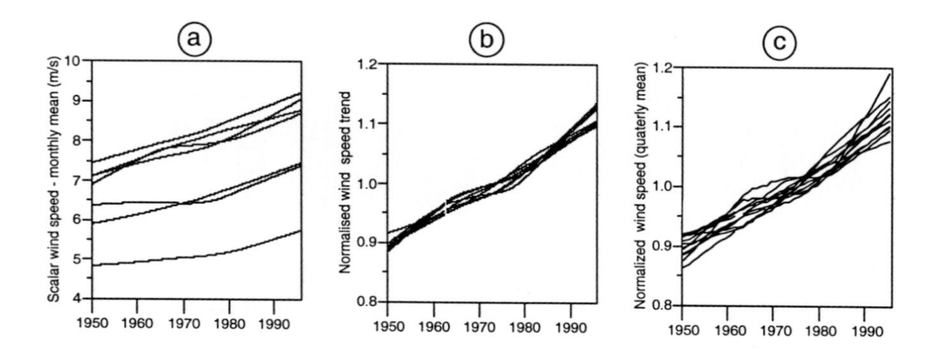

Figure 11-7. a) Scalar wind speed trends ($m \cdot s^{-1}$) over the Canary Current LME and adjacent waters, monthly data from 1950 to 1995 and from 11°N to 43 °N ; b) Same as a) but with normalized wind speed data ; c) Trend of quarterly averaged wind data (1950 to 1995) in selected regions of the Canary Current LME and adjacent waters (15°N, 23°N, 40°N).

The detrended wind data are used to investigate the effect of the interannual variability of the wind on the upwelling strength by looking at the correlation between anomalies of SST and CUI. For each coastal box, quarterly means of SST and CUI anomalies are calculated from the monthly time series of SST and CUI (computed from the detrended wind data). For each coastal box, the correlation between the SST anomalies and the corresponding CUI anomalies are then calculated (Table 11-1). As expected, the global pattern shows significant negative correlation coefficients between SST and CUI anomalies when upwelling is active: a strengthening of the alongshore wind (positive CUI anomalies) enhances the upwelling (negative SST anomalies). In the southern part of the region (between 10°N and 16°N), a significant part of the variability of SST anomalies is explained by the fluctuations of CUI anomalies during the first two quarters. North of 20°N (up to 28°N), upwelling is permanent and the correlation remains relatively high during the four quarters. North of 28°N, upwelling is active in summer and the correlation is high during the third and fourth quarters. Off Spain and Portugal, wind driven processes appear to be of major importance during the spring and summer upwelling seasons but also during the fourth quarter.

Table 11-1. Correlation between quarterly averaged SST anomalies in the Canary Current LME and adjacent waters and the corresponding CUI anomalies, from 1950 to 1995 (* = p<0.05, ** = p<0.01).

Latitude	1st	2nd	3rd	4th
11°N	-0.43 **	-0.60 **	-0.35	-0.27
13°N	-0.49 **	-0.70 **	-0.05	-0.37 *
15°N	-0.49 **	-0.51 **	-0.19	-0.33
17°N	-0.59 **	-0.17	-0.47 **	-0.27
19°N	-0.48 **	-0.14	-0.43 **	-0.57 **
21°N	-0.53 **	-0.45 **	-0.20	-0.61 **
23°N	-0.46 **	-0.47 **	-0.47 **	-0.64 **
25°N	-0.51 **	-0.37 *	-0.43 **	-0.58 **
27°N	-0.42 **	-0.15	-0.57 **	-0.59 **
29°N	0.03	-0.33	0.02	-0.79 **
31°N	-0.25	-0.18	-0.51 **	-0.47 **
33°N	0.03	0.23	-0.47 **	-0.39
35°N	-0.13	0.19	-0.25	-0.50 **
38°N	0.00	-0.53 **	-0.47 **	-0.63 **
40°N	0.00	-0.40 **	-0.44 **	-0.60 **
42°N	-0.19	-0.37 *	-0.28	-0.61 **

TELECONNECTIONS

The North Atlantic Oscillation (NAO) characterizes a meridional oscillation in atmospheric mass with centres of action being the Icelandic low and the Azores high. It is most pronounced in amplitude and areal coverage during winter. NAO is an important contributor to the North Atlantic climatic variability (Hurrel 1995) and has a measurable impact on the North Atlantic ecosystem (Fromentin and Planque 1996).

The link between the NAO and the Canary Current and adjacent waters upwelling is investigated by looking at the correlation between quarterly averaged values of the NAO index and the corresponding SST anomalies over the regions (Table 11-2). The NAO index is based on the difference of normalized sea level pressures between Ponta Delgada, Azores and Stykkisholmur, Iceland from 1865 through 1995. This index is slightly different from the winter version of the index that uses data from

Lisbon. The Ponta Delgada station is chosen instead of Lisbon to adequately capture the NAO during the four quarters. A positive NAO index indicates stronger than average westerlies and anomalously high pressures across the sub-tropical Atlantic.

Negative and statistically significant ($p < 0.01$) correlations between NAO and SST anomalies off the Canary Current occur during the first and second quarters in the central region (between 18°N and 30°N), as well as during the fourth quarter north of 20°N. The inverse relationship indicates that negative (positive) SST anomalies are related to positive (negative) NAO anomalies. In the central region, this correlation suggests that an intensification of the westerlies (positive NAO index) across the sub-tropical Atlantic intensifies the upwelling favourable wind that also

Table 11-2. Correlation between quarterly averaged SST anomalies in the Canary Current LME and adjacent waters and the corresponding NAO anomalies, from 1957 to 1995 (* = $p<0.05$, ** = $p<0.01$).

Latitude	1st quarter	2nd quarter	3rd quarter	4th quarter
11°N	0.02	-0.05	0.03	-0.19
13°N	-0.01	-0.10	0.04	-0.26
15°N	-0.16	-0.16	0.07	-0.38*
17°N	-0.37 *	-0.22	0.13	-0.35 *
19°N	-0.41 **	-0.31	0.04	-0.44 **
21°N	-0.30	-0.39 *	-0.01	-0.44 **
23°N	-0.48 **	-0.44 **	-0.06	-0.49 **
25°N	-0.60 **	-0.40 **	-0.22	-0.41 **
27°N	-0.56 **	-0.34 *	-0.29	-0.54 **
29°N	-0.01	-0.44 **	-0.20	-0.58 **
31°N	-0.15	-0.29	-0.17	-0.40 **
33°N	-0.09	-0.13	-0.04	-0.44 **
35°N	-0.20	-0.14	-0.05	-0.50 **
38°N	0.13	-0.31	-0.19	-0.57 **
40°N	0.26	-0.29	-0.24	-0.50 **
42°N	0.26	-0.24	-0.25	-0.50 **

enhances the upwelling process (negative SST anomalies). In the case of a relaxed meridional oscillation (negative NAO index), the weaker than average atmospheric circulation contributes to a relaxation of the Canary Current and adjacent waters upwelling (positive SST anomalies). In the northern region (north of 30°N), upwelling is not a dominant oceanographic process during the 4th quarter. An alternative mechanism accounting for the correlation between NAO and SST involves the intensification of the westerlies in early winter leading to a premature erosion of the thermocline and to a deepening of the surface mixed layer. Both the erosion of the thermocline and the deepening of the surface mixed layer result in negative SST anomalies.

Table 11-3. Correlation between quarterly averaged SST anomalies in the Canary Current LME and the SOI anomalies during the fourth quarter of the preceding year, from 1957 to 1995 (* = p<0.05, ** = p<0.01).

Latitude	1st quarter	2nd quarter
11°N	-0.41 **	-0.46 **
13°N	-0.48 **	-0.47 **
15°N	-0.48 **	-0.54 **
17°N	-0.40 **	-0.51 **
19°N	-0.38 *	-0.48 **
21°N	-0.42 **	-0.38 *
23°N	-0.36 *	-0.21
25°N	-0.28	-0.15
27°N	-0.17	-0.19
29°N	-0.11	-0.25
31°N	-0.26	-0.28
33°N	-0.28	-0.24
35°N	-0.18	-0.16
38°N	-0.16	-0.09
40°N	-0.15	-0.03
42°N	-0.29	-0.04

Pacific El Niño SouthernOscillation (ENSO) events have a major impact on the world climate. In the tropical Atlantic, the SST and wind fields are regularly affected by Pacific equatorial variability (Hastenrath *et al.* 1987, Nobre and Shukla 1996). Large scale analyses have shown that the North Atlantic warms in response to the Pacific ENSO with a lag of about one or two seasons (3 to 6 months), the effect being stronger in the north-western part of the basin and during the boreal spring and early summer (Enfield and Mayer 1997). Furthermore, it appears that the Atlantic SST response to ENSO is a result of a reduction of the trade wind speeds (Enfield and Mayer 1997, Klein *et al.* 1999). In the Canary Current and adjacent waters, the wind being the driving force of the upwelling, an alteration of the trade wind activity has a pronounced effect on the ecosystem, thus one can expect to observe a connection between ENSO and the upwelling in the Canary Current. This is investigated by looking at the correlation between quarterly SST anomalies during the first semester and the Southern Oscillation Index (SOI) during the fourth quarter of the preceding year. In the southern part of the region (south of 19°N), the correlation is statistically significant (p<0.01) during the first and second quarters (Table 11-3). As expected from the weakening of the trade wind that is associated with ENSO events, there is a negative correlation between SOI and SST anomalies: warm events in the Pacific (negative SOI) lead to the development of positive SST anomalies in the southern part of the Canary Current during late winter and early spring. Moreover, it appears that the correlation holds also for cold events in the Pacific (positive SOI), suggesting that there is a strong and permanent teleconnection between the coastal upwelling activity in the Atlantic and the state of the Pacific ocean. This link between ENSO and the Canary Current upwelling is explored in detail by Roy and Reason (2001). It is thought that the mechanism responsible for this remote forcing involves a tropospheric connection along 10°N-20°N between the Atlantic and the Pacific resulting in an alteration of the Atlantic trade winds by the conditions over the Pacific ocean.

ECOLOGICAL PATTERN IN THE CANARY CURRENT LME

Fisheries and fish population patterns

Annual marine fish catch varied between 1.0 and 2.3 million tonnes from the 1970s to the 1990s in the Canary Current (FAO 1997a). Pelagic fish stocks, which represent approximately 70 percent of the total catch, are not over-exploited and their abundance has been shown to be driven by the strength of the upwelling (Binet *et al.* 1998, Kifani 1998). According to FAO (1997a) most of the demersal stocks are fully exploited in the Canary Current from Mauritania to Guinea-Bissau, and the catches have been decreasing substantially since the end of the 1980s.

Since the 1950s the exploitation of marine resources in the Canary Current has been subject to many changes. In the eastern central Atlantic a total of 25 distant-water fishing nations have exploited the marine resources using long-range fishing fleets (FAO 1997b, Maus 1997). Political changes in the Eastern European countries provoked important changes in the configuration of the exploitation in the Canary Current, particularly in the late 1980s and early1990s. Thus, between 20 percent and 50 percent of the total regional marine production has been harvested by foreign fleets, even though their catches, mainly pelagics, have been declining rapidly over the last few years, following the partial withdrawal of Eastern European and former USSR fleets from the west-African coast (FAO 1997a). In the Canary Current small-scale fisheries are particularly active in Senegal, and to a lesser extent in Mauritania. They have been exploiting both pelagic and demersal resources for several centuries (Chauveau 1991), and have recently improved drastically in their efficiency. In 1992, the small-scale fishery sector in Senegal was composed of 5,700 canoes and employed around 35,000 fishers who landed approximately 270,000 t of pelagic fish and 46,000 t of demersal fish (Ferraris *et al.* 1998). This makes the present Senegalese small-scale fishery far more socially and economically efficient than the industrial fishery sector.

Sardine (*Sardina pilchardus*) is mainly found off Morocco and sardinellas (*Sardinella aurita* and *S. maderensis*) further south off Mauritania and Senegal. Together with the horse mackerels (*Trachurus spp.*) they represent the predominant pelagic fish species. Drastic fluctuations of abundance have been observed for sardines, sardinellas and other pelagics during the last fifty years which have been linked to environmental horse mackerels (*Trachurus spp.*) they represent the predominant pelagic fish species. Drastic fluctuations of abundance have been observed for sardines, sardinellas and other pelagics during the last fifty years which have been linked to environmental changes (Cury and Roy 1991). Off Morocco, decadal patterns in the three fishing areas for sardine (zones A, B and C) seem to be disconnected as the sardine abundance appears to fluctuate in a different manner in the different fishing zones (Figure 11-8). Fluctuations of sardine, linked to the upwelling, have been recorded in

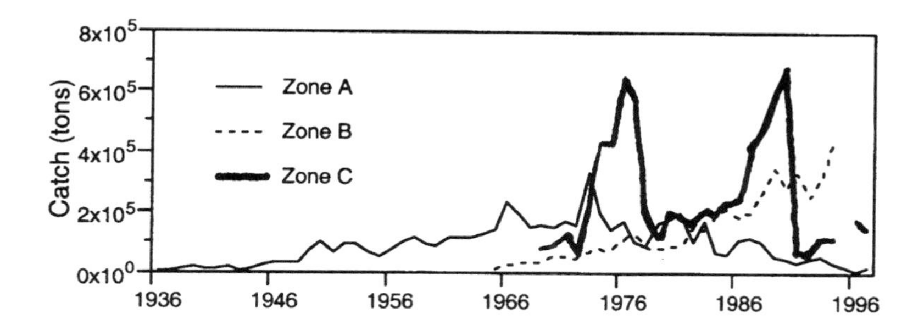

Figure 11-8. Catch of sardine in the three main fishing zones off Morocco (zone A=30°N-33°N, zone B=27°N-30°N, and zone C=20°N-27°N). (Data courtesy of INRH/Morocco)

zone C since the mid-1970s (Binet *et al.* 1998). In this area, periods of high abundance have been associated with a southward expansion of the population as far as the Senegalese coast. Binet (1988) and Binet *et al.* (1998) hypothesized that an acceleration of the trade winds increased offshore and southward surface transport, enhanced the primary versus secondary production rate, and consequently favoured phytoplankton feeders such as sardine. Several hypotheses were recently proposed to explain the change in sardine abundance in areas A and B (Do-Chi and Kiefer 1996). Changes of the migrational tendencies, expansion of the stock, unfavourable environment in the traditional fishery zones affecting fish migration, and a southward shift in the centre of gravity of the sardine population due to climatic changes were the different mechanisms that were put forward to explain the changes observed in the traditional Moroccan sardine fishery.

Sardinellas (*Sardinella aurita* and *Sardinella maderensis*) have been mainly exploited by foreign fleets and by small-scale fisheries off Senegal. Acoustic surveys carried out in the Canary Current showed that sardinellas were very abundant, particularly off Mauritania, with biomass estimated to be about 4 million tonnes in 1992 (FAO 1997a). The two species of sardinella were moderately exploited until 1992. The

Fish population outbursts

In the Canary Current several species, known to be rare, developed such a huge biomass during several years that, in certain cases, they sustained important fisheries. Subsequently, they vanished in a relatively short time period. Off Morocco an outburst of *Micromesistius poutassou* was recorded in the 1960s, which later completely disappeared. In the 1970s *Macrorhamphosus scolopax* and *M. gracilis*, which have very different life-history traits, were found all along the Moroccan coast. In 1976 a first acoustic cruise estimated a potential of about one million tons between Cape Juby and Cape Spartel. The abundance decreased markedly in the 1980s and now the two species are rare.

Between 1972 and 1980 the large trigger fish (*Balistes carolinensis*) expanded its biomass to reach more than 1 million tonnes in the Gulf of Guinea. It also spread geographically from Ghana to Mauritania (Gulland and Garcia 1984): in the early 1970s this species was scarce in the Central Atlantic, the species was at maximum abundance at the end of the 1970s in the Gulf of Guinea and at the beginning of the 1980s in the Canary Current (Senegal). At the end of the 1980s this species almost disappeared from the West African ecosystems. While several authors believe that intensive exploitation drastically reduced the sparid communities and facilitated the outbreak of *Balistes* (Gulland and Garcia 1984), environmental changes could also have affected the trigger fish population dynamics (Caverivière 1991).

The octopus (*Octopus vulgaris*) population significantly increased in abundance in the mid-1960s in the Canary Current (Caddy and Rodhouse 1998). Three stocks of octopus are presently exploited by the Northern African industrial and small-scale fisheries off Dakhla, off Cape Blanc, and off Senegal. The prospective surveys conducted off Mauritania illustrate the rapid increase in biomass during the last thirty years: in 1966, 3.9 percent of octopus were found in the demersal surveys (for a total of 12 percent of cephalopods), compared to 75 percent of octopus in 1971 (90 percent of cephalopods) (Faure *et al.* 2000). In 1968 the octopus represented 10 percent of the commercial catches in Mauritania, 75 percent in 1971, and 84.6 percent in 1989 (approximately 50 percent of the fisheries' value). A substantial small-scale and industrial fishery really started in 1986 in Senegal as the octopus abundance increased in that area. The rapid emergence of the octopus in Mauritania in the late 1960s allowed historical yields between 45, 000 and 50,000 tonnes in the mid-1970s until the end of the 1980s. A rapid expansion of this stock was also observed off Morocco supporting catches that approximated 100,000 tonnes at the beginning of the 1980s. Off Senegal, the octopus stock apparently grew a bit later and was able to sustain an important fishery in the mid-1980s. A peak catch of 17,000 tonnes was recorded in 1986 whereas catches represented only a few hundred tonnes before. Altogether these three fisheries totalled between 40 to 90 000 tonnes from 1985 to 1995 but apparently the catch is rapidly decreasing off Cape Blanc and Dakhla. As in the case of the trigger fish, the outburst of the octopus in the Canary Current appears to be related to a lesser abundance of the sparid community due to strong fishing pressure (Gulland and Garcia 1984) which is supposed to have lowered larval mortality and competition for food (Caddy and Rodhouse 1998). Links between the environment and larval survival off the Banc d'Arguin have been demonstrated, as well as the links between upwelling intensity and the octopus abundance off Senegal and Mauritania (Faure *et al.* 2000; Inejih 2000).

DISCUSSION: CONNECTIONS, TELECONNECTIONS, AND THE LME

Demersal and pelagic fishes migrate seasonally off Mauritania and Senegal according to strong seasonal environmental patterns (Cury and Roy 1988, Fréon 1988). Interannual environmental variability also affects fish abundance and distribution. As noted previously a significant correlation exists between ENSO and upwelling intensity in the Canary Current. Using time series of pelagic fish catch (mainly sardine) in the Eastern Atlantic since 1950 (FAO 1997b) it is possible to quantify the patterns of abundance that occurred during the mid-1970s and the end of the 1980s. Periods of high abundance of sardine appear to be associated with ENSO variability (Figure 11-9). Positive values of SOI are associated with enhanced

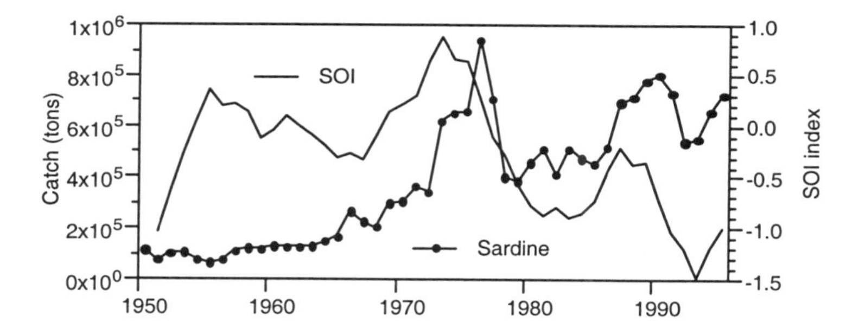

Figure 11-9. Sardine catch in the Canary Current LME from 1950 to 1995 (from FAO 1997b) and smoothed SOI index.

upwelling and coincide with higher catch values. This shows that the interannual variability of both the environment and the fish populations in the Canary Current are related to global environmental signal acting through atmospheric teleconnections. Binet *et al.* (1998) found comparable results where environmental changes can modify fish population abundance and force ecological systems to spread or retract in space. It is noteworthy that the high abundance of Sardinellas at the beginning of the 1990s is associated with a period of higher temperatures, which in turn means that the habitat range for the southern fish species could potentially extend further north as the biomass increases.

During the last several decades the Canary Current was marked by important changes in the fish community which in turn strongly affected fisheries patterns. Species that experienced drastic changes have little in common. They occupy different habitats and have very different ecological requirements and life history traits (Longhurst and Pauly 1987). Many authors believe that the octopus and the trigger fish outbursts were due to a release of predator control on young stages (Longhurst and Pauly 1987) and that current annual fluctuations in recruitment and consequently landings are probably largely environmentally-driven (Caddy and Rodhouse 1998, Faure *et al.* 2000).

African coastal waters fish populations can migrate or spread towards large areas according to their relative abundance and to long-term environmental changes. The most extreme example is the trigger fish population, which apparently expanded from the Gulf of Guinea LME to the Canary LME. Distant ecosystems are connected through environmental teleconnections. The abundance of fish populations in the Canary Current can be related to indices such as the SOI. These teleconnections can potentially induce synchrony among fish populations inhabiting different LMEs. Synchronized patterns in pelagic fish populations have been observed in many distant ecosystems around the world (Schwartzlose *et al.* 1999), which suggests that these emerging patterns of decadal-scale variation are most probably driven by global climatic teleconnections (Klyashtorin 1998, Bakun 1998).

Strong environmental patterns are observed in the Canary Current LME. Drastic decadal patterns of changes in fish populations have also been observed. Even if it is difficult to identify the cause of these changes, it appears necessary to track and examine them at different scales, knowing that causes could be global as well as local.

ACKNOWLEDGEMENTS

Thanks to Dr. S. Kifani (INRH/Morocco) for providing us with the Moroccan sardine catch data. Support for this work was provided by the IDYLE research unit from the Institut de Recherche pour le Développement (IRD - France).

REFERENCES

Arfi, R. 1985. Variabilité inter-annuelle d'un indice d'intensité des remontées d'eaux dans le secteur du Cap-Blanc (Mauritanie). Can. J. Fish. Aquat. Sci., Vol. 42: 1969-1978

Bakun, A. 1973. Daily and weekly upwelling indices, West Coast of north America 1946-71. U.S. Dep. Comm., NOAA Tech. Rep. NMFS SSRF-671, 103p

Bakun, A. 1990. Global climate change and intensification of coastal ocean upwelling. Science, 12 January 1990. 247:198-201

Bakun, A . 1998. Ocean triads and radical interdecadal stock variability: bane and boon for fishery management science. In: Pitcher T.J., P.J.B. Hart and D. Pauly, eds. Reinventing Fisheries Management. Chapman and Hall. 331-358

Barton, E.D. 1998. Eastern boundary of the North Atlantic: Northwest Africa and Iberia. In: A.R. Robinson and K.H. Brink, eds. The Sea, Vol. 11, The Global Coastal Ocean: Regional Studies and Syntheses. Wiley, New-York. 633-657

Bas, C. 1993. Long-term variability in the food chains, biomass yields, and oceanography of the Canary Current ecosystem. In: Sherman, K. L.M. Alexander and B. Gold, eds. Large Marine Ecosystems: Stress, Mitigation, and Sustainability. AAAS Press. 94-103. 376p.

Binet, D. 1988. Rôle possible d'une intensification des alizés sur le changement de répartition des sardines et sardinelles le long de la côte ouest-africaine. Aquat. living Resour. 1:115-132

Binet, D., B. Samb, T.S. Mahfoud, J.J. Levenez and J. Servain. 1998. Sardine and other pelagic fisheries changes associated with multi-year trade wind increases in the Southern Canary current. In: Durand M.H., P. Cury, R. Mendelssohn, C. Roy, A. Bakun and D. Pauly, eds. Global Versus Local Changes in Upwelling Systems. Editions ORSTOM Paris. 211-233

Caddy, J.F. and P.G. Rodhouse. 1998. Cephalopod and groundfish landings: evidence for ecological change in global fisheries. Reviews in Fish Biology and Fisheries 8: 431-444

Cardone, V.J., J.G. Greenwood and M.A. Cane. 1990. On trends in historical marine wind data. J. Climate 3:113-127

Caverivière, A. 1991. L'explosion démographique du baliste (*Balistes carolinensis*) en Afrique de l'Ouest et son évolution en relation avec les tendances climatiques. In Cury P. and C. Roy, eds. Pêcheries Ouest-Africaines: Variabilité, instabilité et changement. Orstom Editions, Paris. 354-367.

Chauveau, J.P. 1991. Les variations spatiales et temporelles de l'environnement soci-économique et l'évolution de la pêche maritime artisanale sur les côtes ouest-africaines. Essai d'analyse en longue période: XVè-XXè siècle. In: Cury P. et C. Roy, eds. Pêcheries Ouest-Africaines: Variabilité, instabilité et changement. Orstom éditions, Paris. 14-25

Cury, P. and C. Roy. 1988. Migration saisonnière du thiof (*Epinephelus aeneus*) au Sénégal: influence des upwellings sénégalais et mauritaniens. Oceanol. Acta, 11(1): 25-36

Cury, P. and C. Roy, eds. 1991. Pêcheries Ouest-Africaines: Variabilité, Instabilité et Changement. ORSTOM Editions, Paris. 525p

da Silva, A., A.C. Young, and S. Levitus, Atlas of Surface Marine Data 1994, Volume 1: Algorithms and Procedures. NOAA Atlas NESDIS 6, U.S. Department of Commerce, Washington, DC, 1994

Do-Chi, T. and D.A. Kiefer 1996. Workshop on the coastal pelagic resources of the upwelling ecosystem of Northwest Africa: research and predictions, Casablanca 15-17 April 1996. FAO:TCP/MOR/4556(A)

Durand, M.H., P. Cury, R. Mendelssohn, C. Roy, A. Bakun and D. Pauly, eds. 1998. From Local to Global Changes in Upwelling Systems. ORSTOM Editions, Paris. 594p

Enfield, D. B. and D. A. Mayer. 1997. Tropical Atlantic SST variability and its relation to El Niño-Southern Oscillation. J. Geophys. Res. 102:929-945

FAO. 1997a. Review of the state of world fishery resources: marine fisheries. FIRM/C920, Fisheries Circular N°920

FAO. 1997b. FISHSTAT PC. Release 5097/97 2 disks. FAO. Rome, Italy

Faure, V., C.A. Inejih, H. Demarcq and P. Cury. 2000. The importance of retention processes in upwelling areas for recruitment of *Octopus vulgaris*: the example of the Arguin Bank (Mauritania). Fisheries Oceanography 9:343-355

Ferraris, J., K. Koranteng and A. Samba. 1998. Comparative study of the dynamics of small-scale marine fisheries in Senegal and Ghana. In: Durand M.H., P. Cury, R. Mendelssohn, C. Roy, A. Bakun et D. Pauly, eds. Global versus local changes in upwelling systems. Editions ORSTOM Paris. 447-464

Fréon, P. 1988. Réponses et adaptations des stocks de clupéidés d'Afrique de l'Ouest à la variabilité du milieu et de l'exploitation. Analyses et réflexion à partir de l'exemple du Sénégal. Orstom Editions, Coll. Études et Thèses, Paris. 287p

Fromentin, J.M. and B. Planque 1996. Calanus and environment in the eastern North Atlantic. 2. Influence of the North Atlantic Oscillation on *Calanus finmarchicus* and *C. helgolandicus*. Mar. Ecol. Prog. Ser. 134:111-118

Gulland, J.A. and S. Garcia. 1984. Observed patterns in multispecies fisheries. In: R.M. May (ed.). Exploitation of Marine Communities. Dahlem Konferenzen. Berlin, Heidelberg, New-York, Tokyo. Springler- Verlag. 155-190

Hastenrath, S., L. C. de Castro and P. Aceituno. 1987. The Southern Oscillation in the Atlantic sector. Contrib. Atmos. Phys. 60:447-463

Hempel, G. 1982.The Canary Current: studies of an upwelling system. Introduction. Rapp. P.-v. Réun. Cons. int. Explor. Mer. 180:7-8

Hurrell, J.W. 1995. Decadal trends in the North Atlantic Oscillation: Regional temperatures and precipitation. Science. 269:676-679

Inejih, C.A. 2000. Dynamique spatio-temporelle et biologie du poulpe (*Octopus vulgaris*) dans les eaux mauritaniennes: modélisation de l'abondance et aménagement des pêches. Thèse de doctorat. Université de Bretagne Occidentale, Brest. 155p

Kifani, S. 1998. Climate dependent fluctuations of the Moroccan sardine and their importance on fisheries. In: Durand M.H., Cury P., Mendelssohn R., Roy C., Bakun A. and D. Pauly, eds. From Local to Global Changes in Upwelling Systems. ORSTOM Editions, Paris. 235-248

Klein, S.A., B.J. Soden and N.-C. Lau. 1999. Remote sea surface temperature variations during ENSO: Evidence for a tropical atmospheric bridge. J. Climate, 12(4):917–932

Klyashtorin, L.B. 1998. Long-term climate change and main commercial fish production in the Atlantic and Pacific. Fisheries Research 37:115-125

Longhurst, A.R. and D. Pauly. 1987 Ecology of tropical oceans. Academic Press, London. 407p

Maus, J. 1997. Sustainable fisheries information management in Mauritania. PhD thesis. Ecosystem analysis and Management Group. University of Warwick, Coventry, UK. 269p

Mendelssohn, R. and C. Roy. 1996. Comprehensive Ocean Data Extraction Users Guide. U.S. Dep. Comm., NOAA Tech. Memo. NOAA-TM-NMFS-SWFSC-228, La Jolla, CA. 67p

Mittelstaedt, E. 1983. The upwelling area off Northwest Africa. A description of phenomena related to coastal upwelling. Prog. Oceanog., 12:307-331

Nobre, P. and J. Shukla. 1996. Variations of sea surface temperature, wind stress, and rainfall over the tropical Atlantic and South America. J. Climate. 9:2464-2479.

Nykjaer, L. and L.V. Van Camp. 1994. Seasonal and interannual variability of coastal upwelling along northwest Africa and Portugal from 1981 to 1991. J. Geophys. Res. 99 (C7):14197-14207

Ould-Dedah, S., W.J. Wiseman Jr. and R.F. Shaw. 1999. Spatial and temporal trends of sea surface temperature in the northwest African region. Oceanol. Acta, 22(3):265-279

Robinson, A.R. and K.H. Brink, eds. 1998. The Global Coastal Ocean: Regional studies and syntheses. The Sea, Vol. 11, Wiley, New-York. 1062p

Roy, C. and Mendelssohn R. 1998. The development and the use of a climatic database for CEOS using the COADS dataset. In: Durand M.H., P. Cury, R. Mendelssohn, C. Roy, A. Bakun et D. Pauly, eds. Global Versus Local Changes in Upwelling Systems. Editions ORSTOM Paris. 27-44

Roy, C. 1989. Fluctuations des vents et variabilité de l'upwelling devant les côtes du Sénégal. Oceanol. Acta., 12(4):361-369

Roy, C. 1991. Les upwellings: le cadre physique des pêcheries côtières ouest-africaines. In: Ph. Cury et C. Roy, eds. Variabilité, Instabilité et Changement dans les Pêcheries Ouest Africaines, ORSTOM, Paris. 146p

Roy, C. and C. Reason. 2001. ENSO related modulation of coastal upwelling in the eastern Atlantic. Prog. Oceanogr.y. 49:245-255

Schwartzlose, R.A., J. Alheit, A. Bakun, T. Baumgartner, R. Cloete, R.J.M. Crawford, W.J. Fletcher, Y. Green-Ruiz, E. Hagen, T. Kawasaki, D. Lluch-Belda, S.E. Lluch-Cota, A.D. MacCall, Y. Matsuura, M.O. Nevarez-Martinez, R.H. Parrish, C. Roy, R. Serra, K.V. Shust, M.N. Ward and J.Z. Zuzunaga. 1999. Worldwide large scale fluctuations of sardine and anchovy populations. S. Af. J. Mar. Sci., 21: 289-347

Ward, M.N. and B.J. Hoskins. 1996. Near surface wind over the global ocean 1949-1988. J. Climate. 9:1877-1895

Woodruff, S.D., R.J. Slutz, R.L. Jenne and P.M. Steurer. 1987. A Comprehensive Ocean-Atmosphere Data-Set. Bull. Amer. Meteor. Soc. 68:1239-1250

Wooster, W. S., A. Bakun and D.R. McLain. 1976. The seasonal upwelling cycle along the eastern boundary of the North Atlantic. J. Mar. Res. 34:131-141

Large Marine Ecosystems of the World
G. Hempel and K. Sherman (Editors)

12

The Humboldt Current - Trends in exploitation, protection and research

Matthias Wolff, Claudia Wosnitza-Mendo, and Jaime Mendo

INTRODUCTION

The south-east Pacific Humboldt Current Large Marine Ecosystem (HCLME) is blessed with the world's most productive coastal waters that provide about 15 percent of the world's catch (FAO 1998). Most of the catch corresponds to the pelagic fish anchovy (*Engraulis ringens*), sardine (*Sardinops sagax*) and horse mackerel (*Trachurus picturatus murphyi*). To a lesser extent mackerel (*Scomber japonicus*), hake (*Merluccius gayi*) and other demersal resources are exploited. Anchovies and a great part of the sardine catch are reduced to fish meal; the other species are used primarily for human consumption. The total annual harvest of the artisanal nearshore fishery is relatively low (about 200,000 t) compared to the pelagic harvest. However, about 170 species of fish and 30 species of invertebrates of high commercial value (Estrella and Guevara 1998) are exploited and this fishery employs thousands of fishermen and sustains as many coastal families.

The high productivity of the HCLME is the result of upwelling processes governed by strong southerly trade winds. These cause offshore Ekman divergence, through which the thermocline is elevated and relatively cold, nutrient-rich deep water is brought to the euphotic zone where the nutrients can be utilized by phytoplankton photosynthesis (Barber *et al.* 1985). The upwelling ecosystem is subjected to considerable interannual climatic variability, with an irregular period and tends to promote large variations in its aquatic populations and their catches. The dominant form of interannual variability occurs when the normal seasonal upwelling is interrupted by the "El Niño-Southern Oscillation" (ENSO), which results in intrusions of warm, clear oceanic waters from the west and north.

In the late sixties and early seventies, the anchovy (*E. ringens*) supported the world's largest fisheries with record catches of 12 million tons in 1972. However, the combined effect of a strong ENSO in 1972/73, overfishing, and possibly other large-

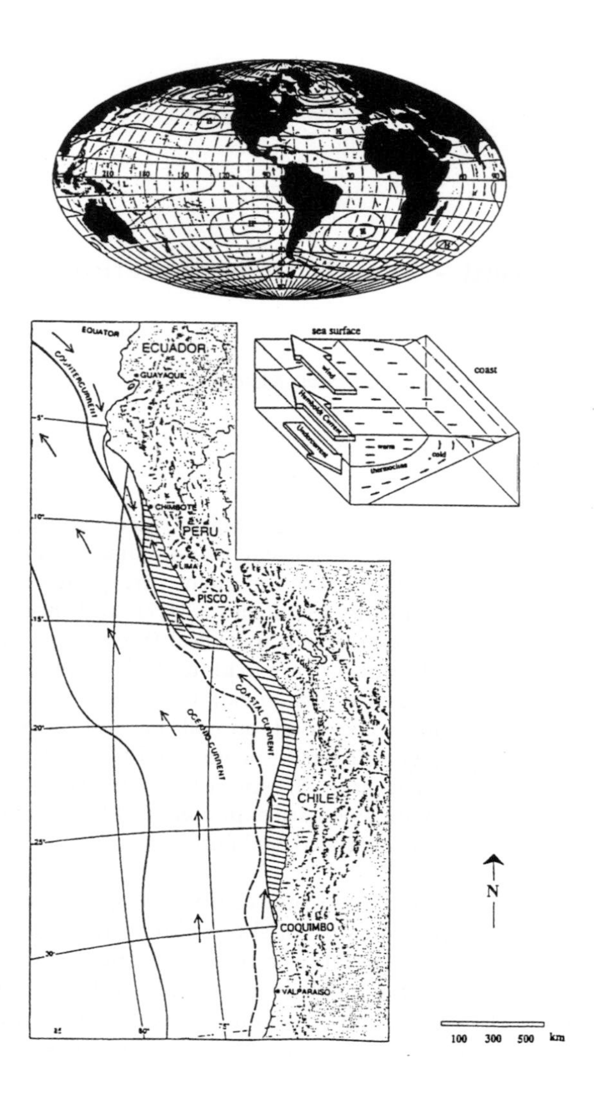

Figure 12-1. Humboldt Current (HCLME); upper part: high and low pressure centres in the world's oceans; middle-right: coastal segment showing currents and water masses of the upwelling system; lower part: main currents and extension of HCLME; the shaded area represents the coastal cold water belt of the central region of the upwelling system.

scale oceanographic changes in the HCLME (see below), led to a collapse of the populations with catastrophic consequences, for the fishing industry and the national economy, that lost several hundreds of millions of US dollars of foreign currency. During the 1970s and the early 1980s pelagic catches remained low, but following the ENSO event in 1983/84, catches rose again to reach, in 1994, the level of the year 1972.

The nature and impacts of ENSO events have been described in numerous articles by physical oceanographers and biologists. In several books the changes of the pelagic fishery have been analyzed (Boje and Tomczak 1978, Pauly and Tsukayama 1987, Pauly *et al.* 1989, Arntz and Fahrbach 1991). More recently, network models have been used to analyze and demonstrate the structural and functional changes that have occurred in the HCS at the decadal timescale or to compare the Humboldt Current ecosystem with other upwelling systems (Jarre-Teichmann and Christensen 1998) and to assess the trophic impact of harvesting small pelagic fish in the Peruvian and other upwelling systems (Mackinson *et al.* 1997). While most researchers have focused their analyses on the changes within the pelagic part of the ecosystem (see Alheit and Bernal 1993 for a review), the demersal shelf community started to receive more attention in the mid-eighties, when discussions came up on the abundance of hake larvae and its relationship to anchovy biomass (Sandoval *et al.* 1989) and when a strong inverse relationship between adult hake and anchovy biomasses from 1953 to 1987 was suggested (Espino and Wosnitza-Mendo 1989). Jarre-Teichmann and Christensen (1998) studied the interannual fluctuations in eastern boundary current systems and included hake and other demersals in their modeling. The nearshore resources came into a stronger management and research focus after the ENSO event of 1982/83, when some species of invertebrates (e.g. the Peruvian scallop) exhibited enormous population proliferations (Arntz and Fahrbach 1991, Wolff 1987, Wolff and Mendo 2000), thus compensating somewhat for the economic losses of the pelagic fishery.

Resource protection and adequate fishery management call for an understanding of the mechanisms that cause the resources in the three sub-systems of the HCLME to vary over time. A central question to be addressed concerns the relative importance of the environmental stochasticity of the system (including ENSO events) compared to (or combined with) top-down mechanisms (Malthusian overfishing, shift in predator biomass) for the observed biomass changes.

The objectives of this chapter are: (1) a brief description of the physical nature of the Humboldt Current Large Marine Ecosystem and El Niño-Southern Oscillation (ENSO), (2) a summary of the changes in exploitation that have occurred in each of the three subsystems over the last decades; (3) review and discussion of research lines along which future investigations should be conducted, (4) suggestion of management policies that incorporate the natural variability observed in each of the subsystems.

THE PHYSICAL NATURE OF THE HUMBOLDT CURRENT LARGE MARINE ECOSYSTEM AND ENSO - A BRIEF SUMMARY

The HCLME has the greatest north-south extent of any of the coastal boundary current systems (from about 5°S to 40°S). Except for the northern part from 6° to 10°S which has a relatively wide shelf (up to 125 km), the shelf of the HCLME is quite narrow with the 200m isobath lying within 10-20 km off the shoreline. Like most other eastern boundary currents, the HCLME is characterized by the equatorward flow of cold, low salinity waters of the Peru or Humboldt Current, which has a complex structure with oceanic and coastal branches (Figure 12-1). Below the Humboldt Current counterflows the Gunther Current or Peru-Chile Undercurrent, with a core at about 100 m depth off Peru. The countercurrent is low in oxygen (< 1ml/l) and high in nutrients. Adjacent to the ocean front of the Humboldt Current, the Peruvian Oceanic Countercurrent (Wyrtki 1967) flows slowly southward ($0.06 m \cdot s^{-1}$). At its oceanic side another surface current (the Chile-Peru Oceanic Current) carries water masses slowly ($0.04\ m \cdot s^{-1}$) northward (Robles *et al.* 1980). Almost year around, nearshore upwelling of the cold and nutrient-rich water occurs as the result of Ekman offshore divergence due to the relatively constant southerly trade winds. The Peruvian and Chilean regions differ in their upwelling characteristics as favourable winds are stronger along the Peruvian coast and subsurface countercurrent water is upwelled here, while off Chile it is subantarctic water of the equatorward flowing coastal current. According to Wyrtki (1963) the boundary between these two source-water regimes lies at about 15°S. In general, the belt of cold upwelling water is much narrower in Peru than in Chile. Near the equator the HCLME is bordered by the oxygen rich Equatorial (Cromwell) Undercurrent that flows eastward below the South Equatorial current in a region of strong water mixing and that can extend far to the south (14°S) during ENSO events.

This current system is related to the large-scale oceanic and atmospheric circulation over the entire tropical Pacific. Rising air over the equatorial region and sinking air over the higher latitudes near 30°S create the high pressure system ("Headly circulation") which is strongest over the winter months. This circulation is modified by rising air over the warm western tropical Pacific and sinking air over the cold upwelled water of the eastern Pacific ("Walker circulation") causing heavy rainfall and low pressures over the western Pacific and sparse rainfall and high pressure over the eastern Pacific. According to Wyrtki (1975), ENSO starts when, after a sustained period of anomalously strong trades, the winds weaken or reverse.

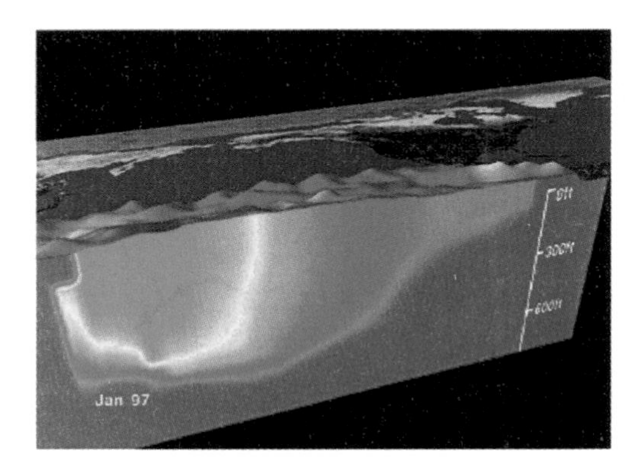

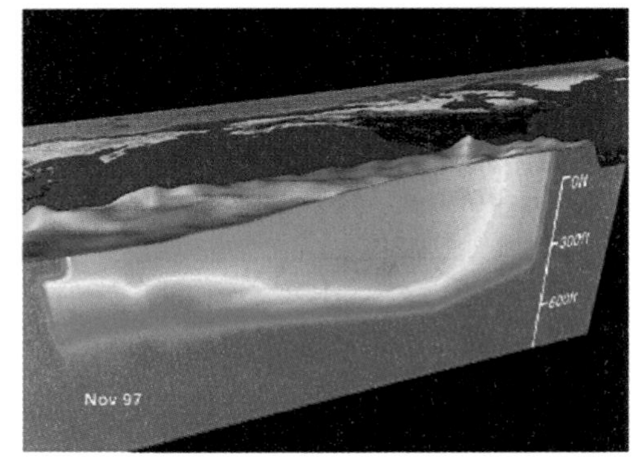

Figure 12-2. Pre-El Niño (Jan 1997, upper part) and El Niño (November 1997, lower part) images of surface topography from NASA's TOPEX satellite, sea surface temperatures (SST) from NOAA's AVHRR satellite sensor, and sea temperature below the surface as measured by NOAA's network of Tropical Atmospheric Ocean Project (TAO) moored buoys. During El Niño, the sea surface rises in the eastern Pacific, the thermocline deepens and the SST increases over a wide area in the eastern Pacific (red is 30°C, blue 8°C; thermocline exists at 20°C, which is the border between the dark blue and the cyan). Images are from NASA Goddard Space Flight Center.

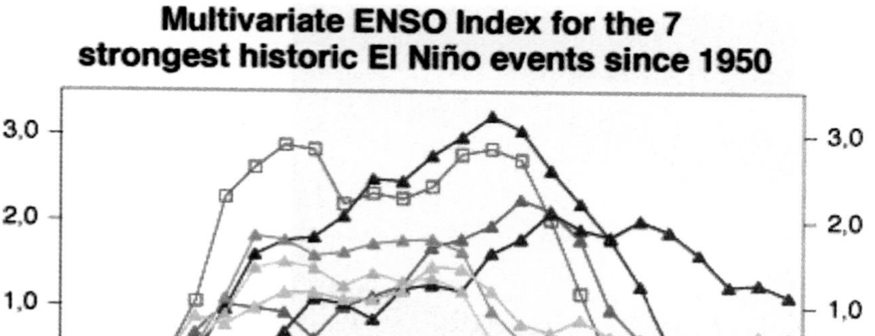

Figure 12-3. Multivariate ENSO Index (MEI) for the seven strongest ENSO events since 1950 (taken from NOAA-University of Colorado, Cooperative Institute for Research in Environmental Sciences (CIRES), Climate Diagnostics Center (CDC), Boulder CO

The period of stronger winds forces an even greater than normal east to west slope of the sea surface, which increases the eastward flow of the subequatorial Cromwell Current and, when the winds slacken or reverse, promotes an eastward flowing Kelvin wave (Enfield and Allen 1980). Upon encountering the South American coast a large intrusion of warm water is observed, which deepens the normally shallow thermocline and causes a rapid rise in sea level and sea surface temperatures along the coast (Figure 12-2).

In terms of sea level, Wyrtki (1977) distinguishes five stages of an ENSO event: (1) a build up of sea level in the western equatorial Pacific; (2) a first relaxation leading to a slow drop in the western and a first peak in the eastern Pacific; (3) a rapid drop of sea level in the west, a filling of the equatorial trough, while it remains above normal in the east; (4) a minimum of sea level in the west and a second peak in the east;

(5) equalization, a very rapid increase of sea level in the west and a return to normal in the east.

The coupled oceanic-atmospheric character of ENSO is being monitored by the multivariate ENSO Index (MEI), which is the weighted average of the main ENSO features contained in the following six variables: sea-level pressure, the east-west and north-south components of the surface wind, sea surface temperature (SST), surface air temperature, and total amount of cloudiness. Positive values of the MEI represent the warm ENSO phase. For comparison, the seven strongest ENSO events since 1950 are shown in Figure 12-3, including 1997/98. The first three events (in 1957/58, 65/66, and 72/73) all featured an early warming in the far eastern Pacific and reached their peaks off Peru before the end of the first year. The more recent ENSO events (1982/83, 86/87, and 91/92) took longer to mature, typically reaching their peaks in the spring of the second year. Early 1983 saw the peak of the strongest El Niño of the century, while 1997/98 featured two peaks just below those of the 1982/83 ENSO, one in July/August 1997, and one in February/March 1998. The ENSO of 1997/98 showed a much higher impact in depth (50-150m) and increases in temperature of 6-10°C within a period of only 50 days (end of March to mid May) (Wosnitza-Mendo and Icochea 1997).

TRENDS IN STOCK SIZES AND EXPLOITATION

The Pelagic Sub-system

Pelagic fisheries of the HCLME began in the early 1950s, reached a peak in catches of > 12 million t just before the collapse in the early 1970s and showed strong fluctuations thereafter. We shall summarize the main trends here only, as the pelagic fisheries system has been analysed in detail in numerous publications (see review by Alheit and Bernal 1993).

Of the four important species anchovy, sardine, horse mackerel and mackerel, anchovy was the dominant species until its collapse in 1972. Fish scale abundance in sediment cores (De Vries and Pearcy 1982) reveal that this was also the case in pre-historic times. The three (northern, central and southern) anchovy stocks in the HCS have risen and fallen simultaneously since 1956. Catches decreased simultaneously in 1972 (ENSO) and fluctuated until the 1980s at low levels. Catches of the stock off Southern Peru and Northern Chile remained relatively high and did not show the dramatic decline of the stock in northern and central Peru. Following the strong ENSO of 1982/83 all anchovy stocks declined to the lowest levels ever registered (about 100,000 t), but rose again considerably thereafter, especially the southern stock. Chilean landings had an historical record of 1.25 million t in 1986. Since that time anchovy landings remained > 1 million t per year in Chile. In Peru, anchovy catches rose thereafter again rather steadily to reach almost 10 million t in 1994. In that year, the combined catches of Peru and Chile exceeded what had ever been caught in the history of the HCLME. While anchovy catches were low after the impact of the last ENSO 1997 (1.5 million t) they rapidly increased thereafter to 6.6 million t in 1999, indicating that stock biomass had suffered much less than during the ENSO of 1983, probably because of favourable recruitment conditions and because fishing was allowed only during very short periods while the stock was constantly monitored (Gutierrez *et al.* 1999).

Sardine catches were insignificant during the 1950s and 1960s. After the ENSO 1972-1973, sardine spawning and catches along the entire coast increased strongly: The stock off Coquimbo (northern Chile) increased significantly in 1974; the stocks off Ecuador, northern, central and southern Peru in 1976, and in 1978 the southern stock off Talcahuano in Chile increased as well.

During the ENSO of 1983-84, sardine landings in southern Peru and northern Chile peaked, while the stocks of northern Peru/Ecuador were at a low, as a result of a southward migration of the sardines due to the high ENSO temperatures. In 1983 Chile landed the highest sardine catches so far recorded, while total pelagic catches in Peru dropped considerably. Two years later, when sardines had migrated northward again, Ecuador had a record year of sardine catches (Serra and Tsukayama 1988). Until 1991 sardine catches in Peru were around 3.0 million t, but steadily decreased thereafter to very low values around 300,000 t in 1999, a trend paralleled in Chile (Figure 12-4).

Horse mackerel showed a very remarkable trend over the last two decades. In Peru catches increased after the ENSO events of 1983-84 and surpassed (low) anchovy landings for the first time. In 1997 catches peaked in Peru at about 650,000 t, but dropped after the ENSO of 1997 to about 80,000 t in 1999. In Chile, catches increased steadily in the late 1980s to > 4 million t in 1994, before they started to decline. Bas *et al.* (1995), based on an extensive data survey, reported a biomass increase of *T. picturatus murphyi* in the South Pacific Ocean from 1978 to 1988-1990, to 15 million tons (Figure 12-4).

Bonito (*Sarda chilensis*) is the top predatory fish in the HCLME, distributed at temperatures < 17°C and salinities between 34.8-35.1. It reached high catches in the 1950s and 1960s (60,000-105,000 t). A sharp decline occurred in the early 1970s, but catches recovered in the mid 1980s and reached a new peak in 1996. The proportion of anchovy in the diet of specimens > 50cm was about 75 percent during periods of high anchovy abundance and declined to about 25 percent after the collapse in 1972.

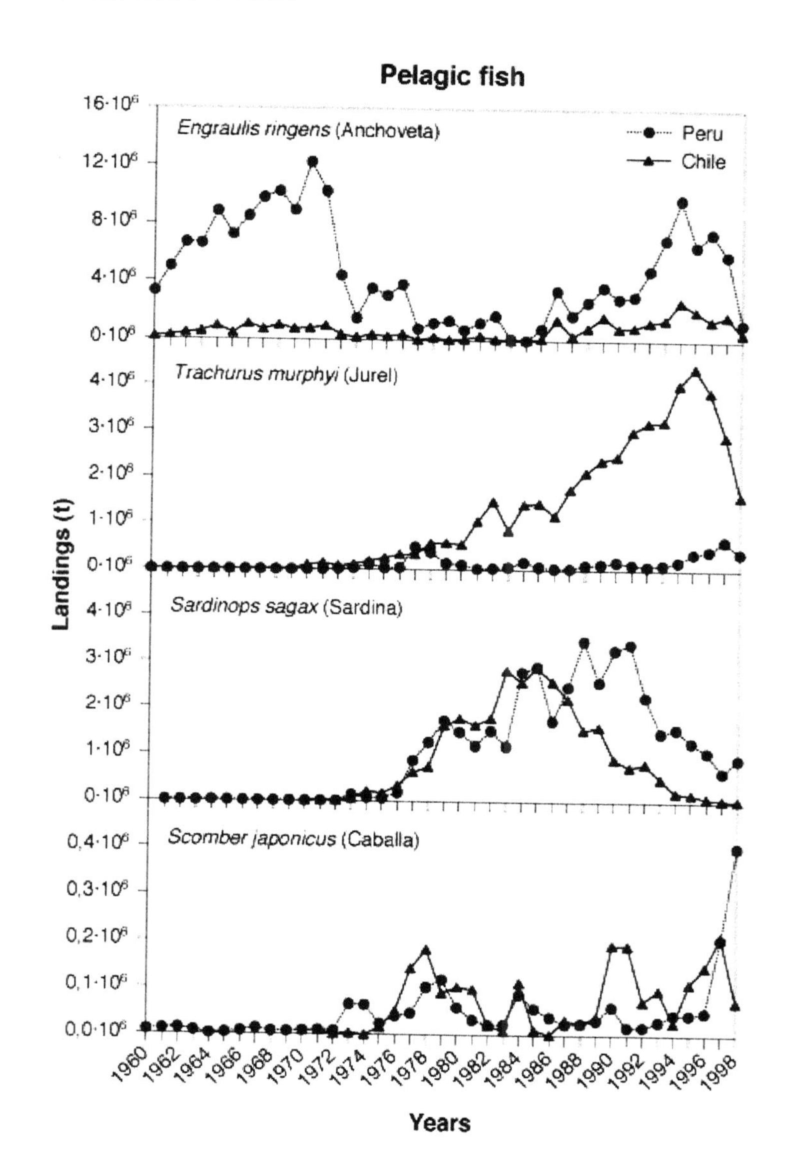

Figure 12-4. Catches of dominant pelagic resources in the HCLME (1960-1999)--from top to bottom: anchovy, horse mackerel, sardine and mackerel

It was estimated that, when populations were large, bonitos devoured approximately 500,000 – 700,000 t of anchovy per year (Pauly *et al.* 1987), suggesting that bonito is strongly related to anchovy biomass in a typical predator-prey way. Seabirds (cormorants, boobies and pelicans) are of similar, or even higher importance as anchovy predators. Their total consumption was estimated around 2 million t before 1955 and dropped thereafter with the decline of the anchovy biomass due to fishing to annual levels of about 30,000 t in the seventies and early eighties (Muck and Pauly 1987). During this decline of anchovy biomass, cormorants, being highly specialized feeders on anchovy, lost their overwhelming dominance over boobies and pelicans. Crawford and Jahncke (1999) showed that numbers and reproductive success of cormorants were significantly related to the biomass of anchovies both in the Peruvian and Benguela upwelling systems. Booby and pelican biomass are more stable but short term decreases are observed for all species following El Niño events.

Looking at the pelagic subsystem would be incomplete without mentioning other large-sized pelagic top predators: Tunas (mainly *Thunnus albacares* and *Thunnus obesus*), and skipjack tuna (barrilete) (*Katsuwonus pelamis*) which are distributed farther offshore in oceanic waters with temperatures > 19°C and salinities > 35.1 between 0-200m depth. Catches of tuna peaked in the 1950s and 1960s with > 10,000 t annually; the same pattern occurred with skipjack (barrilete) which reached > 25,000 t in the 1950s. Thereafter catches of both declined constantly to about 1,000 t annually.

In Peru, a new interesting resource appeared with the giant squid (*Dosidicus gigas*) in 1990, whose catches increased to 210,000 t in 1994 (Peru only harvested part of it, the remainder was taken by Japanese and Korean fleets) parallel to the increase in anchovy catches during that period (Figure 12-5).

The pelagic fishery within the HCLME has thus experienced quite dramatic changes over the last decades in total catch as well as in the catch composition. Most remarkably, anchovy has re-taken its dominant position in the Peruvian part of the HCLME, sardine has decreased in population sizes, first in Chile, then in Peru; horse mackerel, almost insignificant before 1980, has enormously increased its populations to become the main pelagic target in Chile. Combined horse mackerel (mostly in Chile) and anchovy (mostly in Peru) catches provide > 90 percent of total catches of pelagics, which peaked with 14.5 million t in 1994. Remarkably, the anchovy stock did not collapse during the strong ENSO of 1997/98 as has happened during preceeding ENSO events.

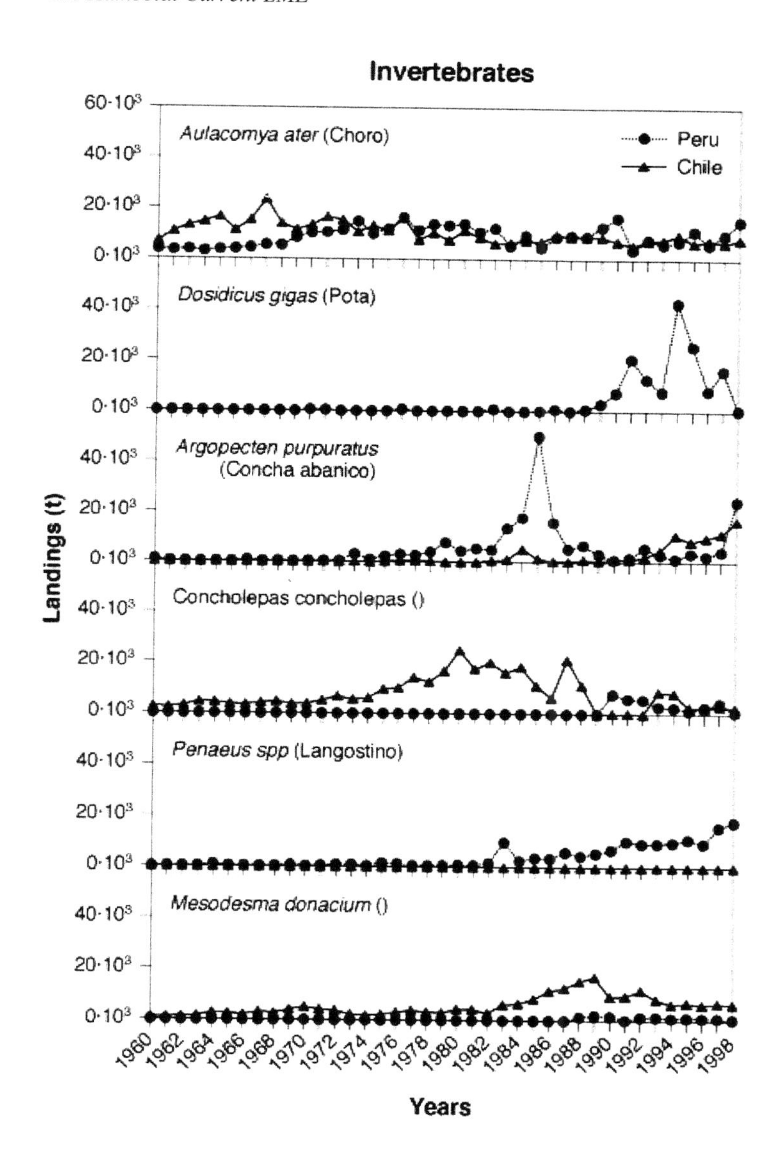

Figure 12-5. Catches of main invertebrate species along the coast of the HCLME—from top to bottom: cholga mussel, jumbo flying squid, Peruvian calico scallop, "loco" snail, Langostino shrimp and macha clam

Presumed Causes for Observed Trends

While it is evident that El Niño has an extremely adverse effect on anchovy (reducing the survival of all life stages, and increasing its vulnerability to the fishery), and a rather positive effect on horse mackerel, mackerel and sardine, Alheit and Bernal (1993) emphasize that other factors (i.e. interspecific, second order interactions and direct effects) also contribute to the observed alternations of the pelagic species. They mention that oceanic offshore transport was significantly lower in the 1960s than in the 1970s-1980s, and cite Mendelssohn (1989) who found SST to have increased from 1971 onwards. Both factors are assumed to negatively affect spawning of anchovies. Bernal et al. (1983) showed that zooplankton biomass dropped at the end of the 1960s; Loeb and Rojas (1988) observed a marked shift of ichthyoplankton composition and abundance from 1965 to 1983, both indicators for biological changes occurring prior to the 1972 ENSO event. The latter authors postulate an atmospherically driven oceanic circulation change, beginning in the late 1960s and possibly involving an onshore presence of subtropical or oceanic waters. They assume that such a circulation change might be responsible for the observed shifts in zooplankton, larval and adult fish in the late 1960s.

In addition to these trends of lower than ENSO frequency in the physical oceanography, predation might play a similar role in modifying the species structure of the pelagic system. According to Alheit (1987) 22 percent of the total egg mortality of anchovy was caused by cannibalism during the first spawning (August-September) in 1981. Interestingly, the author found much less cannibalism in a similar study during the second spawning (February) in 1985, although spawning biomass of adults was 2-3 fold higher. As anchovy eggs seemed to be concentrated rather nearshore in 1981 while distributed over 90 miles offshore in 1985, Alheit and Bernal (1993) suggest that population density (probably higher during first spawning) rather than biomass is decisive for egg cannibalism rates. Mendelssohn and Mendo (1987) who analyzed seasonal recruitment success of the anchovy from 1953 to 1981 confirm these findings as they found successful annual recruitment to be most influenced by the second spawning peak of the respective years.

Other egg predators, such as sardines, juveniles of other species and invertebrates, are assumed to play an important role as well. Adult mortality of anchovy and/or sardine depends on a relative abundance of principal predators that has varied substantially over time. Muck (1989) showed that anchovy mortality due to predation by horse mackerel, mackerel and hake exceeds the consumption of anchovy by birds. Muck and Sanchez (1987) assume that horse mackerel predation on the anchovy stocks throughout the 1950s and in the mid 1970s surpassed catches. As horse mackerel and mackerel are usually distributed farther offshore than anchovy, Muck (1989) assumes that predation mortality should be especially high during years of high SST (i.e. during ENSO events) which bring horse mackerel closer to the coast. The author thus proposes a flexible exploitation scheme according to which, anchovy should be fished particularly when its biomass is high and SST low, and horse mackerel should be exploited heavily when SST is high.

As anchovy recovered its dominant position in the Peruvian upwelling system during the 1990s, it could be assumed that the oceanographical environment had returned to a pre-1970s situation characterized by rather cold waters, strong upwelling and excellent conditions for anchovy recruitment. Gutierrez *et al.* (1999) confirm this for the two years prior to the ENSO of 1997/98. When the warming of ENSO 1997 started, anchovy migrated to the south, onshore and to deeper waters (Gutierrez *et al.* 1999). Although mortality was probably high during this event, stock recuperation was rapid thereafter due to cold conditions again in 1999, the absence of predators and a less destructive fishing regime (Gutierrez *et al.* 1999).

As catches (and possibly available biomass through offshore migration) of horse mackerel and mackerel dropped in Peru along with SST after the ENSO of 1983-1984 (reaching lowest levels in 1986 and 1987), a strong slackening of predation pressure could have favoured the population development of sardines and anchovies, whose catches rose during that time. The subsequent increase in horse mackerel and mackerel catches could be due to an onshore migration of both pelagic predators and thus a better availability to the fishery. The accelerated catch increase of both species from 1992 to the ENSO event of 1997 (Figure 12-4) (from about 120,000 t to 850,000 t) together with the increase of the giant squid to 210,000 t in 1994 and the parallel continuous decline in sardine catches in Peru and Chile could be indicative of a strong predation by these predators upon sardine. It could, however, also be the result of an increase in fishing effort and efficiency in the selective exploitation of these predators as foreign trawlers (Japanese, Korean) joined the fishery during that time. As anchovy catches remained at very high levels in Peru during the 1990s, it is probable that horse mackerel and mackerel (and sardine) were distributed further offshore than anchovy.

The Demersal Sub-system

Hake and other demersal species, mainly of the families *Sciaenidae, Serranidae, Trakidae* and *Triglidae* (the latter increasing during ENSO years, Figure 12-6), are normally distributed over the broad continental shelf from the Peru-Ecuador border (3° S) to at about 10° S. Hake prefers the oxygen rich Cromwell Current and is loosely related to its extension, which defines hake distribution both to the south and in depth. In warm years (ENSO), hake move farther south (14° S in 1972) with the Cromwell Current and in cold years concentrate north of 7° S (1996) and even north of 5° S (1999). The distribution area also amplifies in relation to depth, moving farther offshore in ENSO years. There exists a length gradient with larger and older individuals in the north and offshore, and smaller, younger in the south and onshore. This biological characteristic made it possible to distinguish a different northern stock (sub-population) that moved southward during the strong ENSO 1997 event (Guevara-Carrasco and Wosnitza-Mendo 1997).

In 1963 hake landings in Chile amounted to more than 100,000 t (Figure 12-6). Since then, landings declined to low levels of 50,000 t or less during the 1970s and early 1980s, followed by an increase reaching almost the earlier maximum. A large percentage of hake landings take place in the far south.

In Peru, a limited trawl fishery for hake developed in the mid sixties, converting some purse seiners to trawlers. In 1966, the total Peruvian demersal landings only reached 19,000 t and increased to 46,000 t in 1967. The center of this traditional coastal trawl fleet was and is Paita (5° S); its operational radius is limited to 30 nm offshore, but moves between 4 and 6° S. Annual landings of hake by the Peruvian coastal fleet do not exceed 20,000 t. Since the 1970s, foreign fleets with large factory trawlers up to 2,000 GRT have joined in the hake fisheries, elevating the landings to more than 100,000 t with record landings of about 300,000 t in 1978 (Figure 12-6). The intensive hake fishery was interrupted in the 1982/83 ENSO event, producing a change in distribution and concentration of fish and restraining the normal action of the fleet. Foreign vessels were discouraged from fishing, and only the local fleet continued until the middle of the decade with hake landings of about 20,000 t. During some years between 1986 and 1991 a large fleet of mid-water trawlers, dedicated to horse mackerel as target species, harvested hake as by-catch, raising the landings to about 130,000 t in 1990. The ENSO of 1991-93 ended the intervention of this fleet.

In 1994, an increased and partially modernized coastal trawler fleet and a new national fleet of medium sized, well equipped factory bottom trawlers joined the fishery and landings progressively increased to almost 200,000 t in 1996. One more time, ENSO (1997) restrained the fishing activities and in 1998 landings dropped to 75,000 t.

Hake fishery statistics do not really reflect abundance but rather the fishing effort and success of the fleet--the latter with cyclic changes in the physical environment and the availability of the resource. After 1998 national owners continued fishing, but the increase in fishing effort resulted in the insolvency of the majority of the national trawler fleet in 1999 (economic overfishing). Besides hake, there are other (about 90) species available to the trawler fleet, which amount to about 30 percent of all demersal catches. During normal years, the most important of these are *Mustelus sp., Paralonchurus peruanus, Cynoscion analis, Paralabrax humeralis* and various flatfish species, especially *Hippoglossina macrops.* In ENSO years, *Prionotus stephanophrys* and

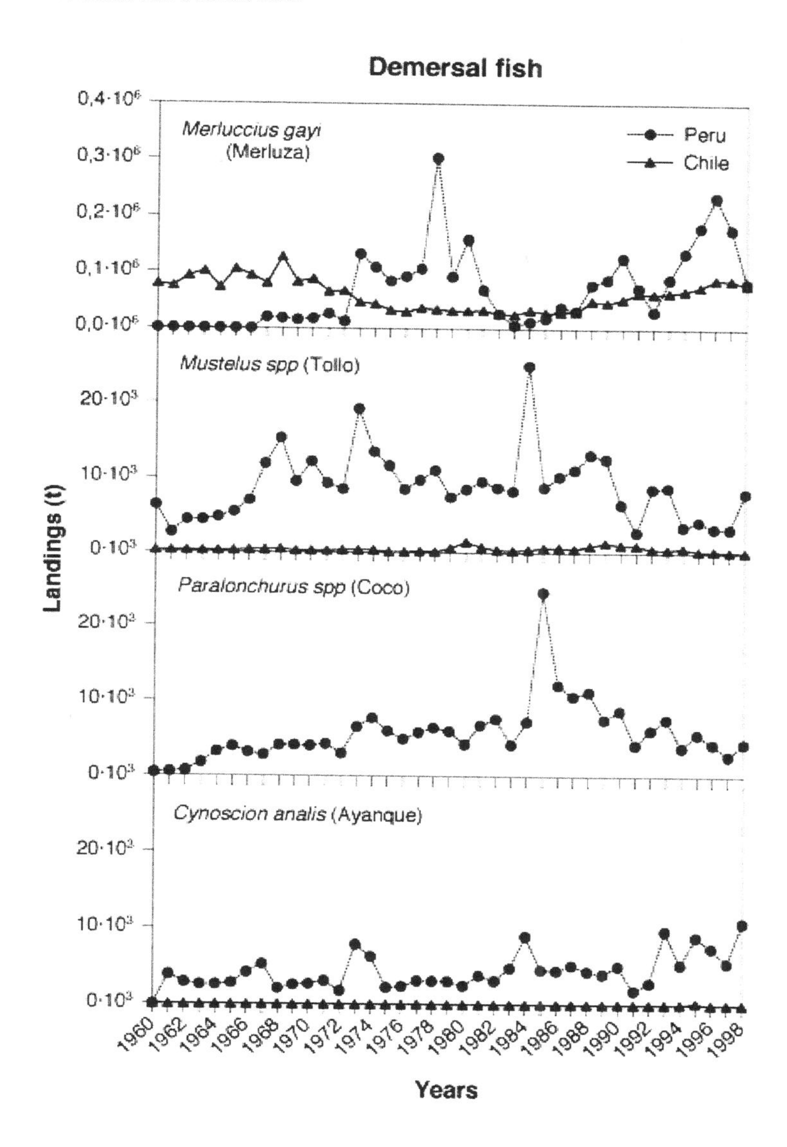

Figure 12-6. Catches of hake and other demersal fish in the HCLME (1960-1998)—from top to bottom: South Pacific hake, smooth-hounds, croaker and Peruvian weakfish

Sciaena deliciosa are favoured. In 1997 the former seem to have displaced hake from its leading position in the demersal subsystem (Elliott and Paredes 1997). Most of the demersal species are traditionally highly appreciated as fresh fish.

Biomass Estimates and Presumed Causes for Observed Trends

In Peruvian waters, hake biomass estimates (using cohort analysis and swept area surveys) in the 1980s were between 400,000-600,000 t (age-group II+), at the same time that anchovy biomass descended due to ENSO 82/83 and sardines began to occupy the niche left by anchovy. This changed at the beginning of the 1990s when after a weaker ENSO in 1992, the subsequent two normal years (1993-1994) and two cold La Niña years (1995-1996) anchovy once more took over the leading role in the Peruvian upwelling system. Sardine and hake biomass decreased to about 100,000-200,000 t (see various technical papers of the Peruvian Fisheries Institute, IMARPE).

In 1992, a sudden decline in mean catch length of about 8 cm for both sexes was observed that cannot be explained solely by increasing fishing pressure but rather by a shift in the distribution of age classes as a response of the stock to the strong fishing mortality (especially on young age groups) and the changes in prey availability due to the decrease in sardine biomass. Other changes observed include an increase in the rate of cannibalism, a change in the diet composition, as well as an earlier maturation (Wosnitza-Mendo and Guevara-Carrasco 2000).

It appears that the combined and simultaneous increase in fishing pressure on anchovy, sardine, pelagic predators such as tunas, barrilete and bonito, hake and other demersals has resulted in an alteration of the species composition with direct influence on the food web of the HCLME. This is evidenced by a shift of the main food items of adult hake over the years from (in rank order) sardines, sciaenids (*Larimus pacificus*) and anchovy in the 1970s and 1980s, to feeding on young hake, *Myctophidae* and other mesopelagic species in the 1990s. Over the last decades, the hake population has changed from a highly structured stock with up to 15 year classes, a mean catch length of about 40 cm, and a length at first maturity of about 35 cm, to one poorly structured, with a mean catch length of about 32 cm and with specimens maturing at smaller sizes.

The Coastal Nearshore Sub-system

Catches of coastal fish fluctuate between 15,000 and 70,000 t annually in Peru; lower amounts are caught in northern Chile. Species of greatest abundance are cojinova [South Pacific bream] (*Seriolella violacea*), Liza [mullet] (*Mugil cephalus*), machete [South Pacific menhaden] (*Ethmidium maculatum*) and lorna [drum] (*Sciaena deliciosa*). During warm ENSO periods most coastal fish species are little affected, some are

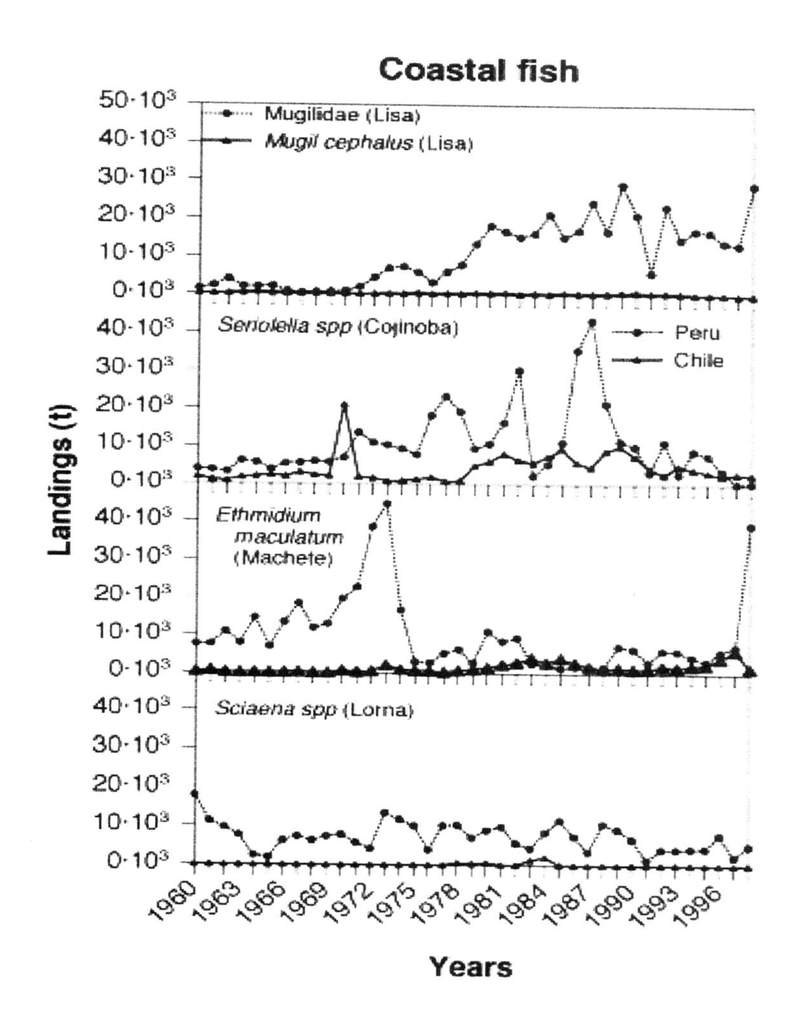

Figure 12-7. Coastal fish landings in Peru and Northern Chile—from top to bottom: mullet, South Pacific bream, Pacific menhaden and drums

clearly favoured (i.e Machete, Figure 12-7). In addition, tropical invaders become important targets of the coastal fishers (i.e. several Lutjanids, skipjack and yellowfin tuna, dorado [dolphin fish](*Coryphaena hippurus*), sierra [Pacific sierra](*Scomberomorus maculatus sierra* and others). Total catch of these species remains low, but the fishery is profitable because of the high market value of these species.

Benthic invertebrates of the HCLME exceed coastal fin fish in catch volumes and, even more in economical importance both in Peru and northern Chile. Among species of highest importance are the scallop (*Argopecten purpuratus*), the mussel (*Aulacomya ater*), the snail (*Concholepas concholepas*), several other bivalve species, shrimps (in northern Peru) and cancrid crabs. The main centres of invertebrate fishing in the HCLME are Independencia Bay in Peru, and Tongoy Bay in Chile. Some invertebrate collection also occurs in the bays of Sechura, Samanco and Tortugas in Northern Peru and of Arica, Iquique and Mejillones in Northern Chile (much invertebrate fishing occurs far south in Chile outside the HCS and is not considered here).

The "loco" snail (*Concholepas concholepas*) and the scallop (*Argopecten purpuratus*) are, in terms of catch volumes and market value, the most important species in Chile and Peru respectively (Figure 12-5) and will be dealt with in more detail here. The loco landings in Chile peaked in the year 1980 at 24,000 t, representing an export value of US$ 20 million, but decreased thereafter due to heavy overfishing. While the "loco fishery" was closed for long periods thereafter, clandestine fishing continued not allowing the stock to recuperate. Castilla and Camus (1992) estimated that 8,000 – 10,000 t of locos were landed in 1991, while the fishery was officially closed. The same problem of overfishing and clandestine fishing occurred for the scallop populations in Chile and Peru despite protective measures taken by the government. The ENSO events of 1982/83 and 1997/98 positively affected the natural scallop populations in Peru (Wolff 1984, Wolff and Mendo 2000) (Figure 12-5), where stocks increased from some 100 t to over 30,000 t within a few months. To a lesser extent, scallops in northern Chile were also favoured (Illanes *et al.* 1985). In addition, octopus stocks proliferated during the ENSO warming events and shrimp stocks extended their distribution from the Ecuadorian border to south of Pisco (14°S). The enormously increased scallop population and the catches of about 25,000 t annually during the years 1983-1986 led to the opening of a Peruvian export market for scallops. The ENSO of 1997/98 had a similar effect on the scallop populations in Peru, but strong growth and recruitment overfishing led to a poor utilization of the stocks' potential and thus to much lower catches (Wolff and Mendo 2000). On the other hand, many species (i.e. mussels, crabs etc.) suffered mass mortalities during both ENSO events (see Arntz and Fahrbach 1991 for a detailed review). Nonetheless, the net effects of the ENSO events on the invertebrate fishery were clearly positive, with increased catches and revenues to the fishers. When the natural scallop stocks were, due to overfishing, almost depleted in Chile in 1988, the species began to be

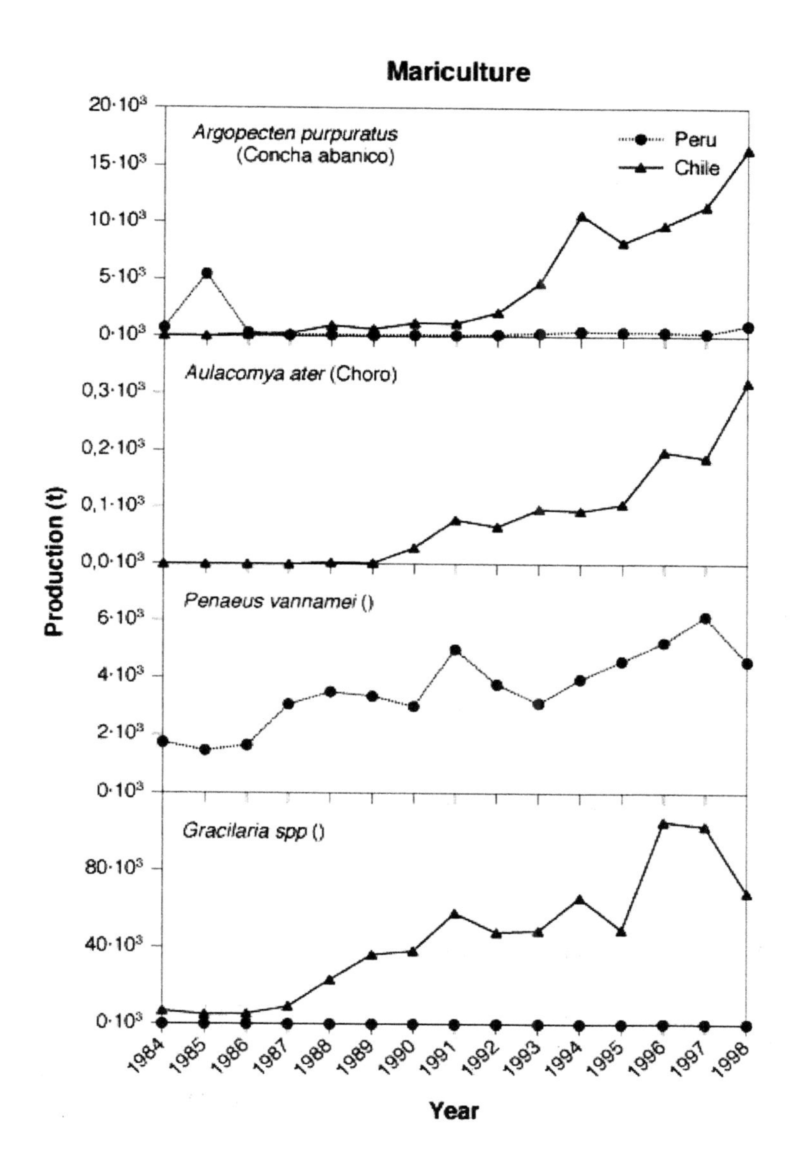

Figure 12-8. Mariculture of seaweed, shrimps and bivalves in the HCLME (1980-1999)—from the top to bottom: Peruvian calico scallop, cholga mussel, white shrimp and red algae

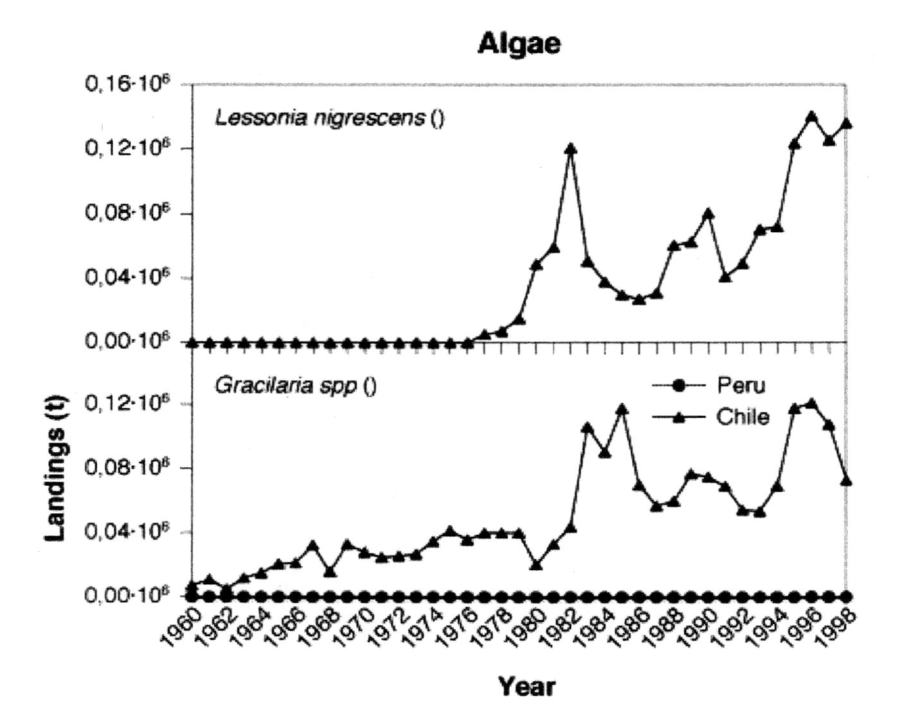

Figure 12-9. Algae landings in Chile (1960-1998)—top to bottom: low growing intertidal kelp and red algae.

cultivated, mainly in Tongoy Bay. Since that time > 98 percent of marketed scallop production in Chile is maricultured (Figure 12-8).

In the early 1980s, the harvest of seaweeds from natural beds started in Chile and strongly increased to levels of > 250 000 t per year during the 1990s (Figure 12-9). The species *Gracilaria* spp. is being cultivated in significant quantities and export volumes have increased tremendously during the last years. In Peru, scallop and seaweed cultures are still at an early stage, but prospects for development are good.

Presumed Causes of Observed Trends

The nearshore fauna of the HCLME is comprised of spatially overlapping species of a tropical/subtropical warm water (Panamanian) and a temperate (Sub-Antarctic) fauna (Arntz and Fahrbach 1991, Waller 1969, Wolff 1994). Depending on the state of the system within an ENSO cycle and its geographic impact (that differs between ENSO events), species of both groups are either favorably or negatively affected. It appears that the functional niches of the system remain occupied over the ENSO cycle, with only the species alternating (Wolff 1994). The process of species alternation at the same functional spot during an El Niño event is rapid (weeks to months) and is achieved by the high colonization capacity of the species involved. The recolonization by the species of the other group following the El Niño event proceeds more slowly and correlates with an average longer life span of these species (4 or more years compared to 1-2 years in the former group). Interestingly, these life spans seem to correlate roughly with the average inter-ENSO and ENSO periods (Wolff 1994).

The causes of the scallop boom during the ENSO events of 1983/84 and 1997/98 are summarized in Wolff and Mendo (2000) as follows: (1) increase in the growth rate of scallops due to elevated temperatures; (2) reduction of larval period and thus mortality rate, (3) decrease in abundance of scallop predators; (4) increase of overall reproductive activity of scallops; (5) decrease in abundance of competing filter feeders. A similar explanation can be given for the strong increase of octopods during the ENSO warming period, that occupied the free niche of top predators of the benthos, being otherwise occupied by a guild of large cancrid crabs – which are favoured during cold (La Niña) years. While ENSO plays a very important role in influencing the variation of invertebrate population sizes and catches in Peru, its effect is much attenuated in Chile. Pronounced interannual catch variations here are due to the fishing regime or to factors other than ENSO.

RESEARCH LINES AND MANAGEMENT POLICIES

The Pelagic Sub-system

After the anchovy collapse in 1972, scientists and fishery managers repeatedly asked the question of the relative importance of bottom up vs. top down processes in determining the distribution and relative population sizes of the important resources of the HCLME. Scientists, believing that the bottom up processes were the more decisive, claimed that a better understanding of the oceanographic and biogeochemical processes (wind and current patterns, intensity, duration and geographical position of upwelling, nutrient concentrations) would ultimately allow them to predict annual population sizes of the main resources in the system. Several descriptors for the hydrographic and oceanographic states of the system were developed and attempts were made to correlate these with the observed biotic changes in the system. Sea surface temperature (SST), depth of the 14° C isotherm, depth of the SST-2°C isotherm and the Southern Oscillation Index (SOI), that measures the difference in atmospheric pressure between the eastern and western tropical Pacific (Brainard and McLain 1987), alongshore windstress (Bakun and Mendelssohn 1989), level of nitrate flux and primary production (Mendo *et. al.* 1989) as well as the above-mentioned multivariate ENSO index were calculated for long time series data. As a result of these studies, the knowledge of the oceanographic features of the HCLME and of the ENSO mechanism has increased significantly since 1972, and nowadays El Niño events are being forecast with a high degree of precision. However, these studies have not yet allowed us to fully understand the factors, processes and interdependencies causing the fish stocks of the HCLME to vary so widely or allowed us to reliably predict catches.

Yañez *et al.* (1998) correlated SST (°C) (as a descriptor for the oceanographic state of the system) with catch per unit of effort of anchovy (as a measure of abundance) and calculated maximum sustainable yield (MSY) levels from a production model as a function of SST (°C). This approach emphasizes that MSY is related to the system's carrying capacity (in terms of food for anchovy) which varies with the oceanographic state of the system. As the future oceanographic state of the system can not be predicted for more than a couple of months in advance (a time span much shorter than the generation time even of the small pelagics), this approach to predictions appears not very useful. Mackinson *et al.* (1997) emphasize other problems with the bottom-up control assumptions: while stocks of small pelagics build up during favourable conditions, community structure will change as will the fishery response. If, for example medium levels of fishing mortality cause a decrease in biomass of small pelagics, predators will decrease subsequently, allowing the catch to be greater. It was shown by using the ECOPATH mass balance simulation model ECOSIM (Walters *et al.* 1997), that predicted F_{MSY} values were more than twice as high under the assumption of bottom–up control as those predicted under the assumption of top-down control (predator limitation). More importantly, the yield curve differed significantly for both scenarios: for top-down control, catch increases to an early maximum and then decreases to zero as F continues to rise, whereas for bottom-up control, yield continues to increase over a wide range of F, predicting that the stock can sustain much higher fishing pressure before it begins to decline. The alternation of sardine and anchovy in the HCLME was evident under assumptions of top-down control mechanisms, but not so evident if only resource competition was considered in the simulations.

It is evident from the accumulated knowledge, that bottom-up and top-down processes are governing the dynamics of the pelagic resources and must be taken into account if management is to be successful. Monitoring of the oceanographic conditions in the HCLME has intensified and strongly improved over the past decade as have the biomass estimates of the pelagic resources by hydroacoustic surveys. Abiotic and biotic changes in the ecosystem can therefore be recognized quickly and the fishery must learn to flexibly adapt to these changes.

A fundamental problem with the pelagic fishery of the HCLME remains its enormous overcapacity of vessels and plants. Csirke and Guny (1996) showed in a bioeconomic analysis, that the 720 Peruvian purse-seiners in the mid-1990s had the capacity to catch around 38.5 million t annually, several times the present landings. Any reduction in this overcapacity could strongly increase the profits.

The Demersal Sub-system

Peruvian hake in the 1970s and 1980s was managed by a total catch quota for adult hake, depending on the estimated total yearly biomass. Although in some years this quota was exceeded, the cyclic appearance of ENSO resulted in catches well below the supposed allowable catch during various years to follow, and until the end of the 1980s this strategy worked well because the hake population with about 600,000 t or even more seemed healthy and safe. This changed in the beginning of the 1990s, when more and more young hake (< 35 cm) appeared in the catches, first attributed to successive good recruitments, but finally, to adaptive response to overfishing and changes in species interactions (Wosnitza-Mendo and Guevara-Carrasco 2000).

Thus far we do not have enough knowledge about the interactions between the species involved, the crucial physical environmental interactions (change in subsurface current system?), and we have little ability to predict what will happen when the present high fishing pressure is maintained. The gradual extinction of large, older females has definitely already resulted in low egg production of the stock, as under present conditions female hake are spawning only once in their lives (at most). This was also reported for the Atlantic cod (Trippel 1995) and should seriously be taken into consideration for management of the fishery. Also, intraspecific diversity might have been reduced through the strong reduction in stock size and age classes, which can lead to genetic change of heritable traits such as growth rate and age at first reproduction. At present, a collapse of the stock does not seem impossible, and it is questionable if a collapsed hake population will ever be able to fully regain its former status in the ecosystem.

At the moment, the market for hake is adjusted to small sizes and catches are far above the 20 percent rate recommended to sustain such fisheries. So far, resilience of Peruvian hake has been enormous and allows the fishery to continue neglecting the effects of fishing on the genetics or stock composition. Even if a regulation of fisheries to the 20 percent rate were to be established, the biomass of the total stock will fluctuate at a long-term lower level. An appropriate model for hake must be developed that incorporates fisheries history, the importance of large spawning females, the role of cannibalism as well as interspecific relationships and environmental changes.

The Near-shore Coastal Sub-system

The severe overexploitation of most of the benthic resources in Chile and Peru has motivated fisheries biologists and ecologists to look for management policies that allow for a more sustainable use and protection of the stocks. Minimum landing sizes and closed seasons were introduced for scallops, snails and other species in the 1970s and 1980s, but (as mentioned above) overexploitation continued due to clandestine fishing and the lack of efficient control by the authorities. Growing shell mounds of scallops and snails along the coast clearly showed the magnitude of clandestine extraction. A recent sad example was the mismanagement of scallops in Independence Bay (Peru) during the ENSO of 1997/98: due to high pressure by the export market, divers collected undersized juvenile scallops and thus depleted the stock before it could grow large and become profitable (Wolff and Mendo 2000).

At the beginning of the 1990s, in its fishery legislation, the Chilean government introduced so called "areas de manjeo" for the loco stocks. These management areas were given to the local fisherfolk with the obligation to protect the loco stock in this area until biomass has reached levels high enough for exploitation. While the increase in resource biomass was significant over the first months, stock biomass surpassed the carrying capacity of the protected area and had naturally declined to low levels when the fishery was finally opened by the local authorities. The result was that the benefit for the fishers was lower than before (under conditions of clandestine fishing). Based on conclusions of Stotz (1997) from this first experience with management areas in Northern Chile and our judgement, the following considerations seem imperative if management areas are to be successful: (1) As the living conditions for most benthic resources are not homogenous along the coastline, only those parts should be chosen that can be considered as "productive"; (2) Each area has a certain carrying capacity for the resource biomass (in terms of available space and production of food organisms), which can roughly be estimated from ecological studies (Stotz and Perez 1992) and which should be monitored throughout the years; (3) In years when natural recruitment is low, spawning biomass in the management areas should be kept above a critical value; (4) The number of fishers allowed to manage and use an area must be determined in relation to the estimated resource productivity in the area; (5) It is advisable to implement a rotation principle, allowing only part of the management areas along the coast to be simultaneously exploited. Thus (closed) high density areas would produce high numbers of offspring (larvae) with an overspill into those (open) areas which are fished and hence of low adult stock density.

The implementation of management areas along the Chilean and Peruvian coastline requires a close cooperation of fishers, fisheries authorities and fisheries scientists who should play complementary roles in the governance of the natural resource. Only by such a co-management, which strongly involves the fishing community in the decision-finding process and calls for constant monitoring of the management areas, can this new management tool can have a chance to be successful.

While recruitment of most coastal invertebrates of the HCLME is probably more bottom –up controlled (as the stock - recruitment relationship seems rather weak (Wolff 1994)), the adult population biomass and thus the harvestable part of the population seems very much related to the trophic (prey-predator) conditions in the management area. For this reason, the trophic modelling approach has become important as research tool for the nearshore - subsystem (Stotz and Perez 1992, Wolff 1994). A simulation exercise with a trophic model constructed for Tongoy bay (Chile) for the period of the early 1990s (Wolff 1994b) allowed the suggestion that the bay had a 10 -15 times higher potential for the production of filter feeder (scallop) biomass, compared to the actual value in the early 1990s. This proved to be right, as scallop cultures successfully increased towards the end of the 1990s by about 10 times.

While the biotic interactions between species of the nearshore coastal subsystem are being studied intensively in Chile and Peru, the drastic scallop proliferation in Peru during the ENSO events 1983/84 and 1997/98 also raised the question whether simple empirical data of environmental conditions could be used to predict (at least the magnitude of) the annual biomass/catches of the stock. While a direct relationship between SST and scallop catches could not be detected, Mendo and Wolff (unpublished) were able to find a significant relationship between the mean SST during the spawning period of the scallop and its subsequent total annual catch. Research will continue along this line in order to find further environmental predictors for the fisheries potential of the nearshore resources.

CONCLUSIONS

- In the Peruvian part of the pelagic sub-system of the HCLME, anchovy has re-taken its dominant position with population sizes and catches of the magnitude before the collapse in 1971. Horse mackerel has become the principal pelagic resource in Chile during the 1980s and early 1990s and combined anchovy and horse mackerel catches in the early 1990s exceeded previous annual catches in the HCLME.
- Due (at least in part) to high fishing pressure, the demersal hake stock off Peru has undergone a dramatic reduction of mean catch size and size at first

maturity. As its food spectrum has also changed from small epipelagic fish to myctophids and other mesopelagic species and as cannibalism has drastically increased, a shift in the food web structure (as yet little understood) seems to have occurred in the demersal part of the Peruvian HCLME.

- Coastal invertebrate resources have received increasing attention in the HCLME, both in Chile and Peru. The ENSO events 1983 and 1997 induced population outbursts of scallops in Peru with the opening of an export market and high revenues of foreign currency. Mariculture of algae (mainly *Gracilaria* spp.) and scallops in northern Chile has increased strongly in the 1990s with very good prospects for the future. In Peru , mariculture is still in an initial state.
- The monitoring and management of the resources of the HCLME has (in general) improved. Biomass and spatial distribution of pelagic resources is being constantly monitored by hydro-acoustic surveys. There is as yet no agreement about adequate yield levels (MSYs) for the pelagics, as their determination heavily depends on the relative importance given to top-down regulation (usually based on multispecies modelling approaches) and bottom-up regulation (usually based on single species approaches) of the pelagic stocks.
- Multispecies trophic mass balance models (ECOPATH and ECOSIM) have been used more recently for modelling the trophic interactions within the pelagic sub-system (including the demersal hake stocks) and have also been applied to the coastal near-shore sub-system. These approaches demonstrated the importance of predation for structuring the communities of the three subsystems of the HCLME and they were also used to explore the impact of different fishing scenarios on the ecosystem. Research along this line should continue together with the intensive monitoring of environmental parameters in order to differentiate between abiotic and biotic factors that influence stock sizes and variability of the resources.
- Knowledge about the physiological ecology of the species should be integrated with real –time environmental information into the forecast process (Sharp 1998). In times when a stock seems naturally endangered (anchovy due to ENSO impact for example), prudent measures should be taken to safeguard the stock.
- There are signs of substantial changes in the structure of the food web of the HCS, that have occurred over the last decades of intensive fishing with a negative effect on energy and material fluxes and on the resilience of the system as a whole. Overfishing seems to be a major threat leading to a loss in the genetic integrity of fish populations (Cury and Anneville 1998). A further system destabilization through an increase in the amplitude of annual stock variations (Upton 1992) could be the consequence.

- A fishery policy should take into account that, while the "fish meal species" are economically of greatest importance to the fishing industry, a far greater proportion of the society heavily depends on a large number of species for fresh fish consumption, fish that require a healthy, equilibrated and diverse ecosystem.
- As the fishery regulation measures for coastal benthic resources (minimum landing sizes, catch quotas and closed seasons) proved not to be effective in preventing overexploitation, the Chilean government introduced so called management areas at the beginning of the 1990s, that were given to the local fisherfolk with the obligation to take care of the stock in this area. The further implementation of these management areas along the Chilean and Peruvian coastline requires a close cooperation of fishers, fisheries authorities and fisheries scientists who should share the governance of the natural resource with complementary roles. Only by such a co-management and the constant monitoring of the management areas, might this new management tools become successful.

REFERENCES

Alheit, J. 1987. Egg cannibalism versus egg predation: their significance in anchovies. S. Afr. J. Mar. Sci. 5:467-470

Alheit, J. and P. Bernal. 1993. Effects of Physical and Biological Changes on the Biomass Yield of the Humboldt Current Ecosystem. In Sherman, K., L.M. Alexander and B.D. Gold, eds. Large Marine Ecosystems: Stress. Mitigation and Sustainability, AAAS Press. 53-68. 376p

Arntz, W. and E. Fahrbach. 1991. El Niño- Klimaexperiment der Natur. Berlin, Birkhäuser Verlag: 264p

Bakun, A. and R. Mendelssohn. 1989. Alongshore wind stress, 1953-1984:correction, reconciliation and update through 1986. In: Pauly, D., P. Muck, J. Mendo and I. Tsukayama, eds. The Peruvian upwelling ecosystem: Dynamics and interactions. ICLARM Conference Proceedings 18: 77-81. 438p

Barber, R.T., F.P. Chavez and J.E. Kogelschatz. 1985. Efectos biologicos del El Niño. Com.Perm.Pacif.Sur Bol.ERFEN.14:3-29

Bas, C., J.J. Castro, J.M. Lorenzo, eds. 1995. Intern. Sysmposium on middle sized pelagic fish, held in Las Palmas de Gran Canaria 1994. Sci. Mar.-Barc., Vol 59, No 3-4

Bernal, P.A., F.L. Robles and O. Rojas. 1983. Variabilidad fisica y biologica en la region meridional del sistema de corrientes Chile-Peru. FAO Fish. Rep. 291:683-711

Brainard, R.E. and D.R. McLain. 1987. Seasonal and interannual subsurface temperature variability off Peru, 1952 to 1984. In: Pauly, D. and I. Tsukayama, eds. The Peruvian anchoveta and its upwelling ecosystem: Three decades of change. ICLARM Studies and Reviews 15: 14-45. 351 p.

Castilla, J.C. and P.A. Camus 1992. The Humboldt-El Niño Scenario: coastal benthic resources and anthropogenic influences, with particular reference to the 1982/83 ENSO. In: A.I.L. Payne, K. H. Brink, K.H. Mann and R. Hilborn, eds. Benguela trophic functioning. Afr. J. Mar. Sci. 12:703-712

Crawford, R.J. and M.J. Jahncke. 1999. Comparison of trends in abundance of guano –producing seabirds in Peru and Southern Africa. S. Afr. J. Mar. Sci. (21):145-156

Csirke, J. and A.A. Guny. 1996. Análisis bioeconómico de la pesquería pelágica peruana dedicada a la producción de harina y aceite de pescado. Bol. Inst. Mar. Perú, 15(2): 25-68

Cury, P. and A. Anneville. 1998. Fisheries resources as diminishing assets: marine diversity threatened by anecdotes. In: Durand, M.-H., P. Cury, R. Mendelssohn, C. Roy, A. Bakun and D. Pauly, eds. Global Versus Local Changes in Upwelling Systems. Editións de l'Orstom. Collections Colloques et séminaires. 537-548. 594p

De Vries, T.J. and W.G. Pearcy. 1982. Fish debris in sediments of the upwelling zone off central Peru: A late quarternary record. Deep Sea Res. 29:87-109

Elliot, W. and F. Paredes. 1997. Estructura del subsistema demersal durante el crucero de evaluación del recurso merluza, BIC Humboldt 9705-06, Callao a Puerto Pizarro. Inf. Inst. Mar Perú 128: 80-114

Enfield, D.B. and J.S. Allen. 1980. On the structure and dynamics of monthly mean sea level anomalies along the Pacific coast of North and South America. J. Phys. Oceanogr. 10(4):557-578

Espino, M. and C. Wosnitza-Mendo. 1989. Biomass of hake (*Merluccius gayi*) off Peru, 1953-1987. In: Pauly, D., P. Muck, J. Mendo and I. Tsukayama, eds. The Peruvian upwelling ecosystem: dynamics and interactions. ICLARM Conference Proceedings 18: 297-305. 43p

Estrella C. and R. Guevara. 1998. Informe estadistico anual de los recursos hidrobiologicos de la pesca artesanal por especies, artes, caletas y meses durante 1997. Inf. Inst. Mar del Peru 132. 422p

FAO, 1998. Yearbook of Fishery Statistics, 1988, Vol. 86

Guevara-Carrasco, R. and C. Wosnitza-Mendo. 1997. Análisis poblacional del recurso merluza (*Merluccius gayi peruanus*) en otoño 1997. Inf. Inst. Mar Perú. 128:25-32

Gutierrez, M., N. Herrera and D. Marin. 1999. Distribucion y abundancia de anchoveta y otras especies pelagicas entre los eventos. Paper presented at VIII Congreso Latino Americanos Sobre Ciencias Del Mar. 17 de octubre de 1999, Trujillo.

Illanes, J.E., S. Akaboshi and E.T. Uribe. 1985. Efectos de la temperatura en la reproduccion del ostion del norte (Argopecten purpuratus) en la bahia de Tongoy durante el fenomeno del Niño 1982-83. Invest. Pesq. (Chile) 32:167-173

Jarre-Teichmann, A., V. Christensen. 1998. Comparative modelling of trophic flows in four large upwelling ecosystems: global versus local effects. In: Durand, M.-H., P. Cury, R. Mendelssohn, C. Roy, A. Bakun and D. Pauly, eds. Global Versus Local Changes in Upwelling Systems. Editións de l'Orstom. Collection Colloques et séminaires. 423-443. 594p

Loeb, V.J., O. Rojas. 1988. Interannual variation of ichthyoplankton composition and abundance relations of northern Chile, 1964-83. Fish. Bull. U.S. 86:1-24

Mackinson, S., M. Vasconcellos, T. Pitcher, K. Walters and K. Sloman. 1997. Ecosystem impacts of harvesting small pelagic fish in upwelling systems: using a dynamic mass-balance model. Proceedings: Forage Fishes in Marine Ecosystems, Alaska Sea Grant College Program, AK-SG-97-01: 731-749

Mathisen, O.A. 1989. Adaptation of the anchoveta (*Engraulis ringens*) to the Peruvian upwelling system. In: Pauly, D., P. Muck, J. Mendo and I. Tsukayama, eds. The Peruvian upwelling ecosystem: dynamics and interactions. ICLARM Conference Proceedings 18:220-234. Instituto del Mar del Peru (IMARPE), Callao, Peru; Deutsche Gesellschaft für Technische Zusammenarbeit (GTZ), Eschborn, Federal Republic of Germany; and International Center for Living Aquatic Resources Management (ICLARM), Manila, Philippines. 438p

Mendelssohn, R. 1989. A re-analysis of recruitment estimates of the Peruvian anchoveta in relationship to other population parameters and the surrounding environment. In: Pauly, D. Mendo, J. and I Tsukayama, eds. The Peruvian upwelling ecosystem: Dynamics and interactions. ICLARM Conference Proceedings 18:364-385. Instituto del Mar del Peru (IMARPE), Callao, Peru; Deutsche Gesellschaft für Technische Zusammenarbeit (GTZ), Eschborn, Federal Republic of Germany; and International Center for Living Aquatic Resources Management (ICLARM), Manila, Philippines. 438p

Mendelssohn, R. and J. Mendo. 1987. Exploratory analysis of anchoveta recruitment off Peru and related environmental series. In: Pauly, D. and I. Tsukayama (eds.) The Peruvian anchoveta and its upwelling ecosystem: Three decades of change. ICLARM Studies and Reviews 15: 294-306. 351p

Mendo. J., M. Bohle-Carbonell and R. Calientes. 1989. Time series of upwelling nitrate and primary production off Peru derived from wind and ancillary data, 1953-1982. In: Pauly D., P. Muck, J. Mendo and I. Tsukayama, eds. The Peruvian upwelling ecosystem: Dynamics and interactions. ICLARM Conference Proceedings 18:64-76. 438p

Muck, P. 1989. Major trends in the pelagic ecosystem off Peru and their implications for management. In: Pauly, D., P. Muck, J. Mendo and I. Tsukayama, eds. The Peruvian upwelling ecosystem: Dynamics and interactions. ICLARM Conference Proceedings 18:168-174. Instituto del Mar del Peru (IMARPE), Callao, Peru; Deutsche Gesellschaft für Technische Zusammenarbeit (GTZ), Eschborn, Federal Republic of Germany; and International Center for Living Aquatic Resources Management (ICLARM), Manila, Philippines.

Muck, P .and D. Pauly. 1987. Monthly anchoveta consumption of guano birds, 1953 to 1982. In: Pauly, D. and I. Tsukayama, eds. The Peruvian anchoveta and its upwelling ecosystem: Three decades of change. ICLARM Studies and Reviews 15: 219-223. 351p

Muck, P. and G. Sanchez. 1987. The importance of mackerel and horse mackerel predation for the Peruvian anchoveta stock (a population and feeding model). In: Pauly, D. and I. Tsukayama, eds. The Peruvian anchoveta and its upwelling ecosystem: Three decades of change. ICLARM Studies and Reviews 15:276-293. 350p

Pauly, D. and I. Tsukayama, eds. 1987. The Peruvian anchoveta and its upwelling ecosystem: Three decades of change. ICLARM Studies and Reviews 15.Instituto del Mar del Peru (IMARPE),Callao, Peru. 351p

Pauly, D., A. CH. de Vildoso, J. Mejia, M. Samamé and M.L. Palomares. 1987. Population dynamics and estimated anchoveta consumption of bonito (*Sarda chiliensis*) off Peru, 1953 to 1982. In: Pauly, D. and I. Tsukayama, eds. The Peruvian anchoveta and its upwelling ecosystem: Three decades of change. ICLARM Studies and Reviews 15: 248-267. 351p

Pauly, D., P. Muck, J. Mendo, and I. Tsukayama, eds. 1989. The Peruvian upwelling ecosystem: dynamics and interactions. ICLARM Conference Proceedings 18, 438p. Instituto del Mar del Peru (IMARPE), Callao, Peru; Deutsche Gesellschaft für Technische Zusammenarbeit (GTZ), Eschborn, Federal Republic of Germany; and International Center for Living Aquatic Resources Management (ICLARM), Manila, Philippines

Robles, F.L., E. Alarcon and A. Ulloa. 1980. Water masses in the northern Chilean zone and their variations in the cold period (1967) and warm periods (1969, 1971-73). In: Proceedings of the Workshop on the phenomenon known as "El Niño," UNESCO. 83-174

Sandoval de Castillo, O., C. Wosnitza-Mendo, P. Muck and S. Carrasco. 1989. Abundance of hake larvae and its relationship to hake and anchoveta biomasses off Perú. In: Pauly,D., P. Muck, J. Mendo and I. Tsukayama, eds. The Peruvian upwelling ecosystem: Dynamics and interactions. ICLARM Conference Proceedings 18: 280-296. 438p

Serra, R., I. Tsukayama. 1988. Sinopsis de datos biologicos y pesqueros de Sardina Sardinops sagax (Jenyns, 1842) en el pacifico suroriental. FAO Sinop. Pesca, 13 (1). 60p

Sharp, G.D. 1998. Dome or U-shaped physiological responses of populations, and ecosystems. In: Durand, M.-H., P. Cury, R. Mendelssohn, C. Roy, A. Bakun and D. Pauly, eds. Global Versus Local Changes in Upwelling Systems. Editións de l'Orstom. Collections Colloques et séminaires. 503-524. 594p

Stotz, W. 1997. The management areas in the fishery law: first experiences and evaluation of its utility as management tool for *Concholepas concholepas*. Estud.Oceanol.16:67-86

Stotz, W. and E. Perez. 1992. Crecimiento y productividad de loco Concholepas concholepas (Brugiere, 1789) como estimador de la capacidad de carga en areas de manejo. Investigaciones Pesqueras (Chile) 37:13-22

Trippel, E.A. 1995. Age and maturity as a stress indicator in fisheries. BioScience 45(11): 759-771

Upton, H.F. 1992. Biodiversity and Conservation of the marine environment. Fisheries 17(3): 20-25

Waller, T.R. 1969. The evolution of the *Argopecten gibbus* stock (Mollusca: Bivalvia), with emphasis on the tertiary and quarternary species of eastern North America. J. Paleontology 43. 125p

Walters, C., V. Christensen and D. Pauly, 1997. Structuring dynamic models of exploited ecosystems from trophic mass -balance assessments. Reviews in Fish Biology and Fisheries 7(2):139-172

Wolff, M. 1984. Impact of the 1982-83 El Niño on the Peruvian scallop *Argopecten purpuratus*. Tropical Ocean-Atmosphere Newsletter, 28: 8-9.

Wolff, M. 1987. Population dynamics of the Peruvian scallop *Argopecten purpuratus* during the El Niño phenomenon of 1983. Can. J. Fish. Aquat. Sci., 44: 1684-91.

Wolff, M. 1994a. Population dynamics, life histories and management of selected invertebrates of the SE Pacific upwelling system. Habilitation Thesis, Universität Bremen, 210p

Wolff, M. 1994b. A trophic model for Tongoy Bay- a system exposed to suspended scallop culture (Northern Chile). J. Exp. Mar. Biol. Ecol.182:149-168

Wolff, M. and J. Mendo 2000. Management of the Peruvian bay scallop (*Argopecten purpuratus*) metapopulation with regard to environmental change. Aquatic Conserv. Mar. Freshw. Ecosyst. 10:117-126

Wosnitza-Mendo, C., L. Icochea. 1997. Crucero de evaluación del stock de merluza en otoño de 1997 (15 de mayo – 8 de junio de 1997). Informe final de los observadores por parte de la Sociedad Nacional de Pesquería. Unpub.ms. 9p

Wosnitza-Mendo, C., R. Guevara-Carrasco. 2000. Adaptive response of Peruvian hake stock to overfishing. NAGA. The ICLARM Quarterly 23(1):24-28

Wyrtki, K. 1963. The horizontal and vertical field of motion in the Peru current. Bull. Scripps Inst. Oceanogr. 8:313-346.

Wyrtki, K. 1967. Circulation and water masses in the eastern equatorial Pacific Ocean. Int. J. Oceanol. Limnol.1:17-147

Wyrtki, K. 1975. El Niño: The dynamical response on the equatorial Pacific Ocean to atmospheric forcing. J. Phys. Oceanogr. 5:572-584

Wyrtki, K. 1977. Sea level during the 1972 El Niño. J. Phys. Oceanogr. 7:779-787

Yanez, E., M. Garcia and M.A. Barbieri. 1998. Pelagic fish stocks and environmental changes in the south-east Pacific. In: Durand,M.-H., P. Cury, R. Mendelssohn, C. Roy, A. Bakun and D. Pauly, eds. Global Versus Local Changes in Upwelling Systems. Editións de l'Orstom. Collection Colloques et séminaires. 275-291. 594 p

III
Tropical LMEs

Large Marine Ecosystems of the World
G. Hempel and K. Sherman (Editors)

13

The Great Barrier Reef: 25 Years of Management as a Large Marine Ecosystem

Jon Brodie

INTRODUCTION

The Great Barrier Reef (GBR) system covers an area of about 350,000 sq km on the north-eastern Australian continental shelf. It is a long, narrow system stretching 2000 km along the coast ranging from 50 km wide in the north to 200 km in the south and bounded by the coast on the west and the Coral Sea on the east (Figure 13-1). It is a relatively shallow system with maximum depths of about 50 m at the shelf break limit of the ecosystems normally considered to form part of the GBR. It encompasses the largest system of coral reefs and related life forms anywhere in the world with approximately 3000 reefs. About 350 species of hard coral are found in the region along with 1500 species of fish, 240 species of seabirds and at least 4000 species of molluscs.

The Great Barrier Reef Marine Park (hereafter 'the Marine Park') is a multiple use marine park established in 1975 by the Australian Federal Government. The overriding objective of the legislation is the conservation of the Great Barrier Reef. The Great Barrier Reef was listed on the World Heritage Register in 1981. The outer boundaries of both the Marine Park and the Great Barrier Reef World Heritage Area lie beyond the shelf break in the east (Figure 13-1) thus enclosing a considerable area of oceanic depth water. The western boundary of the World Heritage Area is the low water mark along the coast with the Marine Park boundary similar except for a few small excluded areas along the coast (Figure 13-1).

The principal habitats of the GBR have only existed in their present form since sea level rose approximately 10,000 years ago, flooding the shelf. Inshore the coastline is dominated by mangroves interspersed with areas of low energy sandy beach and rocky shores. Immediately offshore shallow seagrass beds are common (Lee Long *et*

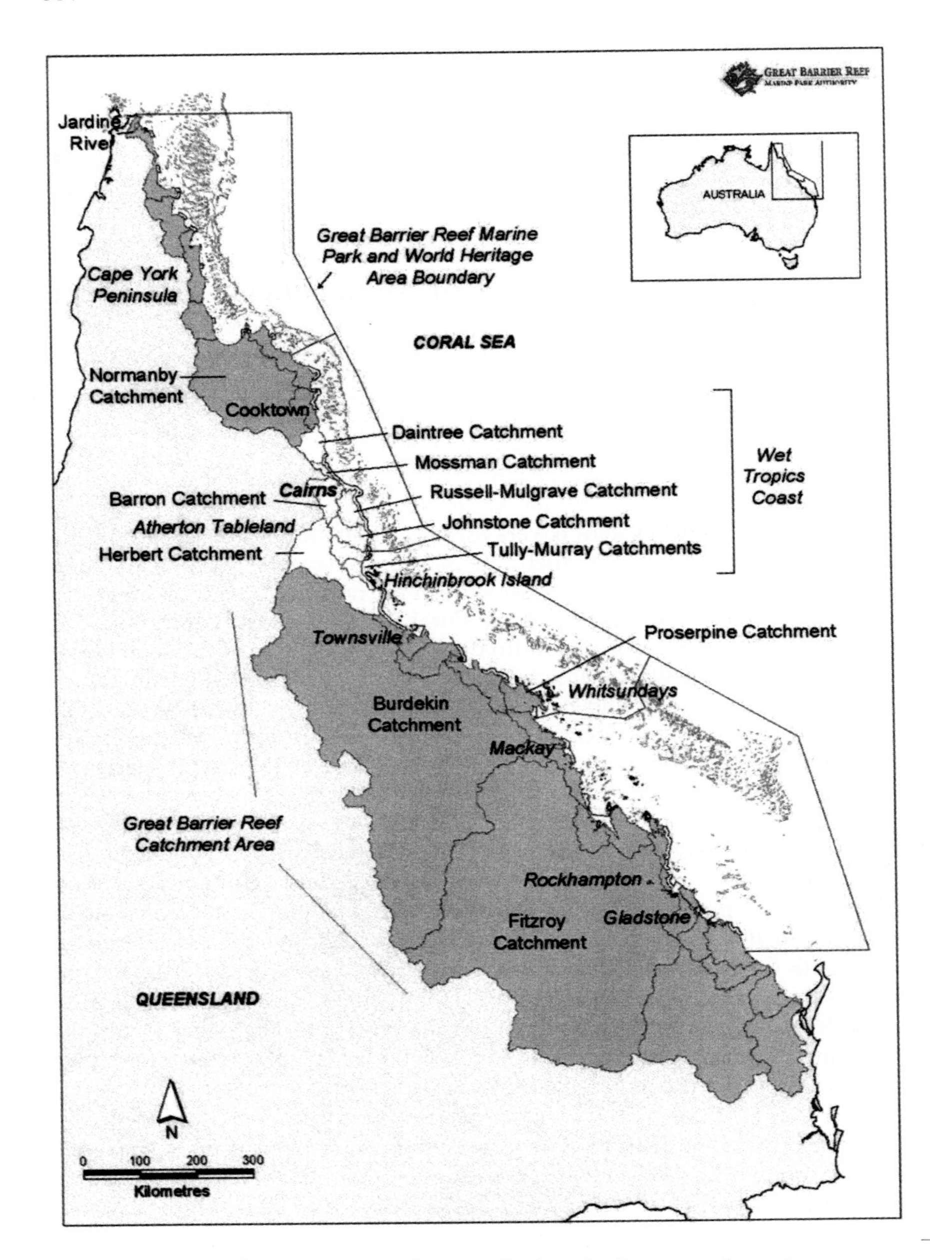

Figure 13-1. The Great Barrier Reef Marine Park and adjacent catchments.

al. 1993) and considerable areas of deepwater (>10m) seagrass are found further offshore (Lee Long *et al.* 2000). The GBR lagoon floor is dominated by soft-bottomed communities of algae, sponges and bryozoans (Birtles and Arnold 1988) interspersed with bare sand. In the north, extensive Halimeda algal beds occupy the deeper off-shore waters, their growth stimulated by nutrient-rich water upwelling from the Coral Sea (Drew and Abel 1988).

The coral reefs of the GBR consist of two main types, the fringing reefs (~ 760 reefs) which occur inshore on the coast and around the high islands, and those of the main reef (~ 2200 reefs) which occupy a band on the outer part of the continental shelf. Optimal conditions for coral reef viability are generally considered to be the clear low nutrient conditions of tropical oceanic waters. However many inshore reefs of the GBR exist in relatively turbid conditions. Sources of nutrients to the GBR include Coral Sea surface water (nutrient poor), upwelling Coral Sea deep water (nutrient rich), terrestrial runoff and atmospheric inputs, including nitrogen fixation by cyanobacteria (Furnas *et al.* 1995). Flushing of the GBR lagoon is limited by the enclosure formed by the main reef (Wolanski 1994).

Individual parts of the GBR are linked together through larval transport. Most GBR biota reproduce using a pelagic larval stage which may be competent from a few days to a few weeks. Reefs (and other ecosystems) may be self-seeded but many receive most of their recruits from 'upstream' reefs (James *et al.* 1990). Thus the reefs exist in a connected system and events on one reef may have significant implications for other 'connected' reefs. The pattern of outbreaks of the crown-of-thorns starfish (*Acanthaster planci)* strikingly illustrates this connectivity (Moran 1986).

As an ecosystem, the GBR is confronted by those large scale human impacts which now affect the earth at a global scale (Vitousek *et al.* 1997a). These impacts began for the GBR in a serious way over one hundred years ago with the overfishing of large marine animals (dugongs, whales, turtles, crocodiles and some species of large fish) (Jackson *et al.* 2001). More recently the GBR now shares in the global oversupply of bioavailable nitrogen (Vitousek *et al.* 1997b); agricultural nonpoint source pollution of surface waters with nitrogen and phosphorus (Carpenter *et al.* 1998; Matson *et al.* 1997); global climate change and the associated coral bleaching phenomenon (Hoegh-Guldberg 1999); increased carbon dioxide concentrations in the surface ocean and associated calcification decline (Kleypas *et al.* 1999a); and increased levels of marine organism diseases apparently associated with the microbialization of the global coastal ocean (Jackson *et al.* 2001). These global scale impacts coexist with the more local scale issues of tourism and recreational use of the Marine Park and together comprise the challenge of management of the GBR as a large marine ecosystem.

MANAGEMENT OF THE GBR

The legislation establishing the Great Barrier Reef Marine Park allowed for the formation of an Authority to manage the Park – the Great Barrier Reef Marine Park Authority (hereafter 'the Authority'). From this legislation the Authority has derived its goal 'to provide for the wise use, protection and understanding of the Great Barrier Reef in perpetuity.'

The GBR lies off the coast of the State of Queensland and the range of activities which occurs within it or affect it are of fundamental significance to the environmental, social, and economic well being of Queensland and Australia. Such is the importance of the GBR in the lives of many Australians (Green *et al.* 1999), community debate over the management of human activity within and adjacent to the GBR continues. A range of these activities is addressed by legislation and regulations under Commonwealth, Queensland State and Local Government jurisdictions as well as by the Great Barrier Reef Marine Park Act itself.

The Marine Park is not a park in the same sense as a terrestrial national park. It is a multiple use protected area and many uses are permitted in parts of the Marine Park. Activities which occur in the Marine Park include commercial and recreational fishing, aquaculture, tourism, shipping traffic, research and defense exercises. Whether such activities are consistent with the GBR also being a World Heritage Area is the focus of considerable debate (Green *et al.* 1999). The one activity which is not permitted in any part of the Marine Park is mining. This proscription arose from the history of the formation of the Marine Park when potential mining was the trigger for its establishment and seen to be an unacceptable threat to the long-term existence of the GBR.

To manage the multiple use area with minimal risk to the GBR ecosystem, many management tools are used by the Authority. Primarily these include: Zoning Plans, Management Plans and Site Plans; environmental impact assessment processes; permits with associated conditions; and a compliance program. Development of policies to influence activities which may impact the Marine Park but which occur outside its boundaries is also important.

Originally the principal policy tool used in the management of the Park was zoning, a form of planning in which activities are allowed or restricted in various zones. Activities in zones in which they are allowed are regulated through a permitting system. The Park is divided into four sections, each of which has a zoning plan which spatially regulates use within the section. Zoning plans characteristically have 5 or 6 zones each with defined objectives and which prescribe 'as-of-right' uses and those requiring a permit. Restrictions are generally as follows: General use--trawling, other fishing, most activities permitted; Habitat protection--trawling not permitted, other fishing, most activities permitted; Conservation park--trawling not permitted, limited line and spear fishing permitted; Marine national park--no extractive uses; Scientific research--scientific research only; and No entry. More recent zoning provisions also place restrictions on the number of reefs within an area permitted to have structures such as tourist pontoons and caged aquaculture facilities.

At the time of the establishment of the Marine Park the concept of large, multiple use managed areas was very new, the Great Barrier Reef region and ecology of coral reefs were not well known scientifically and the urgency for establishment of zoning plans precluded detailed baseline and theoretical scientific studies on management strategies. Nevertheless, the 'GBR model' which subsequently developed is widely regarded as successful (Kelleher1994) and has been applied to other marine protected areas around the world.

Over the past two decades much has been learned about the GBR, the structure and function of coral reefs and marine ecosystems, and planning and management of marine protected areas. Environmental degradation has continued, and the goals of ecologically sustainable development and use and conservation of biodiversity have also been developed as national and international priorities. The Authority has grown from a small staff to a professional agency with 150 staff. Zoning plans have been developed for 360,000 sq km of the Marine Park. Scientific support has increased from a handful of coral reef scientists around Australia to internationally known coral reef research centres including the Australian Institute of Marine Science and James Cook University in Townsville. Human use of the GBR is substantial and continues to increase. The gross value of tourism and fishing is estimated at one billion Australian dollars (Driml 1999).

Strategies for ecological sustainable use and biodiversity conservation are contained in legislation, regulations and policies, and in formal decisions of the Authority, zoning plans and other sources. The Marine Park Act 1975 (with amendments and regulations) protects corals and certain other species, prohibits certain endangering processes (e.g. mining, oil drilling), provides for development of zoning and other management plans and has powers to stop threatening processes. Other Commonwealth and State Acts protect certain Marine Park species and prohibit certain threatening processes. Commonwealth government policies (for example, the Ecologically Sustainable Development strategy), and bilateral and international agreements and conventions also contain goals and objectives for biodiversity management (World Heritage Convention).

With the exception of the protected species identified, the above legislation and policies are very general, and refer to broad goals and concepts ('protection', 'maintenance of ecology', 'ecologically sustainable use'), and have lacked details on the mechanisms by which these goals may be achieved.

The Great Barrier Reef World Heritage Area Strategic Plan (Anon 1994) attempted to provide hierarchical steps and processes for ecological sustainable use and biodiversity management. The Plan identifies the key issues or objectives in the management of the GBRMP as: (1) maintenance of the ecology, (2) management to achieve ecologically sustainable use and, (3) maintenance of traditional, cultural, heritage and historic values. The main 25-year objective is 'to ensure the persistence of the World Heritage Area as a diverse, resilient, and productive ecological system. Several 'broad strategies' are given to achieve this objective (e.g. 'manage use of the Area in accordance with ecological sustainability and the precautionary principle'). Five year objectives and strategies are given, from which many of the Authority's objectives have been developed. However the Strategic Plan fails to provide unambiguous, scientifically-based targets and mechanisms for ecologically sustainable use and biodiversity conservation. The Plan does not attempt to define the processes by which ecological sustainable use may be attained. It makes no attempt to identify threatened species, define limits of acceptable change to habitats, determine proportions of habitat which should be totally protected, or determine the number, size and spatial arrangements of protected areas or specify how representative biological communities can be identified.

The Strategic Plan recognizes that management of potentially damaging activities in the GBR, which may adversely affect conservation values, is spread among a variety of agencies. The Authority maintains a limited level of overall coordination. Catchment land use activities are managed by Queensland Department of Natural Resources; management of fisheries and fishing is by Queensland Department of Primary Industries and Queensland Fisheries Service; shipping activities and oil spill management is by the Australian Maritime Safety Authority and Queensland Department of Transport; and urban and industrial land-use activities management is by the Queensland Environment Protection Agency.

While management of use has apparently been relatively successful, in view of the current favourable assessment of the state of the GBR (Wachenfeld *et al.* 1998), pressure on the overall system has been generally relatively low. Thus management success may be partially illusory and management systems inadequate in the face of pressures from greatly increased use of the system. Only lip service has been paid to the ideals of integration of management under the Strategic Plan. Close integration addressing issues across jurisdictional boundaries is not evident in the eight years since the Plan was developed.

Two key areas where management of human activities is recognized as unsuccessful relate to dugong (*Dugong dugon*) conservation and the management of zoning restrictions on fishing. In the case of the dugong, population numbers in the southern half of the GBR have declined 50% between 1986 and 1994 (Marsh *et al.* 1996). The decline is attributed to high levels of mortality from anthropogenic activities including possible habitat loss, traditional hunting and incidental mortality in commercial gill-nets and in shark nets set for bather protection (Marsh *et al.* 1996). Studies of coral trout (*Plectropomus leopardus*) populations on the reefs of the GBR show little measurable difference in population size, structure or age between zones allowing commercial and recreational fishing and those that do not allow it (Russ *et al.* 1995). The apparent lack of effect of the management regime is attributed to illegal fishing on the closed reefs as well as generally low fishing pressure on the open reefs (Russ *et al.* 1995). The lack of ability to adequately enforce fishing zoning restrictions has apparently negated the benefit of closed areas as refuges, representative areas or stock replenishment sites.

Assessment of Threats to the GBR

The only identifiable catastrophic risk to a significantly sized area of the GBR is that posed by a major oil spill. A major oil spill, either to the east of the GBR or in the inner shipping channel, will cause extensive damage to nearby habitats--mangroves, intertidal seagrass and shallow reefs. There is very limited capacity to deal with such a spill. Measures to minimize the risk of a spill are slowly being introduced but the use of the GBR as an international shipping route prevents many management solutions being easily implemented. Continuing improvements in ship navigation e.g. differential GPS, electronic charts and ship reporting systems appear to be the most promising methods of containing this threat. Research into clean-up technology suitable for use in the GBR environment is limited.

The most important chronic threats to the GBR are believed to be those arising from: increased terrestrial runoff of pollutants associated with agricultural and urban activity; the effects of trawling; the effects of net and line fishing; and localized physical damage from anchoring of tourist, recreational & fishing vessels (Van Woesik 1996). An uncertain risk for the GBR is seen in the cycles of crown-of-thorns starfish outbreaks. These have caused major damage to reefs in the central part of the GBR but it is unclear whether they are a totally natural occurrence, mainly human induced or perhaps natural but with increased frequency caused by human activity. The long term effects of global climate change (including severe coral bleaching and mortality and reduced calcification due to seawater pH changes) may in the end present the gravest threat to the long-term existence of the GBR (Hoegh-Guldberg 1999). The biological level of risk and severity of damage from the impacts mentioned have been hard to quantify against the large inherent natural variability in the system (Done 1998). Many habitats in the GBR are in fair condition reflecting limited effects from present impacts. Some species 'iconic' to the World Heritage Area are showing recent alarming population declines--turtles and dugongs (Marsh *et al.* 1996)--although many populations of large marine vertebrates were lost in earlier times (Jackson *et al.* 2001).

Critical Issue Management

After considerable analysis of the issues facing management of the GBR, based on an extensive research and monitoring program, a restructuring of the management focus of the Authority occurred in 1998. The principal focus after the restructure was a concentration on the critical issues facing the GBR as revealed in the previous risk analysis. The four most critical issues in the Marine Park were identified as: the maintenance of conservation, biodiversity and World Heritage values (particularly the development of a more complete set of protected representative areas); tourism and recreation; fisheries (particularly trawling, inshore gill netting and reef line fishing); and water quality and coastal development (particularly watershed management).

Representative Marine Protected Areas

The need to identify and adequately protect representative examples of all habitats in the Marine Park is now well established as being fundamental to effective long-term management (Day *et al.* in press). Currently worldwide evidence is available to show that no-take and no-go areas help to protect marine biodiversity (Roberts and Hawkins 2000). Many of the GBR biological communities are known to be currently poorly represented in those zones protected from extractive use. The current distribution of highly protected zones reflects an early focus on coral reef habitats as a priority, and pristine reefs located in the remote north. A more comprehensive network of representative natural areas will help ensure protection of the north/south (latitudinal) and east/west (cross-shelf) diversity of a much wider range of marine habitats. The planned representative protected area network within the World Heritage Area will help: maintain options for future users; ensure future economic benefits derived from commercial and recreational fishing and tourism by protecting the resource and offering refuge to some fish populations; guarantee that a healthy marine environment is passed on to future generations; and enhance Australia's international status by implementing a broadscale, biophysically-based network of comprehensive, representative marine protected areas. A first approximation of 'representative habitats' based on physical and oceanographic parameters supplemented with biological data (where available) has now been designed. Public participation in reviewing this GBR bioregionalisation and 'candidate representative areas' has now begun. An assessment of the cultural, social, economic, practical and legal implications of alternative 'candidate' area options will follow. A complete evaluation of existing zoning and management plans will occur to ensure integration with the representative areas program. The result will be a reef-wide re-zoning review.

Dugong protected areas

Many marine mammal populations are in decline world wide as a result of anthropogenic impacts. The seagrass meadows in the Great Barrier Reef Region are feeding grounds for a significant proportion of the world's population of the endangered dugong species *Dugong dugon* (Lee Long *et al.* 2000). The dugong has high biodiversity value and is considered by the World Conservation Union to be vulnerable to extinction. A rapid decline of the dugong population in the southern Great Barrier Reef over the last ten years raised concerns about the survival of the species in that region (Marsh *et al.* 1996; 1999). Reasons for the reported decline in dugong numbers are not fully understood, however, because dugongs have very low rates of population growth any impacts such as mortality by entanglement in nets and degradation and loss of seagrass habitat have the potential to threaten the integrity of dugong populations. A system of 16 Dugong Protection Areas (DPAs) was established in 1997/98 by the Commonwealth and Queensland governments as a key strategy to help the declining dugong population recover in the southern Great Barrier Reef. These areas centre on significant habitat and feeding grounds of dugongs. They were declared to minimise the risk to dugongs from anthropogenic effects such as drowning in fishing nets and collision with boats. The principal restrictions in the DPAs apply to gill-net fishing, boating speed and boating access provisions.

Tourism

Tourism is the boom industry of the GBR region. A forty fold increase in the number of tourists visiting the area in the period 1946 to 1980 has been followed by continuous significant growth. In the last decade visitor nights have doubled (Driml 1994; 1999) and tourism currently attracts around 1.5 million visitors to the area per year (Ilett *et al.* in press). In addition, with advances in the technology of high speed ferries, large numbers of tourists are now able to visit even outer-shelf reefs and a number of reefs now have daily visitation rates of more than 500 persons. The Authority has invested its major management presence in the control of tourism impacts. Monitoring programs have now shown that, with good management, environmental impacts of mass tourism are minimal. Management actions have included the requirement for nutrient-removal tertiary treatment of sewage, supervised installation of well-engineered pontoons, diving and snorkeling behavioural codes and fixed moorings to prevent anchor damage. Further development of codes of practice for tourism operators is now under way as a form of effective self-regulation. The management of access and debate over 'ownership' rights to reef sites are currently subjects of intense interest.

Fishing

Fishing (commercial, recreational and indigenous) is the major extractive activity in the Marine Park. Ten commercial fisheries (trawl, reef line, inshore mesh net and seven harvest fisheries) have a direct economic worth about $200 million per annum (Driml 1999; Cadwallader *et al.* in press). Commercial fishing comprises about 3700 professional fishers and 1400 vessels; up to 24 000 privately registered vessels are used for recreational fishing in the GBR each year involving some 800,000 Queensland resident fishers.

The Offshore Constitutional Settlement between the Commonwealth and the State of Queensland places responsibility for the management of all fish stocks (other than tuna and tuna-like fish) in the World Heritage Area under the Queensland Government, specifically the Queensland Fisheries Service. The Commonwealth has direct responsibility for tuna and tuna-like fish throughout the World Heritage Area through the Australian Fisheries Management Authority. The Commonwealth is also responsible for offshore fisheries in the Australian Fishing Zone adjacent to the World Heritage Area. The Great Barrier Reef Marine Park Authority however, with its aim to protect the natural qualities and ecosystems of the GBR, has the role of ensuring that fishing does not have unacceptable ecological impacts on both the fished areas and the reef system as a whole.

The principal commercial fishery of the GBR is "otter" trawling for prawns and scallops with small amounts of bugs (*Thenus spp.*) and squid. In 2000 some 750 trawlers took about 10,000 tonnes of catch (Cadwallader *et al.* in press). Other commercial fisheries include line fishing for reef and pelagic stocks, crabbing and inshore net fisheries. Recreational fisheries for bottom and pelagic stocks are also important, as are the traditional fisheries of Aboriginal and Torres Strait Islanders. Fish stocks in the GBR are small, a characteristic of many tropical fisheries.

Trawling is potentially the most damaging form of commercial fishing in the Marine Park. The technique is indiscriminate with 70% of the total catch being non-target species (by-catch). There are concerns that trawling may cause long-term modification to the seabed and unacceptable effects on the sessile bottom biota. Research in the GBR and northern Australia shows that bottom communities can be substantially altered by trawling and bottom fish communities tend to be less diverse and smaller in trawled areas (Poiner *et* al. 1998). In the Park 70 percent of the trawlable seabed is trawled at least once a year while 46 percent of the total Park was trawled in 1996, but at varying intensities. Effort in the fishery is steadily rising, increasing from 80,000 days of effort in 1988 to 110,000 days in 1998 (Cadwallader *et al.* in press).

The trawl fishery in the GBR targets tiger prawns (*Penaeus semisulcatus* and *P. esculentus*) and banana prawns (*P. merguiensis*) inshore, and king prawns (*P. longistylus* and *P. latisulcatus*) offshore. Scallops (*Amusium japonicum balloti*) are caught in the southern section of the Park.

At present the Authority's management response to trawling is to exclude trawling from habitat protection zones. The areas closed to trawling represent about 20 percent of the Park but this is only about 5 percent of the trawlable area available in the Park. Recently the Authority required, through a Queensland trawl plan, further management of trawling including: 1. a cap on the level of trawl effort at 1996 effort levels, with a schedule for subsequent reductions in effort, will form part of the management plan for the Queensland east coast trawl fishery; 2. additional closures to limit the areas in which trawling can occur and inclusion of some areas as part of the Authority's Representative Areas Program; 3. By-catch Reduction Devices and Turtle Excluder Devices required on all trawl nets within the World Heritage Area; 4. Vessel Monitoring Systems (satellite tracking devices), mandatory on the commercial prawn trawl fleet in the Marine Park, with progressive introduction on other fishing vessels operating in the Marine Park over the next few years; and 5. increased surveillance and enforcement.

The reef-fish line fishery in the GBR includes both commercial and recreational components and the main target species are the larger reef fish such as coral trout and snappers (*Plectropomus sp, Lutjanus sp,* and *Lethrinus sp*). There are about 250 principal commercial operators and some 1450 commercial operators with more limited licensing arrangements in the fishery (Cadwallader *et al.* in press) while 24,000 recreational boats and more than 150 charter boats also share the fishery. The commercial catch is about 3,500 tonnes with a recreational catch of about 4,000 tonnes and charter catch of 265 tonnes. Smaller fisheries for crabs occur inshore (annual catch about 750 tonnes) while mesh net fishing in coastal and estuarine areas catch about 1,400 tonnes of finfish. Small collection and harvest fisheries target lobsters, aquarium fish, corals and trochus.

The Authority under its zoning plans permits line fishing in the vast majority of the Park with no distinction between commercial and recreational fishing. There is now some concern that fish stocks of the large predatory fish are being overexploited. Research results suggest there has been a 30 percent decline in mean fish size and 50 percent decline in catch per unit effort, coupled with a 25 percent increase in effort in the recreational fishery (Blamey and Hundloe 1991; Williams and Russ 1994). There is a large excess capacity in some commercial and recreational fisheries (latent effort). Increased fishing effort, including technology creep, is leading to increased pressure on both fished and previously unfished areas. Declining catch (or decreased average size of fish caught) is evident in some areas and some fisheries. The inshore barramundi (*Lates calcarifer*) fishery (northern Australia's premium table fish) continues to decline (Williams 1997). The low levels of compliance with State fisheries and Commonwealth Marine Park legislation and the high cost of enforcement and surveillance are continuing problems (Mapstone *et al.* 1996; Davies 2000). Unfortunately current information about stocks of reef fishes is insufficient to clearly demonstrate whether or not current levels of exploitation are sustainable (Wachenfeld *et al.* 1998).

Traditional fishing by indigenous people is confined to areas close to Aboriginal communities. Traditional hunting and fishing are permitted in all zones except Preservation Zones. Fishes, turtles and dugongs are hunted and as turtles and dugong are recognized as threatened species in GBR waters some concern as to management of traditional hunting has been expressed. The Authority has now developed a turtle and dugong strategy. Aboriginal and Islander communities are now involved in the development of management plans for these animals along with the voluntary cessation of hunting rights in some areas.

Traditionally, fisheries management has targeted sustainable take. The Authority considers this is an inappropriate management approach for management of a World Heritage Area as it does not take into account the overall effects of fishing on the whole ecosystem. The Authority now requires fisheries management to adopt a 'whole-of-ecosystem' approach (Cadwallader *et al.* in press).

Watershed management

Terrestrial runoff of sediment, nutrients and pesticides is known to be the major cause of the degradation of water quality in the inshore area of the GBR (Brodie *et al.* 2001; Brodie, in press). Management of land runoff is difficult as the source is in catchments which are adjacent to the GBR but which are outside the legislative boundaries of the Marine Park. The principal agricultural industries on the catchment are cropping (sugar cane, cotton & horticulture) and grazing (Gilbert 2001). Most catchments are small (<10,000 km^2), but the Burdekin (133,000 km^2) and Fitzroy River catchments (143,000 km^2) are among the largest in Australia. The Great Barrier Reef Catchment, of area 400,000 km^2 , has been extensively modified since European settlement by forestry, urbanisation and agriculture. Beef cattle numbers are approximately 4,500,000 and over 80 percent of the catchment area is used for beef cattle grazing. Cropping, mainly of sugarcane, uses considerably less area. The sugarcane cultivation area has increased steadily over the last 100 years with a total of 390,000 ha reached by 1997 (Figure 13-2). The cultivation areas are located near the coast principally on the river floodplains. Fertiliser use is closely linked to sugarcane cultivation as the largest crop on the GBR Catchment. With both continuously increasing cultivation area and increasing rates of fertilization, total

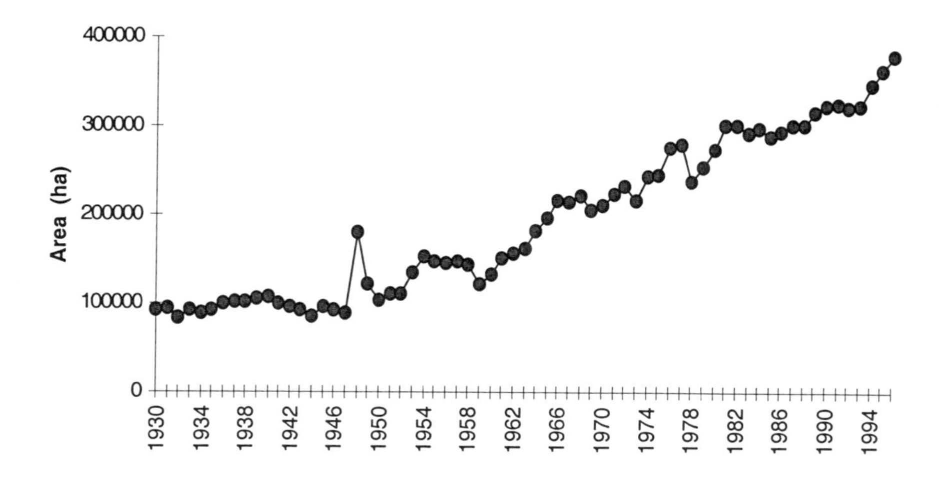

Figure 13-2. Increase in land area used for sugar cultivation from 1930 to 1996 on the GBR Catchment.

fertilizer application has increased rapidly since 1950 (Pulsford 1996) (Figure 13-3) in line with global trends of fertilizer use (Vitousek *et al*, 1997a). The use of pesticides (herbicides, insecticides, and fungicides) is also significant in areas of crop cultivation. Other industries with significant expanding land use (and fertiliser and pesticide use) are cotton and horticulture (particularly bananas). Prawn farming is an expanding industry along the GBR coast and prawn farm effluents may be a considerable local source of nutrients. At present the amounts involved are small with only potential local risks, however, the aquaculture industry along the Queensland coast is very likely to continue to expand.

Historically the environmental performance of agricultural industries has been inadequate. Poor land management has resulted in drainage of wetland systems, inefficient irrigation, oversupply of fertilizer and the clearing and destruction of riparian and wetland vegetation.

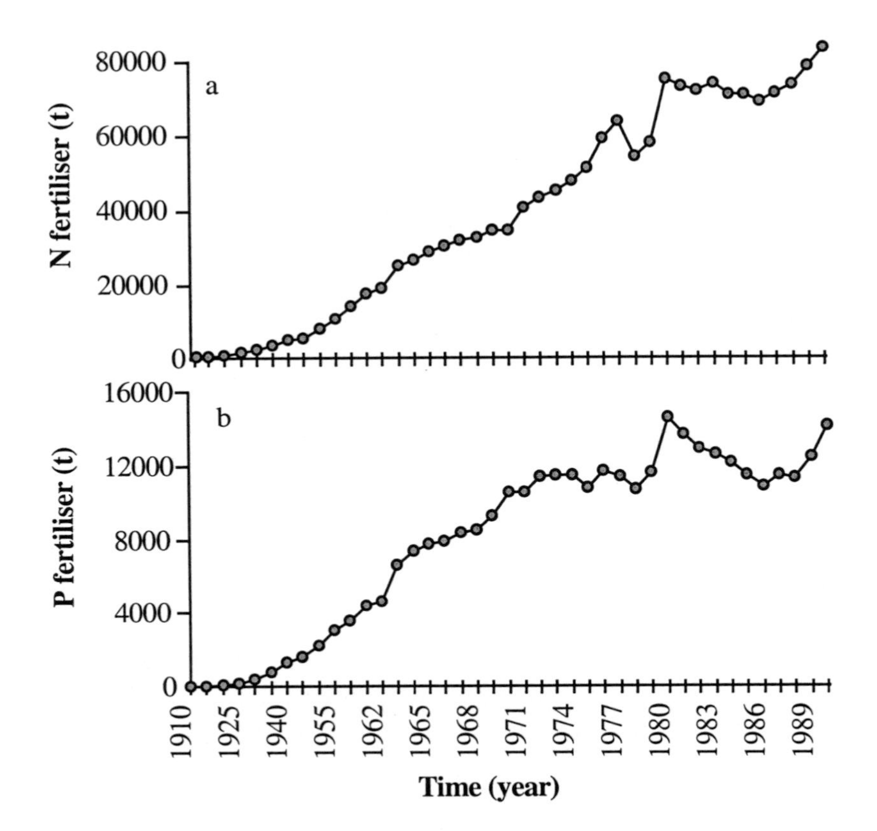

Figure 13-3. Increase in use of nitrogen (a) and phosphorus (b) fertiliser on the GBR Catchment.

It is estimated that 15 million tonnes of sediment, 77 thousand tonnes of nitrogen and 11 thousand tonnes of phosphorus are exported to the coastal waters of the GBR via river discharge annually. Grazing and sugarcane cultivation are the major contributors to the load. Sewage discharges can be significant at local scales but contribute only a few percent of the terrestrial nutrient flux to the coastal waters of the GBR. Sediment and nutrient delivery to the GBR from terrestrial discharge has increased by four times since European settlement of the adjacent coast i.e. the last 130 years (Moss *et al*. 1993; Neil & Yu 1996). The principal nutrients lost from grazing lands are the nutrients naturally present in the soil mobilized through increased soil erosion due to overgrazing. In the case of sugarcane cultivation nutrients are lost from fertilizer addition.

Large quantities of sediment, nitrogen, phosphorus and significant amounts of pesticides lost from agricultural systems are easily measurable in rivers as they discharge into the GBR in flood conditions. For example rivers draining catchments dominated by agriculture typically have dissolved inorganic nitrogen (nitrate and ammonia) concentrations in flood flow 30 times that of rivers with undeveloped catchments (10 – 70 μM compared to 1 – 5 μM; summarized in Brodie (in press). These concentrations are similar to polluted rivers in other parts of the world. Water discharged from these polluted rivers in flood flow covers the inshore part of the GBR, but rarely reaches the outershelf (Devlin *et al.* 2001a). Concentrations of pollutants, such as dissolved inorganic nitrogen, in the river plumes which reach the inshore reefs and seagrass beds are typically 10-50 times ambient concentrations and exceed the 'effects level' for biological impacts on corals, seagrasses and algae. The plumes remain in contact with the inshore ecosystems for periods of days to months (Devlin *et al,* 2001a). The long-term effects of eutrophication on inshore coral reefs are only just becoming evident after a decade of monitoring. In the Whitsundays, coral reef growth has been affected adversely in a eutrophication gradient from the Proserpine River (Van Woesik *et al.* 1999).

Many coral reefs around the world are now in a state of alarming decline due to terrestrial runoff of sediments and nutrients and other factors (Wilkinson 1993; 1999). Deforestation, agriculture on steep slopes and sewage discharges are the causative factors often implicated just as for the GBR. Excess nutrients can have a number of effects on coral and coral reef systems (Koop *et al.* 2001). Either nitrogen or phosphorus can be a limiting nutrient for the growth of phytoplankton, especially in warm, clear tropical waters where light is unlikely to be limiting. Thus phytoplankton flourishes in nutrient enhanced conditions leading to decreased water clarity and reduced light for coral growth on the bottom. The increased phytoplankton crop also encourages the growth of filter-feeding organisms such as sponges, tube worms and barnacles which compete for space with coral as well as weakening the reef physical matrix by boring. In addition nutrients may enhance the growth of algae which overgrow the coral, both competing for space and shading the colonies. Perhaps the most serious direct effect of elevated nutrient loads on corals is the reduction in reproductive success exhibited in these conditions (Koop *et al.* 2001; Ward and Harrison 2000).

Monitoring chlorophyll (an indicator of eutrophic conditions) shows that inshore GBR waters adjacent to developed catchments have twice the concentrations of chlorophyll (0.7 μg/L) compared to waters further offshore (0.3 μg/L) or waters inshore adjacent to the little developed catchments on Cape York (0.3 μg/L) (Devlin *et al.* 2001b).

Heavy metals, pesticide residues and dioxins have been detected in coastal sediments and seagrasses. The herbicide diuron, widely used in minimum tillage crops, has been detected in inshore sediments along a large part of the GBR coast (Haynes *et al.* 2000a). The concentrations found are at or above the levels assessed to cause damage to seagrasses (Haynes *et al.* 2000b).

Terrigenous input reaches only halfway across the shelf. The inner reef area is dominated by terrestrial sediment while the outer area is dominated by carbonate sediment. River plumes are normally constrained close to the coast by hydrodynamic conditions generated by the prevailing south-east wind regime (Devlin *et al.* 2001a). Thus the area of the GBR at risk from land-sourced pollution is a band within about 20 km off the coast adjacent to the developed parts of the catchment (Devlin *et al.* in press).

Several approaches and processes are in place to address agricultural sources (and other land based sources) of pollutants discharging into the GBR. Current approaches rely heavily on industry self-management.

Integrated Catchment Management (ICM) is a consultative approach to land management at the large catchment scale. However, the program has received limited resourcing in Queensland and is a totally voluntary approach with no legislative underpinning. Small scale individual remediation actions have occurred under ICM. However objectives in ICM strategies are not quantitative and are, in most cases, aspirational rather than directed at on-ground implementation. Unfortunately, to date, there is no clearly documented evidence that ICM has been able to prevent the continuing decline of freshwater wetland systems and the loss of riparian vegetation, reduce erosion on a catchment scale, manage fertilizer use to minimize loss from the farm and improve water quality for downstream receiving environments.

Several agricultural industries have developed or are developing Codes of Practice under the Queensland Environmental Protection Act. The Codes are an excellent example of self management. Even though these are voluntary schemes, if fully implemented they should go a considerable way to reducing pollutant discharge to the GBRWHA. Concerns exist about the degree and timing of the implementation of the codes and about effective industry auditing mechanisms that can verify implementation.

Farm Management Plans (also known as Property Management Plans and Land and Water Management Plans) are a valuable tool to institute measures at a farm scale to reduce farm runoff to waterways and protect wetlands and riparian zones. However at present these plans are voluntary and do not generally contain strong environmental performance criteria. Some local government authorities on the GBR coast are now including provisions for regulating 'material change of use' for agricultural purposes in their local government plans. In particular, changes resulting in intensification of agricultural use, such as from beef grazing to fertilised cropping, will be regulated so that areas of riparian vegetation and wetlands will be preserved during such changes.

Management changes in some industries in recent years have led to reduced sediment and nutrient runoff. The most notable example is green cane harvesting and trash retention (blanketing) in sugarcane cultivation where major reductions in soil erosion and hence sediment and phosphorus loss has occurred. In the case of sewage discharges, nutrient removal treatment or land irrigation of sewage effluents on islands in the GBR and on the adjacent mainland, has led to some reduction in direct fluxes to the coastal zone (Brodie 1994).

As a step towards making catchment management strategies contain quantitative objectives for water quality with set implementation timelines, end-of-catchment pollutant load targets have been set by the Authority for rivers on the Great Barrier Reef catchment (Brodie *et al.* 2001). Targets for loads of suspended solids, nitrogen and phosphorus have been set for 26 rivers draining into the GBR. The targets generally require a reduction in load of 30 to 50% over a 10 year period for those rivers with substantial catchment agricultural and urban development. On-ground implementation plans to achieve the targets are yet to be developed.

GLOBAL CLIMATE CHANGE AND CORAL BLEACHING

Since about 1980 coral reefs around the world have experienced an increased frequency of the phenomenon known as coral bleaching. Bleaching involves the loss of the coloured symbiotic algae (zooxanthellae) that live in the coral and upon which corals depend for much of their food supply. While bleaching can be caused by a range of stresses on corals, the mass bleaching of recent years is now known to be caused by seawater temperature increases. Bleaching is often followed by partial or complete coral mortality. Temperatures in tropical oceans have increased by about 1 ^{0}C over the past 100 years and continue to increase at about 1 to 2 ^{0} C per century. In the GBR rates of increase vary from north to south with 0.47 ^{0} C per 100 years in the north, 2.59 ^{0} C in the central and 2.54 ^{0} C in the south (Hoegh-Guldberg 1999). These trends are now believed to be associated with the anthropogenic greenhouse effect. The effects of such increases are now evident in greatly increased global incidents of coral bleaching with subsequent partial coral mortality (Wellington *et al.* 2001).

Mass bleaching of corals has occurred on the GBR on a number of occasions since 1980 but the most widespread and serious incident was in 1998 (Berkelmans and Oliver 1999). The 1998 GBR incident was part of a global sequence of coral reef bleaching events (Wellington *et al.* 2001). It is predicted, using currently accepted climate change models and physiological knowledge of the coral/zooxanthellae relationship, that the incidence of bleaching will continue to increase in the future as atmospheric carbon dioxide increases and seawater temperatures continue to rise threatening the existence of the GBR as a coral reef ecosystem (Hoegh-Guldberg 1999).

In addition to global warming, increased carbon dioxide in the atmosphere will lead to increased concentrations of carbon dioxide in the surface ocean. As coral reef calcification depends on the saturation state of the carbonate mineral aragonite in surface waters it is predicted that calcification of corals will decline by up to 30 percent by the middle of the 21st century (Kleypus *et al.* 1999a). In combination with the increased bleaching scenarios the very existence of many of the existing coral reefs in times of high atmospheric carbon dioxide and high global temperatures has been questioned (Kleypus *et al.* 1999b).

Management of this issue for the GBR is fraught with difficulties ranging from the obvious requirements of a global solution to internal political debate as to Australia's support of the Kyoto agreement. Perhaps more than any other management issue for the GBR, global atmospheric pollution is insoluble at the scale of the large marine ecosystem and the issue least likely to be satisfactorily addressed by the present management regime for the GBR. If the seriousness of the risk to the GBR as a coral reef system is not recognized, then all other management actions for the sustainable use of the GBR may be in vain.

CONCLUSIONS

The Great Barrier Reef large marine ecosystem is fortunate in that it is subject to relatively low level human impact at present, especially when compared to many other coral reef areas (Wilkinson 1993) and marine ecosystems in other parts of the world. The area is also able to be managed as a single unit by virtue of the Marine Park Act although this unity is diluted as many of the human impacts affecting the system originate outside the Park and are not under the direct management control of the Authority. Examples of this boundary and jurisdictional issue at different scales include: 1. watershed management, where management is under control of a different level of government, albeit an Australian jurisdiction; 2. management of turtles where most mortality of GBR stocks occurs outside of Australian waters (primarily in Indonesia and the Pacific island states) and can only be addressed through international arrangements; and 3. global climate change which is only potentially solvable at the international level on a global scale. These examples clearly show that even in the one large marine ecosystem which has a formal integrated management regime, the size of the management unit is still not sufficient to envelop all the issues inherent in managing such an ecosystem. Australia is a relatively rich, developed state with a strong technical base in marine science and management. With the relatively moderate levels of human pressure on the GBR if Australia cannot successfully preserve the system it provides little hope for poorer nations with high populations and heavy pressures on marine resources.

Multiple use management of the Park appears to have been successful so far and provides an example of both ecological and economically sustainable use and development. However loss of dugong populations, ineffective management of fishing and poor land management on the GBR catchment have highlighted the need for better management to address these issues. In the future, successful mitigation of eutrophication associated with terrestrial runoff will test the indirect management power of the Authority. Successful management of both the environmental impacts of fishing and conservation of fish stocks--something rarely achieved in any part of the world--will also be a major test. Measures such as the introduction of dugong protection areas and reductions in traditional hunting of dugongs have not yet been shown to halt the decline in the populations of dugong and turtles. However such measures take long periods to be effective and it may be many years before their success can be measured.

ACKNOWLEDGEMENTS

The assistance of colleagues from the Great Barrier Reef Marine Park Authority, in particular Jon Day, Phil Cadwallader, Annie Ilett and Alison Green is acknowledged and I specifically thank them, and Leon Zann, for providing information, much as yet unpublished, for this paper. I would also like to thank Gilianne Brodie and Gotthilf Hempel for making helpful comments on the manuscript.

REFERENCES

Anon, 1993. A 25 Year Strategic Plan for the Great Barrier Reef World Heritage Area. Great Barrier Reef Marine Park Authority, Townsville, 64p

Berkelmans, R. and J.K. Oliver. 1999. Large-scale bleaching of corals on the Great Barrier Reef. Coral Reefs 18: 55-60

Birtles, R.A. and P.W. Arnold. 1988. Distribution of trophic groups of epifaunal echinoderms and molluscs in the soft-sediment areas of the central Great Barrier Reef shelf. In: Choat *et al.*, eds. Proceedings of the 6th International Coral Reef Symposium, Townsville, 3: 325-332

Blamey, R.K. and T.J.A. Hundloe. 1991. [final report, 1993] Characteristics of recreational boat fishing in the Great Barrier Reef Region. Report to Great Barrier Reef Marine Park Authority, Townsville.

Brodie, J. 1994. Management of sewage discharge in the Great Barrier Reef Marine Park. In: Bellwood *et al.*, eds. Recent Advances in Marine Science and Technology, Proceedings of the 6th Pacific Congress on Marine Science and Technology, PACON, Townsville. 457-465

Brodie, J.E. 2000. Keeping the wolf from the door: managing land-based threats to the Great Barrier Reef. In: Proceedings of the 9th International Coral Reef Symposium. October, 2000, Bali, Indonesia.

Brodie, J.E., C. Christie, M. Devlin, D. Haynes, S. Morris, M. Ramsay, J. Waterhouse and H. Yorkston. 2001. Catchment management and the Great Barrier Reef. Water Science and Technology 43(9): 203-211

Brodie, J., A. Mitchell, M. Furnas, D. Haynes, J. Waterhouse, S. Ghonim, S. Morris, H., Yorkston, and D. Audas. 2001. Developing catchment pollutant load targets for the protection of the Great Barrier Reef. In: Proceedings of the 2nd National Conference on Aquatic Environments: Sustaining our aquatic environments- Implementing solutions. Queensland Department of Natural Resources and Mines, Brisbane, Australia

Cadwallader, P., M. Russell, D. Cameron, M. Bishop and J. Tanzer. In press. Achieving ecologically sustainable fisheries in the Great Barrier Reef World Heritage Area. In: Proceedings of the 9th International Coral Reef Symposium. October, 2000, Bali, Indonesia

Carpenter, S.R., N.F. Caraco, D.L. Correll, R.W. Howarth, A.N. Sharpley, and V.H. Smith. 1998. Nonpoint pollution of surface waters with phosphorus and nitrogen. Ecological Applications 8(3): 559-568

Davies, C.R. 2000. Inter-reef movement of the common coral trout, *Plectropomus leopardus*. . GBRMPA Research Publication No. 61, Great Barrier Reef Marine Park Authority, Townsville, 98p

Day, J., L. Fernandes, A. Lewis, G. De'ath, S. Slegers, B. Barnett, B. Kerrigan, D. Breen, J. Innes, J. Oliver, T. Ward and D. Lowe. In press. The representative areas program for protecting biodiversity in the Great Barrier Reef World Heritage Area. In: Proceedings of the 9th International Coral Reef Symposium. October, 2000, Bali, Indonesia

Devlin, M., J. Waterhouse, J. Taylor and J. Brodie. 2001a. Flood plumes in the Great Barrier Reef: Spatial and temporal patterns in composition and distribution. GBRMPA Research Publication No.68, Great Barrier Reef Marine Park Authority, Townsville

Devlin, M., J. Waterhouse, C. Christie, D. Haynes and J. Brodie. 2001b. Long-term chlorophyll monitoring in the Great Barrier Reef lagoon: Status Report 2, 1993-2000. GBRMPA Research Publication, Great Barrier Reef Marine Park Authority, Townsville

Devlin, M., J. Waterhouse and J. Brodie. In press. Terrestrial discharge into the Great Barrier Reef: Distribution of riverwaters and pollutant concentrations during flood plumes. In: Proceedings of the 9th International Coral Reef Symposium. October, 2000, Bali, Indonesia

Done, T.J. 1998. Ecological criteria for evaluating coral reefs and their implications for managers and researchers. Coral Reefs 14: 183-192

Drew, E.A. and K.M. Abel. 1988. Studies on Halimeda I. The distribution and species composition of Halimeda meadows throughout the Great Barrier Reef Province. Coral Reefs 6: 195-205

Driml, S.M. 1994. Protection for profit. GBRMPA Research Publication No. 35, Great Barrier Reef Marine Park Authority, Townsville

Driml, S. 1999. Dollar values and trends of major direct uses of the Great Barrier Reef Marine Park. GBRMPA Research Publication No. 56, Great Barrier Reef Marine Park Authority, Townsville

Furnas, M.J., A.W. Mitchell and M. Skuza. 1994. Nitrogen and phosphorus budgets for the central Great Barrier Reef shelf. GBRMPA Research Publication No. 36. Great Barrier Reef Marine Park Authority, Townsville. 234p

Gilbert, M., J. Waterhouse, M. Ramsay and J. Brodie. 2001. Population and major land use in the Great Barrier Reef catchment area: spatial and temporal trends. GBRMPA Research Publication, Great Barrier Reef Marine Park Authority, Townsville, Australia

Green, D., G. Moscardo, T. Greenwood, P. Pearce, M. Arthur, A. Clark and B. Woods. 1999. Understanding public perceptions of the Great Barrier Reef and its management. CRC Reef Research Centre, Technical Report No. 29. CRC Reef Research Centre, Townsville. 64p

Haynes, D., J. Müller and S. Carter. 2000a. Pesticide and herbicide residues in sediments and seagrass from the Great Barrier Reef World Heritage Area and Queensland coast. Marine Pollution Bulletin 41(7-12): 279-287

Haynes, D., P. Ralph, J. Prange and W. Dennison. 2000b. The impact of the herbicide diuron on photosynthesis in three species of tropical seagrass. Marine Pollution Bulletin 41(7-12): 288-293

Hoegh-Guldberg, O. 1999. Climate change, coral bleaching and the future of the world's coral reefs. Journal of Marine and Freshwater Research 50: 839-866

Ilett, A., H. Skeat, C. Thomas, V. Bonanno and E. Green. In press. Managing tourism sustainably – lessons learned on the Great Barrier Reef, Australia. In: Proceedings of the 9th International Coral Reef Symposium. October, 2000, Bali, Indonesia

Jackson, J.B.C., M.X. Kirby, W.H. Berger, K.A. Bjorndal, L.W. Botsford, B.J. Bourque, R.H. Bradbury, R. Cooke, J. Erlandson, J.A. Estes, T.P Hughes, S. Kidwell, C.B. Lange, H.S. Lenihan, J. M. Pandolfi, C.H. Peterson, R.S. Steneck, M.J. Tegner and R.R.Warner. 2001. Historical overfishing and the recent collapse of coastal ecosystems. Science 293: 629-637

James, M.K., Dight, I.I. and J.C. Day. 1990. Application of larval dispersal models to zoning in the Great Barrier Reef Marine Park. In: Proceedings Fourth Pacific Congress Marine Science & Technology, Tokyo

Kelleher, G. 1994. Can the Great Barrier Reef model of protected areas save reefs worldwide. In: Ginzberg, R.N. (Compiler). Proceedings of the Colloquium on Global Aspects of Coral Reefs: Health, Hazards and History, 1993. Rosenstiel School of Marine and Atmospheric Science, University of Miami. 346-352

Kleypas, J.A., R.W. Buddemeier, D. Archer, J.P. Gattuso, C. Langdon, B.N. Opdyke. 1999a. Geochemical consequences of increased atmospheric carbon dioxide on coral reefs. Science 284, 118-120

Kleypus, J.A., J.W. McManus and L.A.B. Menez. 1999b. Environmental limits to coral reef development: Where do we draw the line? American Zoologist 39: 146-159

Koop, K.D. A. Booth, J. Broadbent, D. Brodie, D. Bucher, J. Capone, W. Coll, M. Dennison, P. Erdmann, O. Harrison, P. Hoegh-Guldberg, G.B. Hutchings, A.W. Jones, J. Larkum, A. O'Neil, E. Steven, S. Tentori, S. Ward, J. Williamson and D. Yellowlees.2001. "ENCORE: The Effect of Nutrient Enrichment on Coral Reefs. Synthesis of Results and Conclusions," *Marine Pollution Bulletin*, 42[2]:91-120

Lee Long, W.J., J.E. Mellors and R.G. Coles. 1993. Seagrasses between Cape York and Hervey Bay, Queensland, Australia. Australian Journal of Marine & Freshwater Research 44: 19-31

Lee Long, W.J., R.G. Coles and L.J. McKenzie. 2000. Issues for seagrass conservation management in Queensland. Pacific Conservation Biology 5: 321-328

Mapstone, B.D., R.A. Campbell and A.D.M. Smith. 1996. Design of experimental investigations of the effects of line fishing on the Great Barrier Reef. . CRC Reef Research Centre, Technical Report No. 7. CRC Reef Research Centre, Townsville. 86p

Marsh, H., P. Corkeron, I. Lawler, J. Lanyon and A. Preen. 1996. The status of the dugong in the southern Great Barrier Reef Marine Park. GBRMPA Research Publication No. 41, Great Barrier Reef Marine Park Authority, Townsville

Marsh H, C. Eros, P. Corkeron, B. Breen. 1999. A conservation strategy for dugongs: implications of Australian research. Marine and Freshwater Research 50: 979-990

Matson, P.A., W.J. Parton, A.G. Power and M.J. Swift. 1997. Agricultural intensification and ecosystem properties. Science 277: 504-509

Moran, P.J. 1986. The Acanthaster phenomenon. Oceanography and Marine Biology Annual Review 24: 379-480

Moss, A.J., G.E. Rayment, N. Reilly and E.K. Best. 1993. A preliminary assessment of sediment and nutrient exports from Queensland coastal catchments. Queensland Department of Environment & Heritage Technical Report No. 4, Brisbane. 33p

Neil, D. and B. Yu. 1996. Simple climate-driven models for estimating sediment input to the Great Barrier Reef lagoon. In: Great Barrier Reef terrigenous sediment flux and human impacts. CRC Reef Research Centre Technical Report, CRC Reef Research Centre, Townsville.122-128

Poiner, I., J. Glaister, R. Pitcher, C. Burridge, T. Wassenberg, N. Grobble, B. Hill, S. Blaber, D. Milton, D. Brewer and N. Ellis. 1998. Environmental effects of prawn trawling in the Far Northern Section of the Great Barrier Reef: 1991-1996. CSIRO Division of Marine Research, Cleveland, Australia. 2 vols. 500p

Pulsford, J.S. 1996. Historical nutrient usage in coastal Queensland river catchments adjacent to the Great Barrier Reef Marine Park. GBRMPA Research Publication No. 40, Great Barrier Reef Marine Park Authority, Townsville. 98p

Roberts, C.M. and J.P. Hawkins. 2000. Fully-protected marine reserves: a guide. In: WWF Endangered Seas Campaign, Washington, DC and Environment department, University of York, UK. 131p

Russ, G.R., D.C. Lou and B.P. Ferreira. 1995. A long-term study on population structure of the coral trout *Plectropomus leopardus* on reefs open and closed to fishing in the central Great Barrier Reef., Australia. CRC Reef Research Centre, Technical Report No. 3. CRC Reef Research Centre, Townsville. 30p

Van Woesik, R. 1996. Contemporary disturbances to coral communities of the Great Barrier Reef. Journal of Coastal Research Special Issue No. 12: Coastal Hazards, 233-252

Van Woesik, R., T. Tomascik and S. Blake. 1999. Coral assemblages and physico-chemical characteristics of the Whitsunday Islands: evidence of recent community changes. Marine and Freshwater Research, 50: 427-440

Vitousek, P.M., J.D. Aber, R.W. Howarth, G.E. Likens, P.A. Matson, D.W. Schindler, W.H. Schlesinger, D.G. Tilman. 1997a. Human alteration of the global nitrogen cycle: sources and consequences. Ecological Applications 7: 737-750

Vitousek, P.M., H.A. Mooney, J. Lubchenco and J.M. Melillo. 1997b. Human domination of Earth's ecosystems. Science 277: 494-499

Wachenfeld, D., J. Oliver and J. Morrisey. 1998. State of the Great Barrier Reef World Heritage Area. Great Barrier Reef Marine Park Authority, Townsville. 139p

Ward, S. and P. Harrison. 2000. Changes in gametogenesis and fecundity of acroporid corals that were exposed to elevated nitrogen and phosphorus during the ENCORE experiment. Journal Experimental Marine Biology and Ecology 246: 179-221

Wellington, G.M., P.W. Glynn, A.E. Strong, S.A. Nauarrete, E. Wieters and D. Hubbard. 2001. Crisis on coral reefs linked to climate change. EOS 82(1): 1-7

Wilkinson, C.R. 1992. Coral reefs of the world are facing widespread devastation: Can we prevent this through sustainable management practices. In: Proceedings of the 7th International Coral Reef Symposium, Guam, 1992, 1: 11-21

Wilkinson, C.R. 1999. Global and local threats to coral reef functioning and existence: review and predictions. Marine and Freshwater Research 50: 867-878

Williams, L.E. 1997. Queensland's fisheries resources: Current condition and recent trends 1988-1995. Department of Primary Industries, Brisbane, Australia. 500p

Williams, D. McB. and G.R. Russ. 1994. Review of data on fishes of commercial and recreational fishing interest on the Great Barrier Reef. GBRMPA Research Publication No. 33, Great Barrier Reef Marine Park Authority, Townsville. 106p

Wolanski, E. 1994. Physical Oceanographic Processes of the Great Barrier Reef. CRC Press, Boca Raton. 194p

Large Marine Ecosystems of the World
G. Hempel and K. Sherman (Editors)

14

Development of Fisheries in the Gulf of Thailand Large Marine Ecosystem: Analysis of an unplanned experiment

Daniel Pauly and Ratana Chuenpagdee

ABSTRACT

The main ecological features of the Gulf of Thailand (GoT) Large Marine Ecosystem are briefly outlined, to serve as context for a sketch of the development of the fisheries therein. This development took place in the absence of large environmental changes and, except for some squabbles over the incursion of trawlers into the EEZ of Cambodia, in the absence of international conflicts over straddling stocks. This situation may be seen as representing a nearly ideal 'experimental' setting for studying the direct effect of rapidly growing fisheries on the abundance and composition of tropical multispecies resources and on the ecosystem within which these resources are embedded.

This analysis consists of three mutually supporting elements: (1) a narrative, highlighting the sequence of events leading to the present, overcapitalized fisheries and their much depleted resource base; (2) a comparative study, using the multidimensional Rapfish approach, of the impact, in terms of sustainability, of the different gears operating in the GoT; and (3) a demonstration that these fisheries, by fishing down the food webs of the GoT, have fundamentally altered, and continue to alter, the functioning of that ecosystem, albeit in a manner that may still be reversible. These results indicate that a drastic reduction of fishing effort, especially by bottom trawlers, is the only way to halt further ecological degradation of this Large Marine Ecosystem.

INTRODUCTION

The Gulf of Thailand (GoT), with an area of about 350,000 km^2 is a typical Large Marine Ecosystem (LME) in terms of its size (Sherman and Duda 1999). However, it differs from a number of other LMEs in being, to a large extent, part

However, it differs from a number of other LMEs in being, to a large extent, part of the EEZ of a single country (Figure 14-1), thus minimizing straddling stocks issues. Moreover, the GoT appears largely unaffected by the massive regime

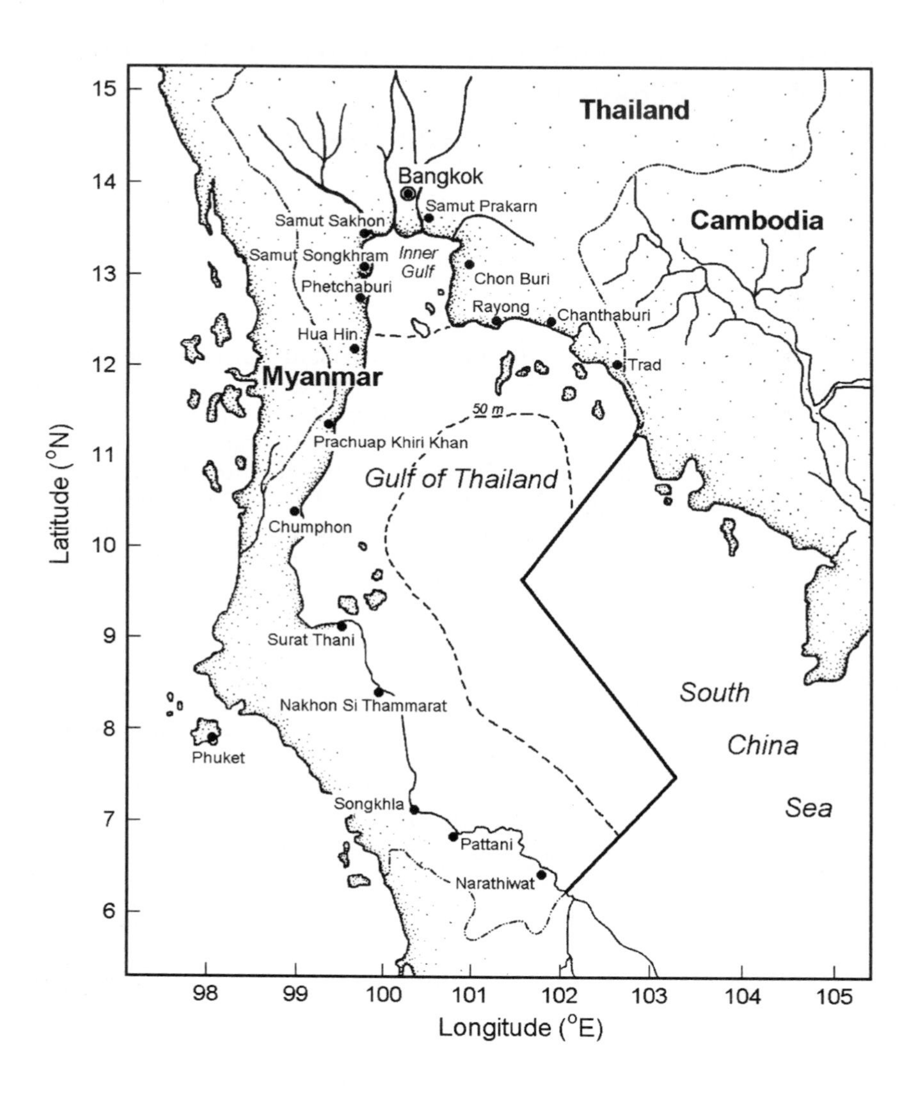

Figure 14-1. Map of the Gulf of Thailand, highlighting features mentioned in the text.

shifts that beset other areas of the Pacific, e.g., the Humboldt Current System (Alheit and Bernal 1993), or the Central/North Pacific (Polovina et al. 1995).

These features lead to a situation where ecosystem effects of fishing can be evaluated in 'pure' form, as it were, without having to account for the confounding effects of, e.g. foreign fleets operating within the GoT, or environmental fluctuations.

In this contribution, we rely on this situation to analyze the impact of fisheries on the GoT LME. In a sense, we thus use the GoT as the site of a giant, if unplanned experiment in fisheries development, conducted in a 'control' LME.

THE PHYSICAL SETTING

The GoT as defined in Figure 14-1 consists of three subsystems:

1) The shallow 'Inner Gulf' (≈10,000 km^2);
2) A band of shallow grounds (down to 50 m), bordering the East and West coasts, reaching to the Cambodian coast on the East, to the border of Peninsular Malaysia on the West (≈ 150,000 km^2), and supporting the bulk of the demersal trawl fishery;
3) A Central Basin (≈ 190,000 km^2), with depths ranging from 50 to 80 m, and a shallow sill (about 50 m) that limits water exchanges with the open South China Sea (Piyakarnchana 1989; Eiamsa-Ard and Amornchairojkul 1997).

The oceanography of the biogeochemical province of which the Gulf of Thailand is a part is discussed in Longhurst (1998), largely based on Wyrtki (1961), and does not need reiterating here. However, we must mention the relatively high primary production prevailing in the GoT, recently boosted by increased nutrients from rivers and shrimp farms, and increasingly leading to harmful algal blooms, oxygen depletion, food poisoning and other pollution effects, particularly in the Inner Gulf (Eiamsa-Ard and Amornchairojkul 1997, Longhurst 1998, Piyakarnchana 1999).

DEVELOPMENT OF THE TRAWL FISHERIES

Until the early 1960s, Thai fisheries had been driven mainly by their own, internal dynamics. This was reflected in an emphasis on small pelagics (mainly Indian mackerels, *Rastrelliger* spp. and anchovies, *Stolephorus* spp.), caught by artisanal fishers operating mainly fixed gears, and supplying local markets.

Most important among these was the supply of anchovies for making 'fish sauce' (*Nam Pla*; Ruddle 1986, Pauly 1996).

In the early 1960s, a Thai-German bilateral project introduced and widely demonstrated the use of the light 'Engels' trawl for catching demersal fish, i.e., a gear much more suited to the bottom type and demersal resources of the GoT than the gear used in a previous, unsuccessful attempt in the area (Tiews 1965, 1972; Butcher 1996).

The subsequent development of the Thai trawl fishery has often been depicted as a showcase of successful transfer of appropriate developed-country technology to a tropical developing country, i.e., 'North-South'. However, detailed analysis (Butcher 1996, 1999) reveals that the technological 'package' that was transferred had been perfected in the Philippines, notably in Manila Bay, at the end of the Second World War (Tiews and Caeces-Borja 1959; Silvestre *et al.* 1987). Thus, the developed-country contribution here consisted mainly of facilitating a 'South-South' technology transfer, i.e., overcoming the isolation of developing-country scientists and managers. There is probably a deep lesson here, as already noted by Tiews (1965).

The rapid build-up of trawling effort in the GoT was fuelled from two sources:

1) Extremely high rates of profit by the first trawl operators, quickly reinvested into more trawlers; and perhaps more importantly,
2) Relatively cheap loans, mainly from the Manila-based Asian Development Bank—loans, for further fisheries development, which markedly reduced the cost of entry into the fishery (Mannan 1997).

A detailed economic analysis of the GoT demersal trawl fishery is given in Panayotou and Jetanavanich (1987), providing details on the various 'market failures' involved here. Also this study analyzes the exacerbating effect of the global fuel price increases of 1973 which, jointly with the declaration of EEZ by neighboring countries, gradually forced the return into the GoT of a large numbers of trawlers that had been operating outside (Butcher 1999).

What we wish to stress here is the ecological impact of the massive increase in trawling effort that occurred from the early 1960s on, and which resulted by the early 1980s in a strong decline in catch per unit effort, from about 300 kg/hour in 1961 to about 50 kg/hour in the 1980s (Figure 14-2), and 20-30 kg/hour in the 1990s (Eiamsa-Ard and Amornchairojkul 1997).

Demersal catches, which had earlier increased in response to the build-up of effort began to stagnate, and to slowly decrease (Figure 14-2), while the catch composition changed both within species (toward smaller individuals; Pope

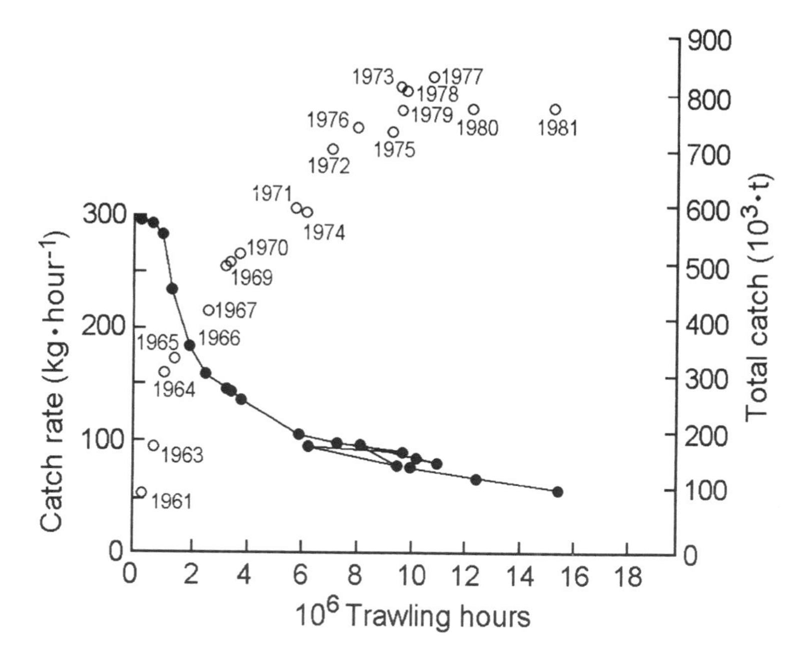

Figure 14-2. Catch and relative abundance of demersal resources vs. effort by Gulf of Thailand trawlers, during the period (early 1960s to early 1980) when the fisheries 'developed' (from Boonyubol and Pramokchutima 1984). Note the concavity of the Catch/effort plot, implying changes in composition of the multispecies catch (see text).

1979; Pauly 1980), and between species (toward a mix consisting predominantly of small, short-lived species; Pauly 1979, Beddington and May 1982).

In addition to the traditional *Nam-Pla* factories, the simultaneous development of the Thai aquaculture industry provided a ready outlet for the small, but increasingly valuable and ill-named 'trashfish' landed by trawlers (Hayase and Meemeskul 1987; Csavas 1993). Sumaila (1999) discusses the now global phenomenon of rapid value increases in small fishes 'subsidizing' the overfishing of larger fishes, a phenomenon which classical economists may view as yet another market failure.

MULTISPECIES AND ECOSYSTEM ANALYSES

The rapid development and scale of the trawl fishery in the GoT soon attracted the attention of stock assessment scientists, who attempted to provide a basis for estimation of 'Maximum Sustainable Yield' (MSY) in tropical multispecies fisheries. The key conceptual steps involved here are briefly recalled (see also Hongskul 1979). The first of these attempts was based on the famous 'Gulland equation', which suggests that Potential Yield = 0.5 x natural mortality x unexploited biomass (Gulland 1971), widely applied through Southeast Asia. This equation, based on the logic of single-species yield per recruit analysis, has been widely criticized (see e.g. Beddington and Cooke 1983).

However tenuous the derivation of Gulland's equation was, this became far worse when a single natural mortality (M) estimate was used to represent all the trophic interactions within a multispecies stock. The high value of M = 1 per year commonly used in Southeast Asia, represented, moreover, only the short-lived components of the multispecies stock (see M estimates for Southeast Asian fishes in www.fishbase.org), and hence led to P_y estimates that were biased upward.

Next were estimates based on what came to be called the Total Biomass Schaefer Model (TBSM), wherein the logic of surplus-production models (Schaefer 1954, i.e., that growth rate of biomass is a parabolic function of biomass itself) was assumed to apply to an ensemble of interacting species (Brown *et al*. 1976, Marr *et al*. 1976, FAO 1978).

However, it soon became obvious that the linear decline of catch per unit effort expected under the Schaefer model did not occur. Rather, as seen in Figure 14-2, the relation between catch/effort and effort in the GoT demersal trawl fishery was strongly non-linear, which led to the TBSM being replaced, for pragmatic reasons, by a model assuming such non-linearity, the 'exponential' surplus-production model of Fox (1970).

The deeper implication of the concavity of catch/effort and effort plots such as in Figure14-2 was not appreciated at the time. We now understand that it was due to rapid depletion, at a low level of fishing mortality, of the less resilient components of the multispecies resources, followed by a less steep decline of its more resilient components, a theme to which we return below.

Pope (1979) was the first to address the multispecies issue in the GoT demersal fishery head on, based on a multispecies Lotka-Volterra model that included trophic interactions. While pointing out that the parameters of this model could not, in practice, be estimated from observed data, he could show that, for any parameter set, multispecies MSY will be always less for an exploited species mix consisting of predators and preys than for the sum of all the individual species MSYs.

Perhaps more importantly, the model also showed that if the species mix is exploited by a non-selective gear such as a bottom trawl (i.e., a gear with fixed ratios of the fishing mortalities it applied to each species), 'true system MSY' cannot be achieved, even if fishing effort is regulated so as to correspond to the apex of a parabolic yield-effort curve. This, of course, results from technological interactions, i.e., from the fact that it is impossible to optimally exploit both large, slow-growing fish and small, fast-growing fish using the same fishing mortality and mesh size.

Another approach for tackling the multispecies issue in the GoT was to construct a dynamic simulation model and to explore its behavior under different management regimes (Larkin and Gazey 1982). Here, the problem was that already faced by Pope (1979), i.e., the parametrization of the model's coupled differential equations. Larkin and Gazey (1982) identified several elegant shortcuts to such parametrization. However, the modeling approach implied therein remained opaque to most practitioners, and had no follow-up.

The application by Christensen (1998) of the now widely-used Ecopath/Ecosim modeling approach to that part of the GoT exploited by the demersal fishery resolved several of the issues addressed by Larkin and Gazey (1982). Thus, the Ecopath software, designed to accommodate the type of data typically collected by fisheries scientists (see Christensen and Pauly 1992, and www.ecopath.org), enabled the rapid parametrization of two food webs for two states ('unexploited', i.e., early 1960s; 'depleted', i.e., early 1980s) of the system in question, based on the vast amount of diet composition, mortality and biomasss information available on GoT fishes (see also www.fishbase.org), and earlier used for a simpler model of the GoT (Pauly and Christensen 1993, Pauly and Christensen 1995, Pauly *et al.* 2000).

The dynamic simulation module of the software, Ecosim (Walters *et al.* 1997), was then used to perform various tests, including some envisaged, but not conducted in Larkin and Gazey (1982). The most interesting of these was to fish the 'unexploited model' with the same rapid increase of effort that occurred in reality. The result after two decades of simulation was a system configuration closely resembling that in the 'depleted' model. Conversely, relaxing fishing effort in the 'depleted model' led to a reestablishment of the food web configuration prevailing in the early 1960s.

As might be seen from Figure 14-3, the species groups most affected by the trawl fishery were crabs/lobsters, rays, sharks, and other large fishes, while penaeid

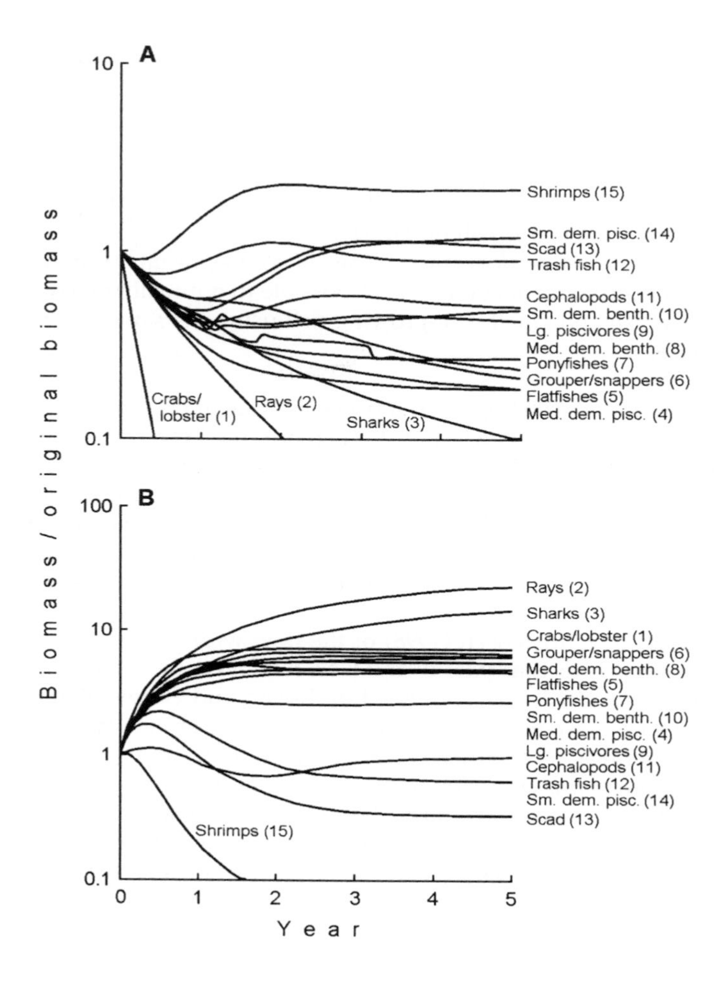

Figure 14-3. Relative changes in the biomass of key components of the Gulf of Thailand Large Marine Ecosystem, as predicted by the Ecosim software of Walters et al. (1997), given the baseline Ecopath models constructed by Christensen (1998). A: Changes from unexploited state (early 1960s), following a 16-fold increase of baseline fishing mortality, and mirroring the changes observed by research trawl surveys; B: Changes from exploited state (early 1980s), following a 16-fold decrease in fishing mortality. The sequence of numbers in A (from crab/lobster (1) to shrimps (15) indicated which groups declined most under fishing. This sequence is reversed in B, suggesting that the effects of fishing pressure would be largely reversible.

shrimps, squids and small fishes (including 'trash fishes') increased. Conversely, crabs/lobsters, rays, sharks, and other large fishes recovered when simulated fishing effort was reduced, while penaeid shrimps, squids and small fishes were reduced by the predatory pressure thus re-established.

These results, which corroborate earlier hypotheses concerning the impact of large predators on shrimps and squids in the GoT (Pauly 1984, 1988), also suggest that the GoT is not presently stuck in some 'alternative stable state' independent of fishing mortality (Beddington and May 1982) but, rather, would rebound if fishing effort were to be reduced. Moreover, the sequence of species impacts suggests that this ecosystem suffers from the 'Fishing down the marine food web' syndrome (Pauly *et al.* 1998), our next topic.

FISHING DOWN THE GULF OF THAILAND FOOD WEBS

Large fish tend to be piscivorous, which puts them near or at the top of marine food webs, and thus gives them high trophic levels (TL). Conversely, small fish and invertebrates such as squids and penaeid shrimps tend to feed on plankton, or bottom organisms and detritus, low in the food web.

Therefore, unselective fishing aimed at maximizing total catch – the 'biomass trash fish production' of James et al. (1991) - will tend to reduce the average TL levels in a system's food web. Also, this should be reflected in the mean TL of aggregate fisheries catches from a given area.

Following demonstration of such decline of mean TL in different parts of the world, and for the world's fisheries as a whole (Pauly et al. 1998), we investigate here the occurrence of this phenomenon in the GoT LME. This was done using three databases:

A) Nominal catches (1950-1976) for Thailand in FAO area 71 (Western Central Pacific), i.e., excluding catches from the Andaman Sea/Indian Ocean coast of Thailand;

B) Catch statistics (1977-1997) of the Thai Department of Fisheries (DOF) for the Gulf of Thailand as defined in Figure 14-1;

C) Species-specific TL estimates from FishBase (see www.fishbase.org).

Figure 14-4 presents the results of this analysis, of which a detailed spreadsheet is available on request. As might be seen from Figure 14-4A, catches from the GoT built up rapidly in the 1960s and 1970s, and fluctuated thereafter. While we are aware that it would be easy to over-interpret these fluctuations (which may be due in part to our heterogeneous database, and to landings originating from

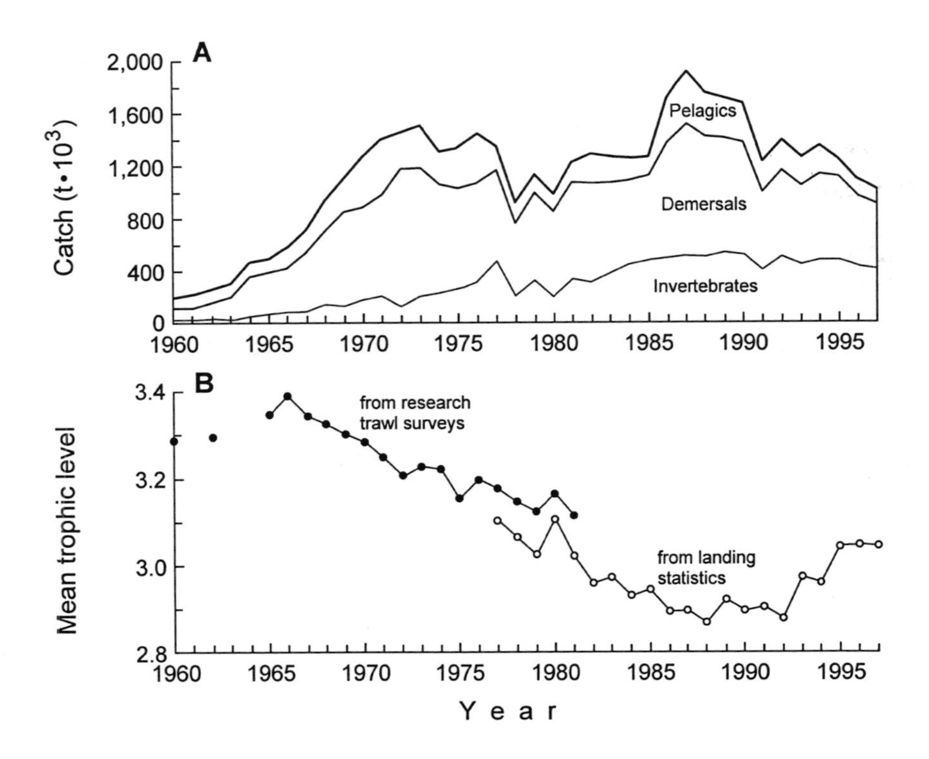

Figure 14-4. Major features of the Gulf of Thailand fisheries and their underlying ecosystem. (A) Catches, by major species groups (excluding tuna and other large pelagics). Note stagnation and decline of demersal catches, following their rapid increase in the 1960s and 1970s. Also note increasing contribution of small and medium pelagics, and overall decline in the 1990s. (B) Trophic level (TL) trends in the catch of research trawlers (reflecting relative abundances in the ecosystems), and in the total landings (both series excluding large pelagics). Lower TL in 1977 to 1997 series are due to inclusion of small pelagics and other low-TL organisms caught by gears other than trawl (see text).

outside the GoT, but not reported as such), we believe they largely reflect a stagnation, then decline of the demersal component of the catch, partly masked by an increased contribution of small and mid-sized pelagic fishes to the overall catch from the GoT.

Figure 14-4B confirms this analysis: the rapid decline of mean TL levels in the survey catch/effort data confirm that the high TL-level species of the GoT were quickly depleted, with small, low-level species soon becoming dominant in the ecosystem. Mean TL levels based on landings data, covering the years 1977 to 1997, show even lower, if fluctuating values, due to the inclusion, in the computation of mean TL, of low–TL fishes (mainly small pelagics) caught by gears other than trawls. We conclude from this that 'fishing down the marine food webs' occurs in the GoT, and that fishing profoundly modified the ecosystem in which these food webs are embedded.

This is important in view of the widespread perception that 'pollution' is the key problem of the GoT. Indeed, it is quite possible (see Parsons 1996) that the fisheries, by removing the upper parts of the natural food webs, have contributed to the observed increase of jellyfish and other consumers of herbivorous zooplankton, i.e., to a trophic cascade leading to phytoplankton blooms and the ensuing oxygen depletions reported by Piyakarnchana (1999).

Thus, the issue for Thai managers is to reduce fishing effort, i.e., to identify and hopefully target those gears and fisheries which most contribute to the unsustainable trends demonstrated here: our next and last topic.

COMPARING THE GULF OF THAILAND FISHERIES USING THE RAPFISH METHOD

Table 14-1 summarizes the key features of the GoT fisheries. Information on each of these fisheries was gathered from the literature cited herein, and in Thailand, from the DOF and other sources. This information was encoded, by choosing from 3-4 options the values of a large number of preset attributes detailed in Pitcher (1999). However, this was done here only in terms of ecological sustainability. Thus, economical, technological, and other dimensions of sustainability (notably compliance to the FAO Code of Conduct) are not reported upon here, although they are important elements of a full Rapfish analysis (Pitcher 1999). The data set resulting from this was analyzed using a form of Multi-Dimensional Scaling, a robust non-parametric method that is largely insensitive to non-linearity and other potentially biasing properties of the input data (Pitcher 1999). Figure 14-5 presents the key result of this analysis, displaying the relative position of each fishery on the sustainability scale.

As might be seen, the fully developed demersal trawl fishery of the 1980s is that closest to the anchor point for low sustainability, followed by the fisheries using Push nets (1980s) and Mackerel purse seines (1990s). All other fisheries have similar, relatively high sustainability scores, although they differ in a number of other attributes leading to differences in the position on the vertical axis (see Figure 14-5).

Table 14-1. Major fisheries, gear and target species in the Gulf of Thailand Large Marine Ecosystem. Figure 14-5 presents an evaluation of the ecological sustainability of these fisheries.

Fishery	Gear	Main target species	Mean catch (t·year^{-1})
Trawl fishery 1960s	Bottom trawl	Fish and shrimps	See Figure 2
Trawl fishery 1980s	Bottom trawl [a]	Threadfin breams	8,434
		Big-eyes	6,321
		Lizardfish	5,586
		Flatfish	4,842
		Croaker	4,007
		'Trash fish'	424,693
		Shrimp & prawns	75,020
	Pair trawl[a]	Threadfin breams	3,443
		Indo-Pacific mackerel	2,960
		Big-eyes	2,135
		Lizardfish	1,580
		Trevallies	1,529
		'Trash fish'	97,716
		Squid & cuttlefish	19,754
Mackerel purse seine 1980s	Purse seine[a]	Sardinellas	21,100
		Indo-Pacific mackerel	12,927
		Indian mackerel	6,144
		Longtail tuna	6,214
		Scads	5,626
	Lighted purse seine[a]	Sardinellas	81,675
		Indo-Pacific mackerel	30,434
		Scads	28,824
		Indian mackerel	23,716
		Big-eye scad	19,524
Mackerel purse seine 1990s	Purse seine	All species (incl. Sardinellas, Indo-Pacific mackerel, Scads, Indian mackerel and Big-eye scad)	[c,d]50,5077
Anchovy purse seine 1980s	Anchovy purse seine	Anchovy	[a]27,033 [b]41,857
Anchovy purse seines 1990s	Anchovy purse seine	Anchovy	[c,d]11,7153
Crab gill net 1980s	Crab gill net[a]	Crabs	7,124
Fish gill net 1980s	Encircling gill net[e]	Indo-pacific mackerel	27,358
		King mackerel	18,112
		Indian mackerel	17,245
Shrimp gill net 1980s	Shrimp gill net	Shrimps	[d,f]8633
Squid cast net 1980s	Squid cast net	Squids	[d,f]1,238
Squid trap 1980s	Squid trap	Squids	[d,f]6,231
Push net 1980s	Push net[a]	Shrimps	1,7216
		'Trash fish'	1,5946
Hook and lines 1980s	Hook and long line		[f]4,674

(a) Mean for 1980 to 1984, Inner GoT;
(b) Mean for 1982 to 1984, Inner GoT;
(c) Mean for 1990 to 1994, including. Outer parts of GoT;
(d) Based on total catches from the gear;
(e) Mean for 1981 to 1984, Inner GoT;
(f) Mean for 1988 to 1989, including the outer GoT.

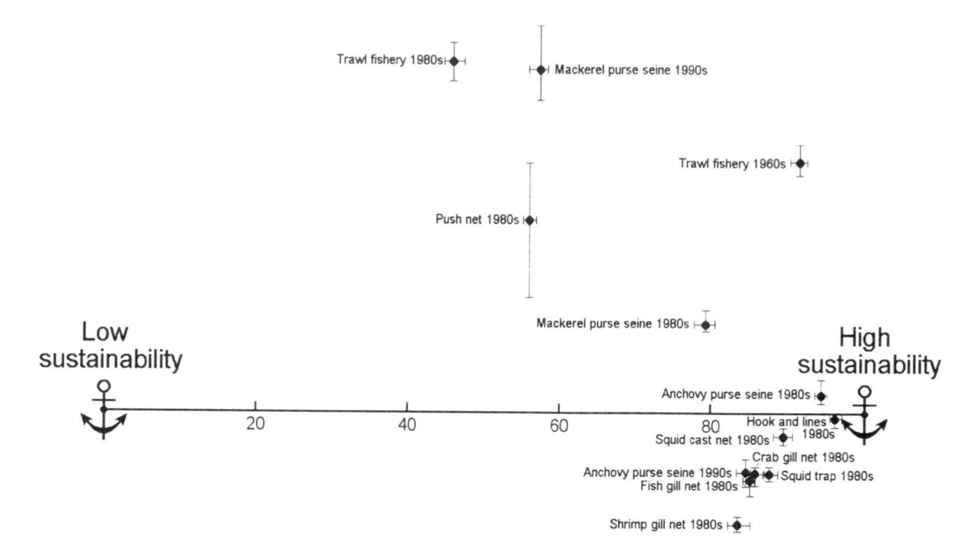

Figure 14-5. Comparison of the ecological sustainability score of the fisheries in Table 1, using the Rapfish method. The anchor point for low sustainability represents a 'model' fishery, given the lowest possible scores on all attributes, and conversely for the other anchor point. This analysis identifies the 'developed' bottom trawl fishery of the 1980s as the worst of all fisheries in Table 14-1 in terms of the horizontal axis, representing ecological sustainability.

CONCLUSIONS

It is now obvious that reducing fishing effort in the GoT and, more precisely, to retire a large fraction of the demersal trawlers operating therein is the only alternative to a continued degradation of that LME. All analyses performed so far converge to the same diagnosis, which is largely free of the potentially confounding effect of large environmental fluctuations.

Needless to say, this does not preclude addressing as well, the numerous other management issues that have led to coastal degradation in and around the Gulf of Thailand (Chuenpagdee 1996). In either case, these are not international issues; they are a matter for the Thai people to resolve, acting in their own best interest. This might involve a shift from the current centralized management regime to a more participatory approach at the community level, as well as an analysis of various policy options (Chuenpagdee et al. 2001), using inputs from scientists, academics, and the public at large.

ACKNOWLEDGMENTS

We wish to thank, in Thailand, the Fisheries Statistics Sub-division, Fisheries Policy and Planning Division, Department of Fisheries, Thailand for supplying catch data, and Dr. K. Juntarashote, at Kasetsart University for his assistance with the Rapfish scoring.

At the Fisheries Centre, UBC, Vancouver, we received crucial assistance from Dr. R. Watson and Mr. D. Preikshot (time series data), Dr. J. Adler (Rapfish and MPS), and Ms E. Buchary (TL level estimates). Ms A. Atanacio, FishBase project, Los Baños, Philippines, prepared the figures, some based on analyses by Drs. V. Christensen and M.L. Palomares (Project). We also wish to thank Drs. G. Hempel and K. Sherman for inviting this contribution.

This work is a product of the Sea Around Us Project, based at the Fisheries Centre, UBC, Vancouver, and supported by the Pew Charitable Trusts, Philadelphia. The senior author also acknowledges support from Canada's Scientific and Engineering Research Council.

REFERENCES:

Alheit, J. and P. Bernal. 1993. Effect of physical and biological changes on the biomass yield of the Humboldt Current Ecosystem. In: Sherman, K., L.M. Alexander and B.D. Gold, eds. Stress, mitigation and sustainability of large marine ecosystems. Amer. Assoc. Advan. Sci., Washington D.C. 53-68

Beddington, J.R. and J.G. Cooke. 1983. The potential yield of fish stocks. FAO Fish. Tech. Pap. (242), 47p

Beddington, J.R. and R.M. May 1982. The harvesting of interacting species in a natural ecosystem. Scientific American 247:42-49

Boonyubol, M. and S. Promokchutima. 1984. Trawl fisheries in the Gulf of Thailand. ICLARM Translations 4. 12p

Brown, B. J. Brennan, M. Grosslein, E. Heyerdahl and R. Hennemuth. 1976. Effects of fishing on the marine fish biomass in the Northwest Atlantic from the Gulf of Maine to Cape Hatteras. ICNAF Res. Bull. 12:49-68.

Butcher, J.G. 1996. The marine fisheries of the Western Archipelago: toward an economic history, 1850 to the 1960s. In: D. Pauly and P. Martosubroto, eds. Baseline Studies of Biodiversity: the Fish Resources of Western Indonesia. ICLARM Studies and Reviews 23:24-39

Butcher, J.G. 1999. Why do Thai trawlers get into so much trouble? MS, presented at a Symposium on "The Indian Ocean: Past, Present and Future", Western Australian Maritime Museum, Freemantle, Western Australia, Nov. 1999. 18p

Christensen, V. 1998. Fishery-induced changes in a marine ecosystem: insights for models of the Gulf of Thailand. J. Fish Biol. 53 (Suppl. A):128-142

Christensen, V. and D. Pauly. 1992. The ECOPATH II - a software for balancing steady-state ecosystem models and calculating network characteristics. Ecological Modelling 61:169-185

Chuenpagdee, R. 1996. Damage schedule - an alternative approach for valuation of coastal resources. NAGA The ICLARM Quarterly 19(4):13-15

Chuenpagdee, R., J.L. Knetsch and T.C. Brown. 2001. Environmental damage schedules: community judgments of importance and assessments of losses. *Land Economics* 77(1):1-11

Csavas, I. 1993. Aquaculture development and environmental issues in the developing countries of Asia. In: R.S.V. Pullin, H. Rosenthal and J. Maclean, eds. Environment and Aquaculture in Developing Countries. ICLARM Conf. Proc. 31: 74-101

Eiamsa-Ard, M. and S. Amornchairojkul. 1997. The marine fisheries of Thailand, with emphasis on the Gulf of Thailand trawl fishery. In: G. Silvestre and D. Pauly, eds. Status and Management of tropical coastal fisheries in Asia. ICLARM Conf. Proc. 53: 85-95

FAO 1978. Some scientific problems of multi-species fisheries. Report of the Expert Consultation on Management of Multi-species Fisheries. FAO Fish. Tech. Pap. 181. 42p

Fox, W.W. 1970. An exponential yield model for optimizing exploited fish populations. Trans. Amer. Fish. Soc. 99:80-88

Gulland, J.A. (Editor). 1971. The fish resources of the Ocean. Fishing News Book, West Byfleet, England. 255p

Hayase, S. and Y. Meemeskul 1987. Fluctuations of trash fish landed by Thai trawlers. Bull. Japan Soc. Fish. Oceanogr. 51(2):124-133

Hongskul, V. 1979. Multispecies systems in fisheries. Thai Fisheries Gazette 32(1):85-95

James, D., S. Garcia, C. Newton and P. Martosubroto 1991. Studies of Thailand, Malaysia, Indonesia, the Philippines and the ASEAN Region. In: Fisheries and aquaculture research capabilities in Asia. The World Bank Technical Paper No. 147, Washington, D.C. 31-70

Larkin, P.A. and W. Gazey 1982. Application of ecological simulation models to management of tropical multispecies fisheries. In: D. Pauly and G.I. Murphy, eds. Theory and management of tropical fisheries. ICLARM Conf. Proc. 9, Manila. 123-140

Longhurst, A.R. 1998. Ecological Geography of the Sea. Academic Press, San Diego. 398p

Mannan, M. A. 1997. Foreword: Asian Development Bank. In: G. Silvestre and D. Pauly (eds.) Status and Management of tropical coastal fisheries in Asia. ICLARM Conf. Proc. 53:*v*

Marr, J.C., G. Campleman and W.R. Murdoch. 1976. An analysis of the present and recommendations for the future development and management policies programmes and institutional arrangements, Kingdom of Thailand. FAO/UNDP South China Sea Fisheries Development and Coordinating Programme, Manila, SCS/76?WP/45.

Panayotou, T. and S. Jetanavanich. 1987. The Economics and Management of Thai Marine Fisheries. ICLARM Studies and Reviews 14. 82p

Parsons, T.R. 1996. The impact of industrial fisheries on the trophic structure of marine ecosystems. In: G.A. Polis and K.O. Winnemiller, eds. Marine Food Webs: Integration of Patterns and Dynamics. Chapman Hall, New York. Chapter 33, 352-357.

Pauly, D. 1979. Theory and management of tropical multispecies stocks: a review, with emphasis on the Southeast Asian demersal fisheries. ICLARM Studies and Reviews 1. 35p

Pauly, D. 1980. A new methodology for rapidly acquiring basic information on tropical fish stocks: growth, mortality and stock-recruitment relationships, In: S. Saila and P. Roedel, eds. Stock Assessment for Tropical Small-Scale Workshop, Sept. 19-21 1979, University of Rhode Island. International Center for Marine Resources Development, Kingston. 154-172

Pauly, D. 1984. Reply to comments on pre-recruit mortality in Gulf of Thailand shrimp stocks. Trans. Amer. Fish. Soc. 113:404-406

Pauly, D. 1988. Fisheries research and the demersal fisheries of Southeast Asia. In: J.A. Gulland, ed. Fish Population Dynamics (2nd ed.). Wiley Interscience, Chichester. 329-348

Pauly, D. 1996. Fleet-operational, economic and cultural determinants of by-catch uses in Southeast Asia. In: Solving By-Catch: Considerations for Today and Tomorrow. University of Alaska, Sea Grant College Program, Report No. 96-03, Fairbanks. 285-288

Pauly, D. and V. Christensen. 1993. Stratified models of large marine ecosystems: a general approach, and an application to the South China Sea. In: K. Sherman. L.M. Alexander and B.D. Gold, eds. Stress, Mitigation and Sustainability of Large Marine Ecosystems, Amer. Assoc. Advan. Sci., Washington D.C. 148-174. 376p

Pauly, D. and V. Christensen. 1995. Primary production required to sustain global fisheries. Nature (374): 255-257

Pauly, D., V. Christensen, J. Dalsgaard, R. Froese and F.C. Torres Jr. 1998. Fishing down marine food webs. Science 279:860-863

Pauly, D., V. Christensen and C. Walters. 2000. Ecopath, Ecosim and Ecospace as tools for evaluating ecosystem impact of fisheries. ICES J. mar. Sci. 57:697-706

Pitcher, T.J. 1999. Rapfish, a rapid appraisal technique for fisheries, and its application to the code of conduct for responsible fisheries. FAO Fisheries Circular No. 947. 47

Piyakarnchana, T. 1989. Yield dynamics as an index of biomass shifts in the Gulf of Thailand. In: K. Sherman and L.M. Alexander, eds. Biomass yield and geography of large marine ecosystems. Westview Press, Boulder. 95-142

Piyakarnchana, T. 1999. Changing state and health of the Gulf of Thailand Large Marine Ecosystem. In: K. Sherman and Qisheng Tang, eds. Large Marine Ecosystems: Assessment, Sustainability and Management. Blackwell Science, Malden. 240-250

Polovina, J.J., G.T. Mitchum, and G.T. Evans. 1995. Decadal and basin-scale variations in mixed-layer depth and the impact on biological production in the Central and North Pacific. Deep-Sea Research 42(10):1701-1716

Pope, J.A. 1979. Stock assessment in multispecies fisheries. South China Sea Fisheries, with specieal reference to the trawl fishery in the Gulf of Thailand. South China Sea Development and Coordinating Programme. SCS/DEV/79/19. Manila. 109p

Ruddle, K. 1986. The supply of marine fish species for fermentation in Southeast Asia. Bull. Nat. Museum of Ethnology. 11(4):997-1036.

Schaefer, M.B. 1954. Some aspects of the dynamics of populations important to the management of the commercial marine fisheries. Bull. Inter. Amer. Trop. Tuna Comm. 1(2):27-56.

Sherman, K and A.M. Duda. 1999. An ecosystem approach to global assessment and management of coastal waters. Marine Ecology Progress Series. 190:271-287

Silvestre, G.T., R. Federizon, J. Munoz and D. Pauly. 1987. Overexploitation of the demersal resources of Manila Bay and adjacent areas. In: Proceedings of the 22nd Session of Indo-Pacific Fisheries Commission, Darwin, Australia, 6-26 February 1987, RAPA. Bangkok. 269-287

Sumaila, U. R. 1999. Pricing down marine food webs. In: D. Pauly, V. Christensen and L. Coelho, eds. Proceedings of the EXPO '98 Conference on Ocean Food Webs and Economic Productivity, Lisbon, 1-3 July 1998. ACP-EU Fish. Res. Rep. (5):13-15

Tiews, K. 1965. Bottom fish resources investigations in the Gulf of Thailand and an outlook on further possibilities to develop the marine fisheries of South East Asia. Archiv. FischWiss. 16(1): 67-108

Tiews, K. 1972. Fishery development and management in Thailand. Archiv. FischWiss. 24(1/3):271-300

Tiews, K. and P. Caeces-Borja 1959. On the availability of fish of the Family Leiognathidae Lacepède in Manila Bay and San Miguel Bay and their accessibility to controversial fishing gears. Philipp. J. Fish. 7(1):59-85

Walters, C. V. Christensen and D. Pauly. 1997. Structuring dynamic models of exploited ecosystems from trophic mass-balance assessments. Rev. Fish Biol. Fish. 7(2):139-172

Wyrtki, K. 1961. Physical Oceanography of the Southeast Asian Waters. Naga Report: Scientific Results of Marine Investigations of the South China Sea and the Gulf of Thailand. Vol. 2. Scripps Institution of Oceanography, La Jolla. 195p

Large Marine Ecosystems of the World
G. Hempel and K. Sherman (Editors)

15

A Review and Re-Definition of the Large Marine Ecosystems of Brazil

Werner Ekau and Bastiaan Knoppers

ABSTRACT

The geographical definition of the current two Brazilian LMEs (Northeast Brazil Shelf and Brazil Current) goes back to the early seventies. It can be argued that the limits of these LMEs were mainly a result of the paucity of information of Brazil's coastal-shelf systems, particularly those of the north-eastern and eastern geographical regions. Since then, many national and international studies have been performed and reviews of different fields of research have emerged, thus permitting a more integrated analysis of the structure and functioning of Brazil's systems. The present synthesis shows that the north-eastern and eastern marine habitats are very similar and may, within the scope of LME criteria, be considered as one unique system. In our opinion, a redefinition of the Brazilian LMEs has thus become necessary and three instead of two LMEs are proposed, namely the North Brazil Shelf, the East Brazil and the South Brazil Shelf LMEs. The East Brazil LME is located between the North and South Brazil LMEs from the Parnaíba estuary in the North to Cape São Tomé in the South (Figure 15-1-right). It thus incorporates parts of both old LMEs. The East Brazil LME is a typical oligotrophic system dominated by oceanic boundary currents and a diverse food web with low production. It corresponds basically to the geographically defined Northeast and East Brazil Sectors (Table 15-1). In contrast, the North and South Brazil Shelf LME's are more controlled by shelf topography and external material sources, and sustain less diverse food webs and higher production.

INTRODUCTION

The concept of Large Marine Ecosystems (LME) defines large regions of similar bathymetry, hydrography, productivity and trophically dependent populations as distinct ecosystems, existing within and beyond the territorial waters of the nations. Two of these LMEs have been defined to lie within the Brazilian Exclusive Economic Zone (Bakun 1993): 1) The Northeast Brazil Shelf, along the northern coast from Cape São Roque to the West, and 2) The Brazil Current from Cape São Roque to the South along the Brazilian coast up to the La Plata estuary (Figure 15-1).

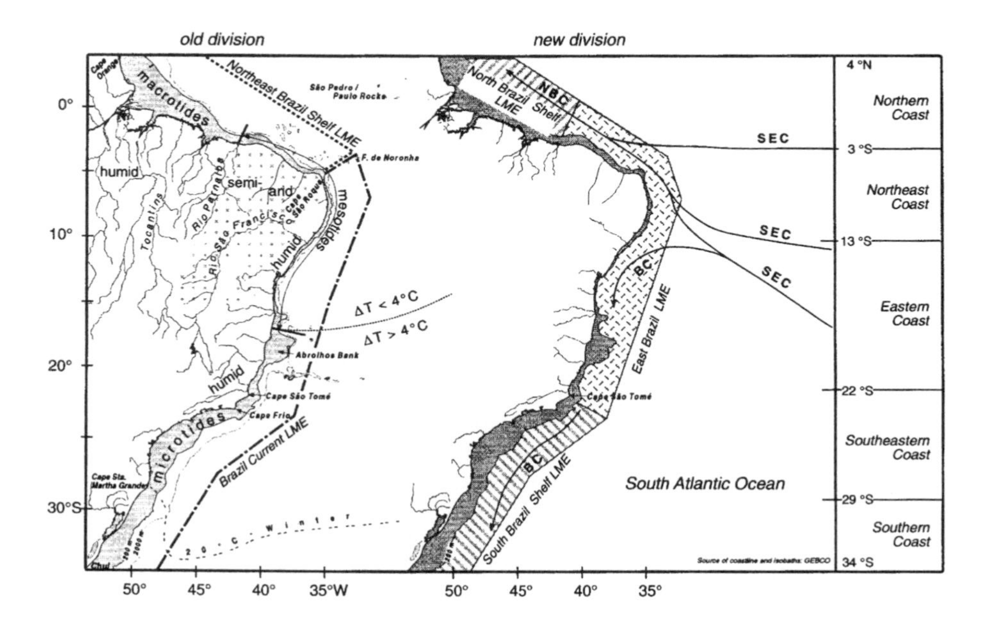

Figure 15-1: Boundaries of the Large Marine Ecosystems along the Brazilian coast: *Left*: Setting proposed by Sherman (1993); *Right*: Suggested new division of the Brazilian coastal waters based on results from Joint Oceanographic Projects II (JOPS-II) and other studies discussed in this review. The Brazilian coast and its characteristics in geomorphology, tides, currents, climate, and vegetation are shown after: Kjerfvre and Lacerda 1990; Yesaki 1974; Paiva 1997; Kelletat 1995; ANEEL 1998; Knoppers and Barbiéri, unpubl.; Dominguez and Bittencourt 1996; Diretoria de Hidrografia e Navegação, Niteroí, Brazil: Tidetables; Brandini 1990; Knoppers and Pollehne 1991; Gaeta *et al.* 1999; Smith and Demaster 1996; Ciotti *et al.* 1995; Stuhr 1996; Brandini *et al.* 1997; Knoppers and Kjerfvre B. (1999); Platt *et al.* 1983; Medeiros *et al.* 1999; Seeliger *et al.* 1997; Castro and Miranda 1998; Teixeira and Gaeta 1991. NBC = North Brazil Current. SEC = South Equatorial Current. BC = Brazil Current.

The tropical and subtropical shelf of Brazil is most productive at the Amazon shelf in the North and along the South Brazilian Bight in the South. The latter accounts for over 50 percent of Brazil´s commercial fisheries yield (Table 15-1; Valentin 1988, Bakun 1993, Ciotti *et al.* 1995; Smith and DeMaster 1996; Matsuura, 1995; Paiva 1997, Odebrecht and Garcia 1997). Less has been known on the productivity of the remaining east and north-east Brazilian shelves, governed by western boundary currents. Studies have shown that primary production, plankton biomass and fisheries yields are low on the east and northeast shelves (Aidar-Aragao *et al.* 1980; Vieira and Teixeira 1981, Matsuura 1995, Brandini *et al.* 1997, Paiva 1997, Ekau and Knoppers 1999, Knoppers *et al.* 1999a). More is known on the composition and abundance of plankton, particularly zooplankton in the Northeast, but synoptic assessments of all plankton components have been scarce in the Northeast and East.

Much of the information has been obtained from the ongoing Brazilian programmes on coastal zone management (GERCO-PNMA 1996) and the assessment of the sustainable living resources of its exclusive economic zone (REVIZEE), and from results of the recently performed oceanographic expedition JOPS-II—results which suggest a redefinition of the limits of the LMEs for Brazil. Three instead of two LMEs are proposed: The North Brazil Shelf, the East Brazil, and the South Brazil Shelf LMEs (Figure 15-1). The North and South Brazil Shelf LMEs are controlled particularly by shelf topography and external material sources, and sustain less diverse food webs and higher production. The East Brazil LME is a typical oligotrophic system dominated by oceanic boundary currents and a diverse food web with low production. It corresponds basically to the geographically defined Northeast and east Brazilian Sectors (Table 15-1). It includes the region from the Parnaíba estuary in the North to the Cape São Tomé in the South.

Five topics are envisaged in support of the redefinition of the LMEs along the Brazilian coastline: coast topography, water masses and currents, sediments, plankton characteristics, and fisheries yield and composition.

Coast topography

The Brazilian coastline stretches from Cape Orange at Lat. 4°N to Chui at Lat. 34°S (Figure 15-1). The latest estimated length by GERCO-PNMA (1996) of 7089 km is adopted here. The Brazilian coast is classified by geologic, geographic, climatic, hydrographic and sediment criteria (see reviews by Ekau and Knoppers 1999; Knoppers *et al.* 1999a). Figure 15-1 presents an overview of the climate (humid vs. arid), the drainage pattern and the tidal range along the coastal zone, the extension of the shelf and the governing oceanic boundary currents. Geographical criteria distinguish five main regions, including the North, Northeast, East,

Southeast and South. The corresponding physiographic data, fresh water input, and shelf production data are summarized in Table 15-1; physical oceanographic criteria are according to Castro and Miranda (1998). The northern, eastern and south-eastern coasts are governed by a tropical humid climate (Köppen type Af), the southern sector by a tropical warm climate (Köppen type Caf), and part of the north-eastern coast is tropical dry with typical semi-arid conditions (Köppen type Bs; Figure 15-1).

The mean annual fresh water input to the coastal regions is summarized in Table 15-1. The total input along the northern sector is in the order of 135,000 $m^3 s^{-1}$. The northern sector fresh water input includes the contribution of the Amazon River, Tocantins River and smaller sized rivers of the Amapá and western Pará coastal plains. The input at the north-eastern sub-sector, from the Parnaíba river delta to Cape São Roque, lies in the order of 2,000 $m^3 s^{-1}$. The Northeast from Cape São Roque to the São Francisco river delta has an input of about 3,400 $m^3 s^{-1}$, including the contribution of the São Francisco river itself with 2,850 $m^3 s^{-1}$. Along the entire eastern sector, the fresh water input increases gradually from North to South and amounts to a total of 3,600 $m^3 s^{-1}$. About 35 rivers discharge at the Northeast and East Brazilian coastline. The larger Jequetinhonha, Mucuri, Doce and Paraíba do Sul rivers contribute about 80 percent of this total amount. The Southeast has a fresh water input of 1,100 $m^3 s^{-1}$ and the South about 3,000 $m^3 s^{-1}$.

The geomorphologic configuration of the shelf is diverse. The largest shelves are encountered in the North, with a maximum width of about 320 km, and in the Southeast and South, up to 220 km wide. In contrast, the north-eastern and part of the eastern shelf regions are narrow, usually varying between 20 and 50 km and sporadically up to 90 km in width. The exception is found at the Abrolhos Bank, located in the middle of the eastern shelf. The Abrolhos Bank widens up to a maximum of about 220 km and lies directly within the pathway of the southward flowing Brazil Current (Figure 15-1), inducing fundamental changes and spatial variability in physical, chemical and biological features in the region (see below).

Water masses and currents

Summary descriptions of the current system and characteristic water masses off Brazil are presented in Emílsson (1959, 1961), Evans and Signorini (1985), Johns *et al.* (1990), Geyer *et al.* (1991), Peterson and Stramma (1991), and Castro and Miranda (1998). An overview of the main oceanic boundary currents affecting the north-east and east shelves is presented in Figure 15-1.

The Brazilian continental shelf lies within the direct pathway of the South Equatorial Current (SEC), which reaches the NE Brazilian shelf between 11 and 15 °S. In it originate both the North Brazil Current (NBC) and the Brazil Current

Table 15-1. Characteristic features of the Brazilian coast and shelf. Sources:1) Kjerfvre and Lacerda, 1990; 2) Yesaki, 1974; 3) Paiva, 1997; 4) Castro and Miranda, 1998; 5) ANEEL, 1998; 6) Knoppers and Barbiéri, unpubl.; 7) Knoppers and Kjerfvre,(1999)

Ekau/Knoppers: LME in Northeast and East Brazil.....Tab. 1 25.1.2001

New LME - regions	North Brazil Shelf LME			East Brazil LME									South Brazil Shelf LME				
Old LME - regions	Northeast Brazil Shelf LME					Brazil Current LME											
Geographical Regions	North			Northeast							East		Southeast				South
River Discharge 5,6,7) ($m^3 \times sec^{-1}$)	135 000			≈ 5 400							3 620		1 100				3 000
Coastline 1) (km)	1 820			1 775							1 324		1 887				620
Shelf Area (km^2)	285 383			76 844							92 842		180 754				122 443
Water Regime 4)	Amazon Estuary and NBC			NBC+SEC							BC		BC and SACW				BC and SACW
Demersal Fish Biomass 2) (t)	494 - 715 000			32 - 52 000							13 - 20 000		168 - 218 000				409 - 567 000
($t \times km^{-2}$)	1,73 - 2.51			0.42 - 0.68							0.14 - 0.22		0.93 - 1.21				3,3 - 4,6
Demersal Fisheries Prod. 3) (t)	94 000			46 000							34 000		350 000				75 000
States	Amapá	Pará	Maran-hão	Piauí	Ceará	R.G. do Norte	Paraíba	Pernam-buco	Ala-goas	Sergipe	Bahia	Espírito Santo	Rio de Janeiro	São Paulo	Paraná	Santa Catarina	R.G. do Sul
Total catch per km coastline (t)	*4,88*	*33,26*	*78,50*	*32,18*	*40,20*	*19,33*	*5,39*	*17,40*	*13,46*	*22,17*	*20,44*	*24,50*	*158,67*	*122,74*	*21,52*	*226,52*	*121,10*
Total catch composition 1980-94 (%);Taxa with more than 5% 3)																	
Mullets	7,7	8,0	9,3				14,1				7,7						5,2
Pink shrimp	8,0	11,1	11,4												7,7		
Catfish (*Arius couma*)	13,9	5,9	17,0														
Weakfishes	7,3		8,5														5,2
Catfish (*Brachyplatystoma flavicans*)	24,9	8,6															
Gillbaker sea catfish	22,7	7,1															
Angel sharks		5,2													6,3		8,6
Whitemouth croaker			8,5														16,7
Mangrove crabs		18,3		24,4			20,4			15,9	8,2						
Shrimps				33,9	7,8	7,0	13,2		38,3	27,9	14,2						
Lobsters					23,8	19,1	12,2	12,2									
Southern red snapper					19,1												
Fourwing flyingfish						10,2											
Broadband anchovy							6,4										
Swimcrabs											8,1						
Silk snapper											5,8						
Triggerfish												22,2					
Yellowtail snapper												5,8					
Atlantic seabob												13,3		5,6	27,2		
Sardine													48,5	37,7		46,6	
Chub mackerel													10,9				
White shrimp															8,9		
Drum																	10,7
Hake																	12,1
Others	15,5	35,8	45,3	41,7	49,3	63,7	41,0	87,8	61,7	56,2	56,0	58,7	40,6	56,7	49,9	53,4	39,5

(BC). A total of 12 Sv (Sverdrup units) or more of SEC give origin to the strong and continuous (mean speed 75 cm s^{-1}, max. 100 cm s) north-westward flowing NBC. From July to December, the NBC undergoes offshore retroflection at about 50° W. However, the flow pattern over the continental shelf is mostly undisturbed during the whole year (Richardson *et al.* 1994).

About 4 Sv SEC give origin to the southward flowing Brazil Current (BC). Other branches of SEC, driven towards the East Shelf sector at higher latitudes, feed waters to BC at a rate of about 5 percent per 100 km (BC, 500 m depth limit; Peterson and Stramma 1991; Gordon and Greengrove 1986). The Abrolhos Bank (Figure 15-1) forms, together with the Vitória-Trinidade-Ridge, a topographical barrier to the BC. Eddies along the eastern shelf edge and cyclonic vortices at the southern edge are generated, such as the Vitória Eddy (Schmid *et al.* 1995). The latter moves towards the shelf and creates upwelling of nutrient rich South Atlantic Central Water (SACW) (Gaeta *et al.* 1999). Coastal upwelling of SACW occurs at the shelf south of the Abrolhos Bank (Gaeta *et al.* 1999), but not as close to the shore as off Cape Frio (Figure 15-1).

The impact of the BC on the shelf system from Cape São Tomé to the South gradually diminishes in comparison to the East. The Southeast and South are subject to more intense shelf edge and wind-driven coastal upwelling of nutrient rich waters from SACW in summer (Hubold 1980a,b; Bakun and Parrish 1991; Garcia 1997). This region, also called the South Brazilian Bight, has a similar upwelling regime to that of the Southern California Bight (Bakun 1993; Bakun and Parrish 1991). The South is also influenced by the outflow of the Patos and La Plata estuaries.

In all, the high velocity of NBC on the narrow shelf (Table 15-1), the steep continental slope (Summerhayes *et al.* 1976, Martins and Coutinho 1981) and low freshwater discharge per km coastline (Table 15-1) along the northeastern coast, are the factors responsible for the nearshore impact of the oligotrophic oceanic waters. Similar conditions are found upon the shelf and nearshore in the eastern region, governed by the BC (Knoppers *et al.* 1999b), except for south of the Abrolhos Bank (Gaeta *et al.* 1999).

Sediments

The north-eastern region represents one of the few areas in the world where an open, passive margin is almost completely covered by biogenic carbonate sediments (Summerhayes *et al.* 1975). The interaction between terrigenous and carbonate sedimentation is clearly reflected by the sediment distribution. The higher the fresh water input, the lower the content of carbonate sediment. Sediments in the Northeast and in the East up to the Abrolhos bank are largely composed of carbonates (> 75 percent). In contrast, south of the Abrolhos bank up to Cape São Tomé the sediments are terrigenous, in part relict (Figure 15-2, Melo *et al.* 1975; Summerhayes *et al.* 1975). The carbonate sediments vary from coarse sand to gravel, with the latter predominating at the outer shelf. The carbonate is derived from branching algae and *Halimeda*. Corals, bryozoans and encrusting algae form the reefs of the Abrolhos region (Leão 1996). Relict terrigenous sediments cover some isolated sections of the inner and mid-shelf off Ceara in the Northeast. The São Francisco, Jequitinhonha, Doce, Mucuri and Paraíba do Sul rivers in the East, have modern terrigenous sediment off their mouths (Summerhayes *et al.* 1975, 1976; Tintelnot 1995). Most of the modern and relict terrigenous sediments are also relatively poor in organic matter and phosphorites, even where upwelling persists (Summerhayes *et al.* 1976).

Plankton characteristics

The most productive regions of the Brazilian shelf are the North (Amazon), Southeast and South, and the least productive are the Northeast and sections of the East (see review by Ekau and Knoppers 1999). The composite plot of chlorophyll *a* concentrations from the SEAWIFS satellite image (Figure 15-2) clearly indicate the lower potential for primary production in the Northeast and East. However, studies on primary productivity in these regions have been scant and lack integrated estimates of the water column. They mainly cover oceanic waters, such as the NBC in the North (Teixeira and Gaeta 1991) and the BC in the East (Aidar-Aragao *et al.* 1980; Panouse and Susini 1987). From these limited productivity data it may be derived that SEC and the western boundary currents lie within the range of 0.02 to 0.2 $gC \times m^{-2} \times d^{-1}$, which are plausible values for tropical oceanic waters (Platt *et al.* 1983). Knoppers and Pollehne (1991) estimated 0.2 $gC\ m^{-2}\ d^{-1}$ in the BC off Cape Frio, the transition zone between the east and south-east sectors.

Primary production of the East is given by Gaeta *et al.* (1999). Primary production rates were characterized by a marked spatial variability. At the Royal Charlotte Bank towards the north of the Abrolhos Bank, rates were within the range of 0.1 to 0.5 $gC \times m^{-2} \times d^{-1}$, over the Abrolhos Bank within 0.3 and 0.8 $gC \times m^{-2} \times d^{-1}$ and south of the Abrolhos Bank within 0.3 and 1.1 $gC \times m^{-2} \times d^{-1}$. Highest rates were recorded at the inner shelf where coastal upwelling waters of SACW proliferated along the bottom at 50 m depth (Gaeta *et al.* 1999). Medium to high rates were also found offshore south of the Abrolhos Bank in the cyclonic Vitória Eddy embedded in the BC. The eddy moving towards the shelf (Schmid *et al.* 1995) creates upwelling of nutrient-rich SACW enhancing productivity in sub-surface waters (Gaeta *et al.* 1999). This could represent an example of an offshore-shelf fertilization mechanism. In all, the lower southern part of the East seems to be a distinct functional region, where primary production may temporarily be sustained by three different material sources: the medium sized Paraíba do Sul and Doce rivers, coastal upwelling and likely also the Vitória Eddy, when it reaches the shelf.

The offshore phytoplankton in the Northeast is dominated by picoplankton (Cyanobacteria), an indicator for oligotrophic conditions (Bröckel and Meyerhöfer 1999). Similar dominance of picoplankton was found along the east with a gradual increase of nanoplankton from the Abrolhos Bank towards the south, showing concentrations typical for shelf and oceanic habitats (Susini-Ribeiro, 1999).

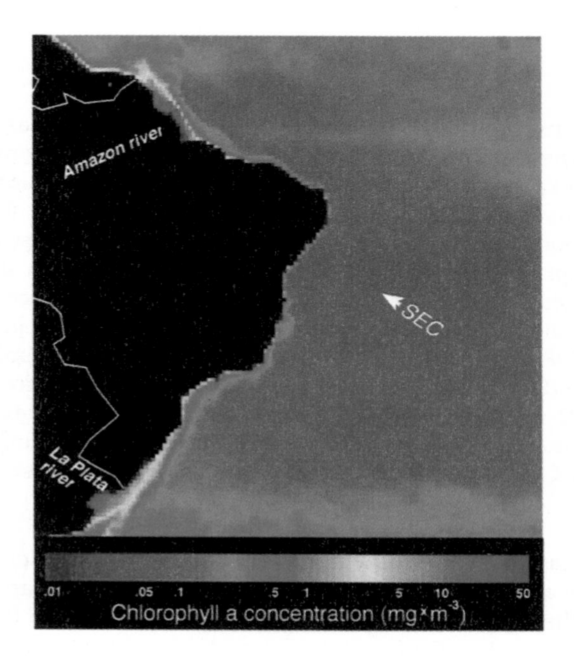

Figure 15-2: Satellite view of the Brazilian Shelf chlorophyll-a distribution, showing the Amazon and La Plata Rivers with the South Equatorial Current indicated

The macrozooplankton in the Northeast and East is dominated by calanoid and cyclopoid copepods (Ekau 1999, Neumann-Leitão *et al.* 1999). The distribution patterns of the different taxonomic groups and the microplankton as well follow the current system. The high number of oceanic indicator species such as several

Scolecithrix danae, Lucicutia flavicornis, Euchaeta marina for Tropical Water (TW) after Boltovskoy 1981, Brenning 1985) confirms the strong influence of the oligotrophic ESW, that reaches the area between the 10 and 20 m isobath, representing 5 to 15 km from shore. Neritic species dominated only in the very nearshore area, demonstrating the low fertilisation effect of the various mangrove estuaries located along the north-eastern and eastern coast, especially from Cape São Roque to the South. Nutrients and organisms are exported from these estuaries (Medeiros *et al.* 1999, Ekau *et al.* 1999, Schwamborn *et al.* 1999a), but the fertilization effect by nutrient and particulate organic matter (POM) export was found to be restricted to the 5 to 15 km band mentioned above (Schwamborn *et al.* 1999b, Medeiros 1991).

The fish larvae community in the Northeast was found to be diverse without the presence of dominant taxa. Most of the taxa could be classified as shelf and demersal (Ekau *et al.* 1999). Regional differences were found between the shelf of Pernambuco and Ceará. The nearshore-offshore gradient was less pronounced in the North, as inferred from the community analyses of the macrozooplankton (Schwamborn *et al.* 1999a) and the fish larvae (Ekau *et al.* 1999). However, both studies showed that conditions in the vicinity of the Jaguaribe river were similar to those off the Pernambuco coast.

In the southern part of the East coast, between the southern edge of Abrolhos Bank and Cape São Tomé/Cape Frio, a transition zone is observed, following in its structure the large and small-scale current pattern (Nogueira and Oliveira 1991; Bonecker *et al.* 1991, Freire 1991, Valentin and Monteiro-Ribas 1993). Oceanic and neritic species are found in alternating dominance indicating influence from a northern tropical warm water community and a southern colder upwelling community with changes in the trophic web structure (Ekau 1999).

In contrast to the situation in the Northeast and East regions, in the North, the Amazon River is the main nutrient source, and in the South, multiple terrigenous sources sustain production, such as the Patos Lagoon and La Plata systems, as well as coastal, shelf-edge and offshore (cyclonic vortices) upwelling (Smith and DeMaster 1996; Garcia 1997, Ciotti *et al.* 1995). Primary production rates in the turbid nutrient-rich Amazon mid-plume vary from 0.8 to 4.0 $gCm^{-2}d^{-1}$ (mean 2.18 $gCm^{-2}d^{-1}$), occasionally up to 8 $gC\,m^{-2}d^{-1}$ in the transition zone, and in offshore waters from 0.3 to 0.8 $gCm^{-2}d^{-1}$ (Smith and DeMaster 1996). Primary production in the Southeast and South is marked by spatial and seasonal variability. Rates are higher during summer, when upwelling of SACW is frequent. Primary production rates in the Southeast in coastal and upwelling waters off Rio de Janeiro vary between 0.3 and 1.3 $gCm^{-2}d^{-1}$ (Knoppers and Pollehne 1991), in the mid- and outer shelf waters off the coast of São Paulo (SP) and Paraná (PR) between 0.1 and 0.5 $gCm^{-2}d^{-1}$ (Brandini 1990, Brandini *et al.* 1997), and in the South off Rio Grande (RG) between 0.3 and 2.9 $gCm^{-2}d^{-1}$ (i.e. 160 $gCm^{-2}yr^{-1}$; Ciotti *et al.* 1995, Odebrecht and Garcia 1997).

Fisheries yield and composition

The oligotrophic character of the north-eastern and eastern shelf systems and their diverse food web structures are in clear contrast to the system of the Southeast Brazilian Bight (Pires-Vanin *et al.* 1993, Matsuura 1998). In the offshore ichthyoplankton community, mesopelagic species dominate at stations with water depth > 200 m (Ekau *et al.* 1996, 1999). They form part of the oligotrophic pelagic oceanic system and serve as food for the higher trophic level carnivorous fish. At nearshore stations and on the Abrolhos Bank, demersal or coastal species dominate the system. Mostly herbivorous fish are found, possibly relying on the primary production of benthic algae. Thus the food web here, with a high diversity in herbivorous fish, is in sharp contrast to the south-east Brazilian system, where diversity is low at the herbivorous level with only three species: sardine (*Sardinella brasiliensis*), anchovy (*Engraulis anchoita*) and *Maurolicus* spp. (Matsuura, pers. comm.). The mechanisms of bentho-pelagic coupling seem to be of importance in these areas for sustaining the demersal fish stocks, although this is not reflected in high fishery yields.

Fisheries yield in Brazilian waters varied widely during the last 15 years. The yield decreased from about 830,000 tonnes in the late 1980s to nearly 700,000 tonnes in 1995, and increased again to about 760,000 tonnes in 1998 (FAO 2001), a quarter of it coming from inland fisheries. Information on catch composition in marine fisheries (580,000 tonnes) is limited. More than 50 percent of the nominal catches are not given by species. About 165,000 tonnes are given as "unidentified marine fish", of which 100,000 tonnes are attributed to subsistence or recreational fisheries (FAO 2000). Fish catches in the Northeast and East regions are low with 46 000 t and 34 000 t per year (1980-94), respectively. The fishery is dominated by artisanal fishing methods. Industrial fisheries are significant only in Ceará and Espirito Santo State, on lobster and Southern red snapper in the first, and on triggerfish, Atlantic seabob and snapper in the latter case (Table 15- 1). The annual catch per km coastline varies between 5.4 and 40.2 t (Figure 15-3).

In the Northern states Amapá, Pará and Maranhão the catch varies between 4.9 and 78.5 t per km coastline, and in the Southeast and South between 121 and 227 t (except Paraná with an unusually low value). While in the Northern region the artisanal fisheries produce the high landings, especially in Maranhão, in the South the industrial fisheries are the most important branch and contribute 66 to 89 percent of the total catches. Older estimates of pelagic fish biomass indicate a major percentage for the south-eastern (22 percent) and southern (33 percent) areas, while 27 percent of the total pelagic biomass off the Brazilian coast was attributed to the Northern region. Thus only 17 percent of the pelagic fish biomass is distributed off the Northeast and East regions (Jacobsen *et al.* 1978). Hempel (1972) estimated the potential catch of pelagic fish to be 900,000 t for the whole coastline, about 150,000 t of this (17 percent) are for Northeast and East. Dias Neto and Mesquita (1988) calculated the same total pelagic resource for the entire coast. For the Northeast region, however, they estimated only 100,000 tonnes of potential

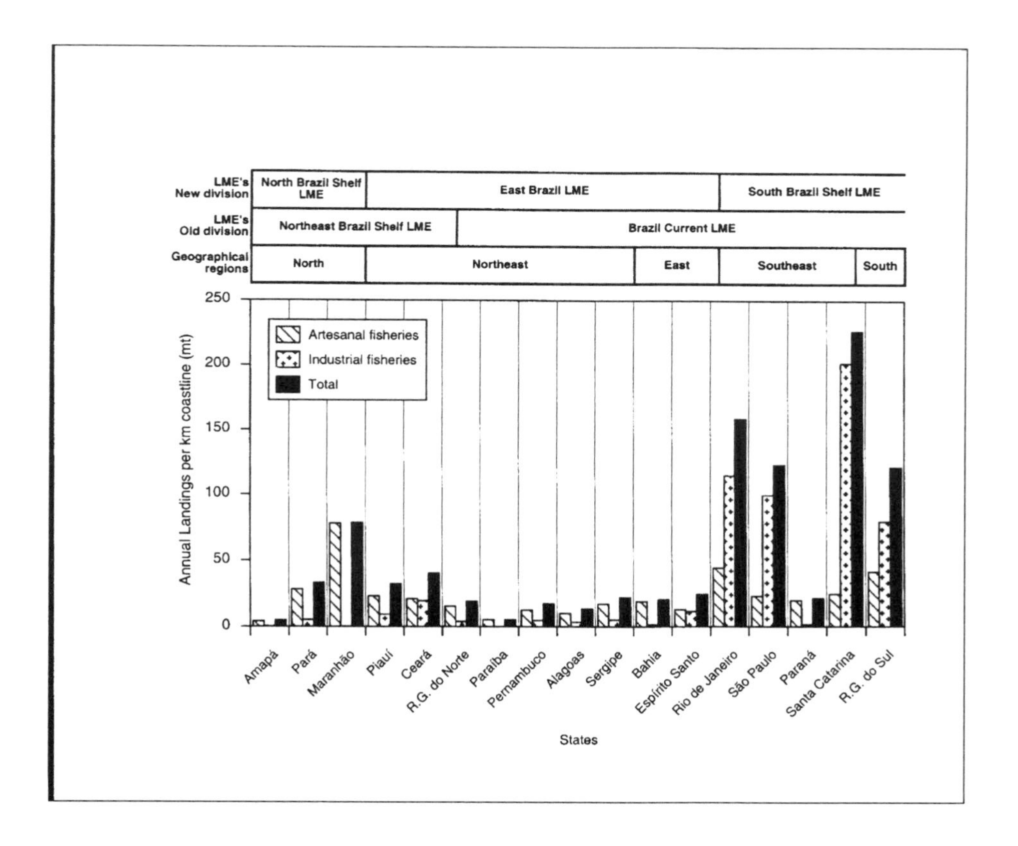

Figure 15-3. Resources of the marine and estuarine fisheries in the different states along the Brazilian coastline, separated after artesanal and industrial catches. Data represent the mean yearly catches for the period 1980 to 1994, calulated per km coastline.(After Paiva 1997)

resources, while the South was believed to have a higher potential than Hempel (1972) estimated. Taking into account the very low primary production (see above), it seems likely that pelagic fish resources in the Northeast are as low as estimated by Dias-Neto and Mesquita (1988) or even lower.

Demersal fish biomass estimates show similar amounts for the North and South with 400,000 to 700,000 tonnes each. The Southeast, however, with a much lower biomass (200,000 tonnes) accounts for over 50 percent or 350,000 t (0.9 t x km^{-2}) of Brazil's commercial fisheries yield, the South only for 10 percent (75,000 t; 3.3 t x km^{-2}), and the North for 13 percent (94,000 t; 1.7 t x km^{-2}) (Matsuura 1995, Paiva 1997). In the Northeast and East, demersal fish biomass, density and catch are at a minimum (Table 15-1). Hempel (1972) gives an estimation for the potential catches of demersal fishes of about 825,000 tonnes with a concentration in the North (250,000 t) and the Southeast and South (425,000 t). The estimates of Dias-Neto and Mesquita (1988) are more conservative reaching 800,000 t only as the upper confidence limit. The minimum is found in the Southeast, showing the pelagic dominance in that area. But also the Northeast and East have low demersal fish potential with similar values as the pelagic resources.

Fish catch composition in the Northeast and East, from Piauí to Bahia, differs significantly from the North and Southeast/South. In the Northeast/East it consists of eight main taxa, e.g. mangrove crabs, lobsters, shrimps, snappers and sardines. Mullets were of some importance in the Northeast, where an offshore fishery on minke whales was predominant during the eighties (the latter was excluded from Table 15-1 and Figure 15-3) showing some shift from Bahia to the south, mainly because of the Abrolhos Bank, where snappers, triggerfish, swimcrabs and Atlantic seabobs are dominant in the landings. Tunas are caught offshore around seamounts and banks.

In the Southeast, sardines (*Sardinella brasiliensis*) and mackerels (*Scomber japonicus*) are most important by weight, reflecting the pelagic oriented fish production in the area. In the South, demersal fish biomass is higher than in the Southeast, the important fisheries species are the demersal hake (*Merluccius hubbsi*) and sciaenids (*Umbrina canosai, Micropogonias furnieri*), but also shrimps (*Xiphopenaeus kroyeri*) and pelagic anchovies (*Engraulis anchoita*), and sardines (*Sardinella brasiliensis*) (Paiva 1997, Haimovici *et al.* 1997). Mullets, catfishes (*Arius* spp.), crabs (*Ucides cordatus*) and shrimps (*Penaeus brasiliensis*) are harvested in the North (Paiva 1997).

Information on the exploitation of Brazilian fish stocks is not available for all areas and species. Fisheries along the Brazilian coastline are managed by 4 regional departments of the Brazilian Institute of Environment and Natural Renewable Resources (IBAMA): CEPNORTE is responsible for the northern coast from Cape Orange to Rio Paranaíba; CEPENE is responsible for the north-eastern and eastern coast from Rio Paranaíba to north of Abrolhos, CEPLESTE covers the area of Abrolhos until Cape Frio and CEPSUL the coastline from Cape Frio to the

Uruguayan border (Chuí). Formerly, several working groups have been responsible for providing advice on the management of fisheries in Brazil, but they have been mostly inactive in recent years (FAO 1997). Information on the level of exploitation is available only for few species. The Brazilian sardine (*Sardinella brasiliensis*) has been overfished for several years and has also been affected by some adverse environmental conditions. Catches declined from up to 230,000 t per year in the 1980s to 32,000 t by 1990. Although catches have recovered to 82,000 t in 1998 (FAO 2000), the stock still seems to be in poor condition (FAO 1997). Haimovici (1998) investigated the fishery in the South from 1975 to 1994. He stated that 41 percent of the catches came from depleted or overexploited stocks, another 24 percent from heavily exploited species. Catches of tunas and other large pelagics are increasing slowly (FAO 1997). The most important offshore species is the skipjack tuna with an annual catch of 24,000 tonnes (FAO 2000) taken mainly along the shelf edge from Rio de Janeiro to Rio Grande do Sul (BDT 2001). The MSY was estimated as 27,000 tonnes by Jablonski and Matsuura (1985) and Alencar-Vilela and Castello (1993) for that area. Other large pelagics are found mainly in the offshore areas off Northeast and East Brazil, contributing c. 18,000 tonnes to the annual landing in Brazil (FAO 2000).

REFERENCES

Aidar-Aragão, E, C. Teixeira, and A.A.H. Vieira. 1980. Produção primária e concentração de clorofila-a na costa brasileira (Lat. 22°31'S - Long. 41°52'W a Lat. 28°43'S - Long. 47°57'W). Bol. Inst. oceanogr. S. Paulo. 29:9-14

Alencar-Vilela, M.J. and J.P. Castello. 1993. Population dynamics of the skipjack tuna (*Katsuwonus pelamis*) fishery off south and southern Brazil in the period 1980-1986. Frente-Marit 14:111-124

ANEEL, Agencia Nacional de Energia Eletrica. 1998. Internet source http:www.aneel.gov.br

Bakun, A. 1993. The California Current, Benguela Current, and Southwestern Atlantic shelf ecosystems: A comparative approach to identifying factors regulating biomass yields. In: Sherman, K., L.M. Alexander, and B.D. Gold, eds. Large Marine Ecosystems: Stress, Mitigation and Sustainability of Large Marine Ecosystems. AAAS Publication, 92-39S. 199-224.

Bakun, A. and R.H. Parrish. 1991. Comparative studies of coastal pelagic fish reproductive habitats: the anchovy (*Engraulis anchoita*) of the southwestern Atlantic. ICES J. Mar. Sci., 48:343-361

Boltovskoy, D. 1981. Atlas del zooplancton del Atlantico Sudoccidental. Publicación especial del INIDEP, Mar del Plata, Argentina. 936p

Bonecker, S.L.C., A.C.T. Bonecker, C.R. Nogueira, and M.V. Reynier. 1991. Zooplâncton do litoral norte do Espírito Santo - Brasil: estrutura espaço-temporal. Anais do IV Encontro Brasileiro do Plâncton, Recife, UFPE: 369-391

Brandini, F.P. 1990: Produção primária e características fotosintéticas do fitoplâncton na região sueste do Brasil. Bolm. Inst. oceanogr., S. Paulo, 38(2):147-159.

Brandini, F.P., R.M. Lopes, K.S. Gutseit, H.L. Spach, and R. Sassi. 1997. Planctonologia na plataforma continental do Brasil - Diagnose e revisão bibliográfica. Ministério do Meio Ambiente, dos resursos hídricos e da Amazônia legal -MMA-, Comissão Interministerial para os recursos do Mar -CIRM-, Fundação de Estudos do Mar -FEMAR-: 196p

Brenning, U. 1985. Structure and development of calanoid populations (*Crustacea, Copepoda*) in the upwelling regions off North West and South West Africa. Beitr. Meeresk. 52:3-33

Bröckel, K.v. and M. Meyerhöfer. 1999. Impact of the rocks of São Pedro and São Paulo upon the quantity and quality of suspended particulate organic matter. Arch. Fish. Mar. Res. 47(2/3):223-238

BDT. 2001. Necton: Grandes Peixes Pelagicos. Base de Dados Tropical, Fundação André Tosello, Campinas, São Paulo, Brazil. www.bdt.org.br

Castro, B.M. and L.B. Miranda. 1998. Physical oceanography of the western Atlantic continental shelf located between 4°N and 34°S. In: Robinson, A.R. and K.H. Brink, eds. The Sea. New York, John Wiley & Sons. Vol. 11, 209-252

Ciotti, A.M, C. Odebrecht, G. Fillmann, and O.O. Möller Jr. 1995. Freshwater outflow and subtropical convergence influence on phytoplankton biomass on the southern Brazilian continental shelf. Cont. Shelf Res. 15(14):1737-1756

Dias-Neto, J. and J.X. Mesquita. 1988. Potencialidade e explotação dos recursos pesqueiros do Brasil. Ciência e Cultura, São Paulo. 40(5):427-441

Diretoria de Hidrografia e Navegação, Niteroí, Brazil: Tidetables

Dominguez, J.M.L. and A. C. Bittencourt. 1996. Regional Assessment of Long-term trends of coastal erosion in northeastern Brazil. An. Acad. bras. Ci., 68(3):355-371

Ekau, W. 1999. Topographical and hydrographical impacts on macrozooplankton community structure in the Abrolhos Bank region, East Brazil. Arch. Fish. Mar. Res. 47(2/3):307-320

Ekau, W. and B. Knoppers. 1999. An introduction to the pelagic system of the North-East and East Brazilian shelf. Arch. Fish. Mar. Res. 47(2/3):113-132

Ekau, W., Y. Matsuura and S. Torbohm-Albrecht. 1996. Diversity and distribution of macrozooplankton in the eastern continental shelf waters off East Brazil. In: Ekau, W. and B.A. Knoppers, eds. Sedimentation processes and productivity in the continental shelf waters off east and northeast Brazil. Cruise Report and first results of the Brazilian German project JOPS-II (Joint Oceanographic Projects). Center for Tropical Marine Ecology, Bremen: 139-147

Ekau, W., P. Westhaus-Ekau, and C. Medeiros. 1999. Large scale distribution of fish larvae in the Continental Shelf waters off NE-Brazil. Arch. Fish. Mar. Res. 47(2/3):183-200

Emílsson, I. 1959. Alguns aspectos físicos e químicos das águas marinhas brasileiras. Ciência e Cultura, S. Paulo, 11(2):44-54

Emílsson, I. 1961. The shelf and coastal waters off southern Brazil. Bolm. Inst. oceanogr., S. Paulo 11(2):101-112.

Evans, D.L. and S.R. Signorini. 1985. Vertical structure of the Brazil Current. Nature 315:48-50

FAO, 1997. Review of the state of world fishery resources: Marine fisheries. 6. Southwest Atlantic. FAO Fisheries Circular No. 920 FIRM/C920, Rome

FAO. 2000. Fishery statistics. FAO yearbook vol 86/1 1998, FAO, Rome

FAO. 2001. Summary tables of fishery statistics. http://www.fao.org/fi/statist/summtab/pr_a1c.asp

Freire, A.S. 1991. Variação espaço-temporal do zooplâncton e das espécies de Euphausiacea (Crustacea) ao largo da Costa L'este do Brasil (23 - 18 °S, 41 - 38°W). Dissertação de Mestrado, Instituto Oceanográfico, Universidade de São Paulo. 75p

Gaeta, S.A., J.A. Lorenzetti, L.B. Miranda, S.M.M. Susini-Ribeiro, M. Pompeu, and C.E.S. De Araujo. 1999. The Victoria Eddy and its relation to the phytoplankton biomass and primary productivity during the austral fall of 1995. Arch. Fish. Mar. Res. 47(2/3): 253-270

Garcia, C.A.E. 1997. Water masses and transports. In: Seeliger, U., C. Odebrecht, and J.P. Castello, eds. Subtropical Convergence Environments. Springer Verlag, Heidelberg. 308p

GERCO-PNMA. 1996. Macrodiagnóstico da zona costeira do Brasil na escala da união. - Gerenciamento Costeiro, Programa Nacional do Meio Ambiente, MMA (Ministério do Meio Ambiente, UFRJ, SUJD, LAGEJ, Brasilia. 280p

Geyer, W.R.; R.C. Beardsley, J. Candela, B.M. Castro, R.V. Legeckis, S.J. Lentz, R. Limeburner, L.B. Miranda, J.H. Trowbridge, 1991. The physical oceanography of the Amazon outflow. Oceanography 4(1):8-14

Gordon, A.L. and C.L. Greengrove. 1986. Geostrophic circulation of the Brazil-Falkland confluence. Deep Sea Res. 3:573-585

Haimovici, M., J.P. Castello, C.M. Vooren. 1997. Fisheries. In: Seeliger, U., C. Odebrecht, J.P. Castello, eds. Subtropical Convergence Environments. Springer Verlag, Heidelberg. 308p

Haimovici,M. 1998. Present state and perspectives for the southern Brazil shelf demersal fisheries. Fish. Manage. Ecol., 5(4):277-289

Hempel, G. 1972. Southwest Atlantic. In: Gulland, J.A., ed. The Fish Resources of the Ocean. Fishing News Ltd., Surrey, England. 146-152

Hubold, G. 1980a. Hydrography and plankton off southern Brazil and Rio de la Plata, August - November 1977. Atlântica 4:1-22

Hubold, G. 1980b. Second report on hydrography and plankton off southern Brazil and Rio de la Plata, Autumn cruise: April - June 1978. Atlântica 4:23-42

Jablonski, S. and Y. Matsuura. 1985. Estimate of exploitation rates and population size of skipjack tuna off the southeastern coast of Brazil. Bol. Inst. Oceanogr. São-Paulo 33(1):29-38

Jacobsen, K.H., H.v. Westernhagen, and M. Zurck. 1978. Die Küstenfischerei in Brasilien und Möglichkeiten ihrer Förderung insbes. im Nordosten. Gutachten, unpubl. 39p

Johns, W.E., T.N. Lee, F.A. Schott, R.J. Zantopp, and R.H. Evans. 1990. The North Brazil Current retroflection: seasonal structure and eddy variability. J. Geophys. Res., 95(C12):22103-22120

Kelletat, D.H. 1995. Atlas of coastal Geomorphology and Zonality. J. Coastal Res., Spec. Iss. 13. 286p

Kjerfvre, B. and L.D. Lacerda. 1990. Mangroves of Brazil. In: Lacerda, L.D., ed. Conservation and sustainable utilization of mangrove forests in Latin America and Africa regions. Part I - Latin America. ITTO/ISMA Project PD 114/90:245-271

Knoppers, B and B. Kjerfvre. 1999. Coastal lagoons of southeast Brazil: Physical and biogeochemical characteristics. In: Perillo, G.M.E, M.C. Piccolo and M. Pino-Quivira, eds. Estuaries of South America. Springer Verlag, Berlin. 35-62

Knoppers, B., W. Ekau, and A.G. Figueiredo. 1999a. The coast and shelf of east and northeast Brazil and material transport. Geo-Marine Letters 19:171-178

Knoppers, B., M. Meyerhöfer, E. Marone, J. Dutz, R. Lopes, T. Leipe, and R. Camargo. 1999b. Compartments of the pelagic system and material exchange at the Abrolhos Bank coral reefs, Brazil. Arch. Fish. Mar. Res. 47(2/3):285-306

Knoppers, B.A. and F. Pollehne. 1991. The transport of carbon, nitrogen and heavy metals to the offshore sediments by plankton sedimentation. In: Ekau, W., ed. JOPS 90/91 Cruise Report, Alfred Wegener Institute for Polar and Marine Research, Bremerhaven. 25-30

Leão, Z.M.A.N. 1996. The coral reefs of Bahia: Morphology, destruction and the major environmental impacts. An. Acad. bras. Ci. 68(3):439-452.

Martins, L.R. and P.N. Coutinho. 1981. The Brazilian continental margin. Earth Science Review 17:87-107

Matsuura, Y. 1995. Exploração pesqueira. In: Os ecossistemas brasileiras e os principais macrovetores de desenvolvimento: subsídios ao planejamento da gestão ambiental. Programma nacional do meio ambiente, Brasília: 77-89.

Matsuura, Y. 1998. Brazilian sardine (*Sardinella brasiliensis*) spawning in the southeast Brazilian Bight over the period 1976-1993. Rev. bras. oceanogr. 46(1):33-43

Medeiros, C.Q. 1991. Circulation and mixing processes in the Itamaracá estuarine system, Brazil. Ph.D. thesis. Feitosa, F.A.N., M.L. Koening. 1999. Hydrology and phytoplankton biomass of the northeastern Brazilian waters. Arch. Fish. Mar. Res. 47(2/3):133-151

Melo, U., C.P. Summerhayes and J.P. Ellis. 1975. Salvador to Vitória, Southeastern Brazil. In: Upper Continental Margin sedimentation off Brazil. Stuttgart: Contribution to Sedimentology 4: 78-116

Neumann-Leitão, S., L.M.deO. Gusmão, T.A. Silva, D.A. do Nascimento-Vieira, and A.P. Silva. 1999. Mesoplankton biomass and diversity of coastal and oceanic waters off Northeastern Brazil. Arch. Fish. Mar. Res. 47(2/3):153-165

Nogueira, C.R. and S.R. Oliveira Jr. 1991. Siphonophora from the coast of Brazil (17°Sto 24°S). Bolm Inst. oceanogr., S. Paulo 39(1):61-69

Odebrecht, C. and V.M.T. Garcia. 1997. Phytoplankton. In: Seeliger, U., C. Odebrecht and J.P. Castello, eds. Subtropical Convergence Environments. Springer Verlag, Heidelberg. 105-109

Paiva, M.P. (ed). 1997. Recursos pesqueiros estuarinos e marinhos do Brasil. Avaliação do potencial sustentável de recursos vivos na zona econômica exclusiva. Universidade Federal do Ceará, Fortaleza. 286p

Panouse, M. and S.M. Susini. 1987. Production primaire du phytoplancton. In: Guille, A. and J.M. Ramos, eds. Terres Australis et Antarctiques Françaises. Mission de recherche. Les rapports des campagnes a la mer a bord du "Marion Dufresne." 91-107

Peterson, R.G. and L. Stramma. 1991. Upper-level circulation in the South Atlantic Ocean. Prog. Oceanogr. 26:1-73.

Pires-Vanin, A.M.S, C.L.DE.B. Rossi-Wongtschowski, E. Aidar, H. de S.L. Mesquita, L.S.H. Soares, M. Katsuragawa and Y. Matsuura. 1993. Estrutura e função do ecosistema de plataforma continental do Atlântico Sul brasileiro: síntese dos resultados. Publção esp. Inst. oceanogr., S. Paulo 10: 217-231

Platt, T., D. V. Subha Rao, B. Irwin. 1983. Photosynthesis of picoplankton in the oligotrophic ocean. Nature 301:702-704.

Richardson, P., G.E. Hufford and R. Limeburner. 1994. North Brazil current retroflection eddies. J. Geophysical Res. 99:5081-5093

Schmid, C., H. Schäfer, G. Podestá and W. Zenk. 1995. The Vitória Eddy and its relation to the Brazil Current. J. Phys. Oceanogr. 25(11):2532-2546

Schwamborn, R., W. Ekau, A. Pinto, T. Silva and U. Saint-Paul. 1999a. The contribution of estuarine decapod larvae to marine macrozooplankton communities in Northeast Brazil. Arch. Fish. Mar. Res. 47(2/3):167-182

Schwamborn, R., M. Voss, W. Ekau and U. Saint-Paul. 1999b. Stable isotope composition of particulate organic matter and zooplankton in northeast Brazilian shelf waters. Arch. Fish. Mar. Res. 47(2/3):201-222

Seeliger, U., C. Odebrecht and J.P. Castello, eds. 1997. Subtropical Convergence Environments. Springer Verlag, Heidelberg. 308p

Smith, W.O. and D.J. Demaster. 1996. Phytoplankton biomass and productivity in the Amazon river plume: correlation with seasonal river discharge. Cont. Shelf Res. 16(3): 291-319

Stuhr, A. 1996. Phytoplankton production measurements. In: Ekau, W. and B.A. Knoppers, eds. Sedimentation processes and productivity in the continental shelf waters off east and Northeast Brazil. Cruise Report and first results of the Brazilian German project JOPS-II (Joint Oceanographic Projects). Center for Tropical Marine Ecology, Bremen. 83-85.

Summerhayes, C.P., P.N. Coutinho, A.M.C. França, J.P. Ellis. 1975. Salvador to Fortaleza, Northeastern Brazil. In: Upper Continental Margin Sedimentation off Brazil. Stuttgart: Contribution to Sedimentology 4: 44-77

Summerhayes, C.P., U. De Melo and H.T. Barretto. 1976. The influence of upwelling on suspended matter and shelf sediments off southeastern Brazil. J. Sedimentary Petrology 46(4):819-828

Susini-Ribeiro, S.M.M. 1999. Biomass distribution of pico-, nano- and microplankton on the continental shelf of Abrolhos, East Brazil. Arch. Fish. Mar. Res. 47(2/3):271-284

Teixeira, C. and S.A. Gaeta. 1991. Contribution of picoplankton to primary production in estuarine, coastal and equatorial waters of Brazil. Hydrobiologia 209:117-122

Tintelnot, M. 1995. Transport and deposition of fine-grained sediments on the Brazilian continental shelf as revealed by clay mineral distribution. Unpublished PhD thesis. Heidelberg. University of Heidelberg. 294p

Valentin, J.L. 1988. A dinâmica do plâncton na ressurgência de Cabo Frio-RJ. Memórias do III Encrontro Brasileiro de Plâncton. Caiobá, 5-9 December 1988. 25-35.

Valentin, J.L. W.M. Monteiro-Ribas. 1993. Zooplankton community structure on the east-southeast Brazilian continental shelf (18-23°S latitude). Cont. Shelf Res. 13(4):407-424

Vieira, A.A.H. and C. Teixeira. 1981. Excreção de matéria orgânica dissolvida por populações fitoplanctônicas da costa leste e sudeste do Brasil. Bolm. Inst. oceanogr., S. Paulo, 30(1): 9-25.

Yesaki, M. 1974. Os recursos de peixes de arrasto ao largo da costa do Brasil. Programa de Pesquisa e Desenvolvimento Pesqueiro/ Série Documentos Técnicos, Rio de Janeiro, 8: I-II and 1-47.

IV
Mapping Natural Ocean Regions and Large Marine Ecosystems

Large Marine Ecosystems of the World
G. Hempel and K. Sherman (Editors)

16

Mapping Fisheries onto Marine Ecosystems for Regional, Oceanic and Global Integrations

Reg Watson, Daniel Pauly, Villy Christensen, Rainer Froese, Alan Longhurst, Trevor Platt, Shubha Sathyendranath, Kenneth Sherman, John O'Reilly, and Peter Celone

ABSTRACT

Research on ecosystem-based fisheries management, marine biodiversity conservation, and other marine fields requires appropriate maps of the major natural regions of the oceans, and their ecosystems. A global ocean classification system proposed by T. Platt and S. Sathyendranath and implemented by A.R. Longhurst, defined largely by physical parameters that subdivide the oceans into four 'biomes' and 57 'biogeochemical provinces' (BGCPs), is merged with the system of 64 Large Marine Ecosystems (LMEs) identified by K. Sherman and colleagues as transboundary geographic coastal and watershed units. This arrangement enhances each of the systems, and renders them mutually compatible. LMEs are ecologically defined to serve as a framework for the assessment and management of coastal fisheries and environments including watersheds, while the BGCPs have physical definitions, including borders defined by natural features, and extend over open ocean regions. The combined mapping will, for example, allow the computation of GIS-derived properties such as temperature, primary production, and their analysis in relation to fishery abundance data for any study area in the combined system. A further useful aspect of the integration is that it allows for the quantification, even within the EEZs of various countries, of the distribution of marine features (e.g. primary production, coral reef areas) so far not straightforwardly associated with different coastal states. Applications to shelf, coral reef and oceanic fisheries, and to the mapping of marine biodiversity are briefly discussed.

INTRODUCTION

There is broad consensus in the scientific community that fisheries management should be ecosystem-based, but very little agreement as to what this means (NRC 1999). Also, there is a need to analyze biodiversity data at larger scales than have generally been done so far, as demonstrated by e.g., Sala *et al.* (2000), for terrestrial and freshwater biomes. Clearly, when dealing with such complex issues, the first task, as in all science-based approaches to a problem, is to define the object(s) of concern, and to develop a consistent method to show how these objects are interrelated. Here, the objects are the marine ecosystems within which fisheries and biodiversity are to be analyzed, and marine life in general, is embedded.

Fortunately, reaching a consensus on the classification of marine ecosystems may be relatively easy, given the compatibility, originally noted in Pauly *et al.* (2000), of two classification schemes proposed in recent years. Both of these integrate an enormous amount of empirical data, and are sensitive to previous analyses of marine ecology. The two schemes are: (1) the global system of 57 'biogeochemical provinces' (BGCPs) developed by Platt and Sathyendranath (1988, 1993), Platt *et al.* (1991, 1992), Sathyendranath *et al.* (1989), Sathyendranath and Platt (1993), implemented by Longhurst (1995, 1998), and defined at scales appropriate for understanding physical forcing of ocean primary production and related processes; and (2) the 64 coastal Large Marine Ecosystems (LMEs), incrementally defined by Sherman and co-workers (Sherman and Alexander, 1986, 1989; Sherman *et al.* 1990, 1993; Sherman and Duda 2001; IOC 2002), whose ecologically-based definition, size, coastal locations and ecologically-based definitions make them particularly suitable for addressing management issues, notably those pertaining to fisheries on continental shelves, and coastal area management (Sherman and Duda 1999a; 1999b, Duda and Sherman 2002).

After reviewing selected features of these two schemes, we describe how the 64 Large Marine Ecosystems (LMEs) relate to the 4 biomes and their 57 biogeochemical sub-provinces (BGCPs). The joint classification which then emerges is presented in the form of a spatial hierarchy, and is presented as maps, each emphasizing a key feature of the classification. Overall, the integrated scheme allows for explicit consideration of different scales, as discussed e.g. by Levin (1990).

THE BIOMES

In this outline of geographic areas, the four biomes are the largest units. In Figure 16-1 biomes are defined by the dominant oceanographic process that determine the vertical density structure of the water column, which is what principally constrains the vertical flux of nutrients from the interior of the ocean. In the **Polar biome**, vertical density structure is largely determined by the flux of fresh or low-salinity water derived from ice-melt each spring and which forms a prominent halocline in polar and sub-polar oceans. In oceanographic terms, this occurs in each hemisphere polewards of the Oceanic Polar Front, whose location in each ocean is determined by the characteristic circulation of each. Though looming large on Mercator maps, the

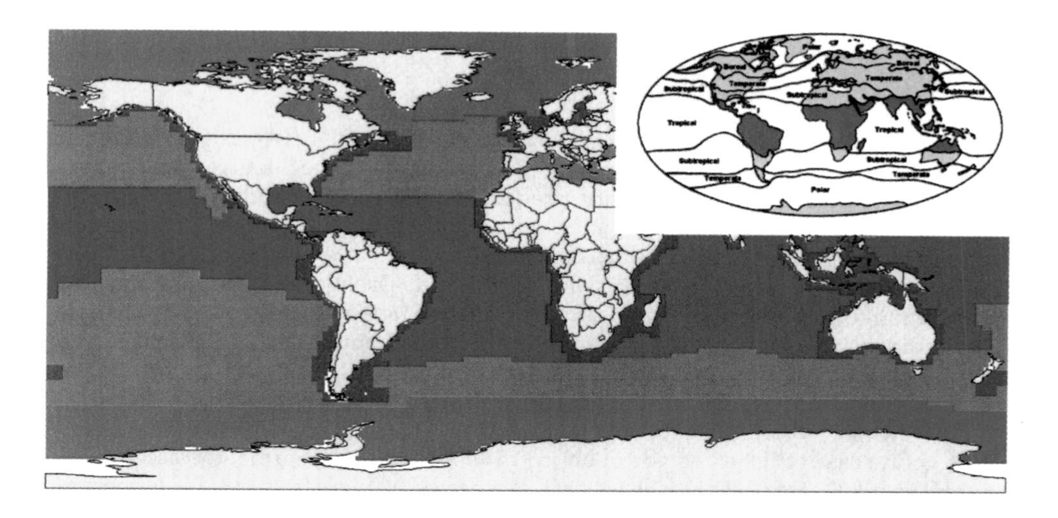

Figure 16-1. Map of the 4 world ocean's biomes: Polar (aqua), Westerlies (green), Trade Winds (red) and Coastal Boundary (blue). Biomes are the largest category in the proposed classification of the world oceans. Note its overall similarity to a conventional map of the atmospheric climate (inset, adapted from Anon. 1991)

Polar biome occupies only about 6 percent of the ocean's surface. Between the Polar fronts and the Subtropical Convergence in each ocean lies the **Westerlies biome**. Here, large seasonal differences in mixed-layer depth are forced by seasonality in surface irradiance and wind stress. Biological processes consequently may have sufficiently strong seasonality so that a spring bloom characterizes the plankton calendar. Across the equatorial regions, between the boreal and austral Subtropical convergences lies the **Trade-wind biome**. Here, the conjunction between low values for the Coriolis parameter, a strong density gradient across the permanent pycnocline and weak seasonality in both wind stress and surface irradiance result in relatively uniform levels of primary production throughout the year. Upper continental slopes, continental shelves and marginal seas comprise the **Coastal Boundary biome**. This is constrained between the coastline itself and (usually) the oceanographic front characteristically found at the shelf-edge. The single generalization that characterizes this biome is that nutrient flux in the water column is forced by a great variety of processes, including, for example: coastal upwelling, tidal friction, and fresh-water outflow from river mouths. In the partitions discussed above, subdivision of this biome into provinces was not carried as far as might be useful for some purposes. One of the objectives of the present study is to do just that, to clarify the linkages among biogeochemical provinces and Large Marine Ecosystems.

The boundaries between the biomes vary seasonally and between years, as can readily be inferred from satellite images. Dynamic boundaries that respond to this variability are discussed for primary production and related studies by Platt and Sathyendranath, (1999). However, such dynamic schemes are neither practical nor necessarily useful for biodiversity and fisheries studies. For example, one of the tasks facing biodiversity investigations is the creation of global maps documenting the distribution of hundreds of thousands of marine species. Requiring that these distributions are assigned to habitats having variable boundaries would make even simple, first-order assignments of species extremely difficult and postpone the delivery of products whose need is already keenly felt by students of biodiversity.

Thus, in the case of fishes, of which about 15,000 species are marine, the assignment within FishBase (see www.fishbase.org) of species to climate type (as defined in Figure 16-1, inset), required us to distinguish tropical from non-tropical species (see Pauly 1998), and this task alone required several person-months to complete. Moreover, there are numerous types of floral or faunal assemblages whose location does not vary, though their habitat is part of, or affected by a surrounding or overlying pelagic ecosystem. Thus, the reef fishes of the Galapagos do not change their location when an El Niño event strikes the archipelago. Rather, it is their abundance which is affected (Grove 1985, Grove and Lavenberg 1997). A similar argument applies to benthic communities, whose boundaries will tend to reflect the long-term average location of the boundaries of the overlying pelagic systems, rather than tracking their changing location (Ekman 1967).

The ecosystem classification scheme presented here is thus deliberately fixed in space. On the other hand, we anticipate that its use by various authors will quickly lead to the identification and quantification of changes in species compositions, thus reintroducing the dynamic element required at various spatial and temporal scales (Levin 1990).

Oceanographic conditions within the four biomes are obviously not uniform, and each can be subdivided further using the same set of principles as those that determined the biomes themselves. For example, in both the westerlies and trades biomes there are definable ocean regions where heavy tropical rainfall or excessive continental fresh water runoff lead to the existence of a quasi-permanent low salinity 'barrier- layer' occupying the upper portion of the thermally-stratified surface layer. This has important biological consequences and suggests that these regions should be recognized as individual partitions. Using such methods, based on close examination of regional physical oceanography, the four primary biomes can be further partitioned into 57 provinces, the BGCPs discussed above. Figure 16-2 illustrates these BGCPs, as defined by Longhurst *et al.* (1995). This schema has been used to stratify the world ocean in two studies, pertaining to the global distribution of primary production (Longhurst *et al.* 1995, Platt and Sathyendranath 1999, Pauly 1999) and tuna catches (Fonteneau 1998).

BIOGEOCHEMICAL PROVINCES

The next largest units in the hierarchy are the 57 Biogeochemical Provinces (BGCPs), based on Platt and Sathyendranath (1988) who proposed this recognition of natural regions of the ocean, each region having characteristic physical forcing to which there is a characteristic response of the pelagic ecosystem. These regions are dynamic biogeochemical provinces because their boundaries respond to annual and seasonal changes in physical forcing and are 'biogeochemical' because, within each, the biota respond to those characteristic geochemical processes which determine nutrient delivery to the euphotic zone. The concept has been used to partition both global and basin-scale analyses of primary productivity, though the 'dynamic' boundary aspect of the system remains to be exploited. So far, most applications of the partition have assumed that boundaries between provinces were fixed at locations representing average conditions.

The central principle in locating boundaries between provinces is that of the critical depth model of Sverdrup (1953), which remains the most useful formulation to relate phytoplankton growth to surface illumination, and to the vertical density structure of the water column. It successfully predicts, for example, the timing of the North Atlantic spring bloom. A proposed partition of the North Atlantic into 18 BGCPs (Platt *et al.* 1995) was followed by a partition of all oceans and adjacent seas into 57 provinces (Figure 16-2, Longhurst *et al.* 1995, Longhurst 1998).

After examination of 26,000 archived chlorophyll profiles to determine Gaussian parameters describing the regional/seasonal characteristic profiles, surface chlorophyll from 43,000 grid-points from monthly Coastal Zone Colour Scanner images, and about 23,000 monthly mean mixed-layer depths, together with other oceanographic variables, a two-level partition was created to adequately represent regional differences in the expression of the Sverdrup model. The first partition is into the 4 biomes, following the usage of this term by terrestrial ecologists to mean a region of relatively uniform dominant vegetation type, with its associated flora and fauna: grassland, tundra, steppe, humid forest and so on (Golley 1993). Secondly, these biomes are each partitioned into a number of regional entities, the biogeochemical provinces. It is at this geographic areal level that the LME's exist.

Table 16-1. Countries Participating in GEF/Large Marine Ecosystem Projects

Approved GEF Projects

LME	Countries
Gulf of Guinea (6)	Benin, Cameroon, Côte d'Ivoire, Ghana, Nigeria, Togo[a]
Yellow Sea (2)	China, Korea
Patagonia Shelf/Maritime Front (2)	Argentina, Uruguay
Baltic (9)	Denmark, Estonia, Finland, Germany, Latvia, Lithuania, Poland, Russia, Sweden
Benguela Current (3)	Angola,[b] Namibia, South Africa[b]
South China Sea (7)	Cambodia, China, Indonesia, Malaysia, Philippines, Thailand, Vietnam
Black Sea (6)	Bulgaria, Georgia, Romania, Russian Federation, Turkey,[b] Ukraine
Mediterranean (19)	Albania, Algeria, Bosnia-Herzegovina, Croatia, Egypt,[b] France, Greece, Israel, Italy, Lebanon, Libya, Morocco,[b] Slovenia, Spain, Syria, Tunisia, Turkey, Yugoslavia, Portugal
Red Sea (7)	Djibouti, Egypt, Jordan, Saudi Arabia, Somalia, Sudan, Yemen
Western Pacific Warm Water Pool-SIDS (13)	Cook Islands, Micronesia, Fuji, Kiribati, Marshall Islands, Nauru, Niue, Papua New Guinea, Samoa, Solomon Islands, Tonga, Tuvalu, Vanuatu
	Total number of countries: 72[c]

GEF Projects in the Preparation Stage

LME	Countries
Canary Current (7)	Cape Verde, Gambia, Guinea,[b] Guinea-Bissau,[b] Mauritania, Morocco, Senegal
Bay of Bengal (8)	Bangladesh, India, Indonesia, Malaysia, Maldives, Myanmar, Sri Lanka, Thailand
Humboldt Current (2)	Chile, Peru
Guinea Current (16)	Angola, Benin, Cameroon, Congo, Democratic Republic of the Congo, Côte d'Ivoire, Gabon, Ghana, Equatorial Guinea, Guinea, Guinea-Bissau, Liberia, Nigeria, Sao Tome and Principe, Sierra Leone, Togo
Gulf of Mexico (3)	Cuba,[b] Mexico,[b] United States
Agulhus/Somali Currents (8)	Comoros, Kenya, Madagascar, Mauritius, Mozambique, Seychelles, South Africa, Tanzania
Caribbean LME (23)	Antigua and Barbuda, The Bahamas, Barbados, Belize, Columbia, Costa Rica, Cuba, Grenada, Dominica, Dominican Republic, Guatemala, Haiti, Honduras, Jamaica, Mexico, Nicaragua, Panama, Puerto Rico, Saint Kitts and Nevis, Saint Lucia, Saint Vincent and the Grenadines, Trinidad and Tobago, Venezuela
	Total number of countries: 54[c]

[a]The six countries participating in the Gulf of Guinea project also appear in a GEF/LME project in the preparatory phase
[b]Countries that are participating in more than one GEF/LME project
[c]Adjusted for multiple listings

LARGE MARINE ECOSYSTEMS

The term 'Large Marine Ecosystem' (LME) is used to distinguish regions of ocean space encompassing coastal areas out to the seaward boundary of continental shelves and the outer margins of coastal current systems. As such, LMEs are regions of the order of 200,000 km^2 or greater, characterized by distinct bathymetry, hydrography, productivity and trophic patterns (Sherman *et al.* 1991; 1996; 1999a; 1999b; Sherman 1994; Sherman and Duda 1999a; 1999a; 1999b; 1999c; IOC 2002). The 64 LMEs are the source of more than 90 percent of the world's annual marine fisheries yields. Also, most of the global ocean pollution, overexploitation, and coastal habitat alteration occur within these 64 LMEs. They provide, therefore, a convenient framework for addressing issues of natural resources management. Moreover, given that most of them border developing countries, LMEs also provide a framework for addressing issues related to economic development.

Also, as part of the collaboration between the Sea Around Us Project (details at www. saup.fisheries.ubc.ca) and the FishBase project (Froese and Pauly 1999), the world's marine fishes (about 15,000 species; see above) have been assigned to BGCPs and LMEs, if somewhat tentatively in a few cases. We note that this work, which relied on a large number of local ichthyofaunal lists, required about 12 person-months to complete. However, it would have required much longer had it been necessary first to compile a global list of fish species, and to assign them directly to the BGCP, without prior assignment to FAO areas, countries, and oceanic islands, as is provided by FishBase (Froese and Pauly 1999; www.fishbase.org). This point is even more important with regard to invertebrate groups, whose global distribution will have to be mapped, in the long term, in a manner compatible to that used for fishes. This should, for example, be an important component of the Ocean Biogeographic Information System currently under consideration in the USA.

Various development agencies, notably the Global Environment Facility (GEF), the United Nations Development Programme, the UN Environment Programme, the UN Industrial Development Organization, and the World Bank have endorsed the LME concept as framework for several of their international assistance projects, for example in the Gulf of Guinea, the Yellow Sea, and the Benguela Current, with additional projects forthcoming (Sherman and Duda 1999). Table 16-1 shows the numbers of countries currently represented in transboundary LME projects being funded by or in preparation for the GEF. Given this considerable amount of interest, it is fortunate that a number of BGCPs, in the coastal domain, are nearly congruent with the 64 LMEs. Thus, Figure 16-3 illustrates the areal congruities between 19

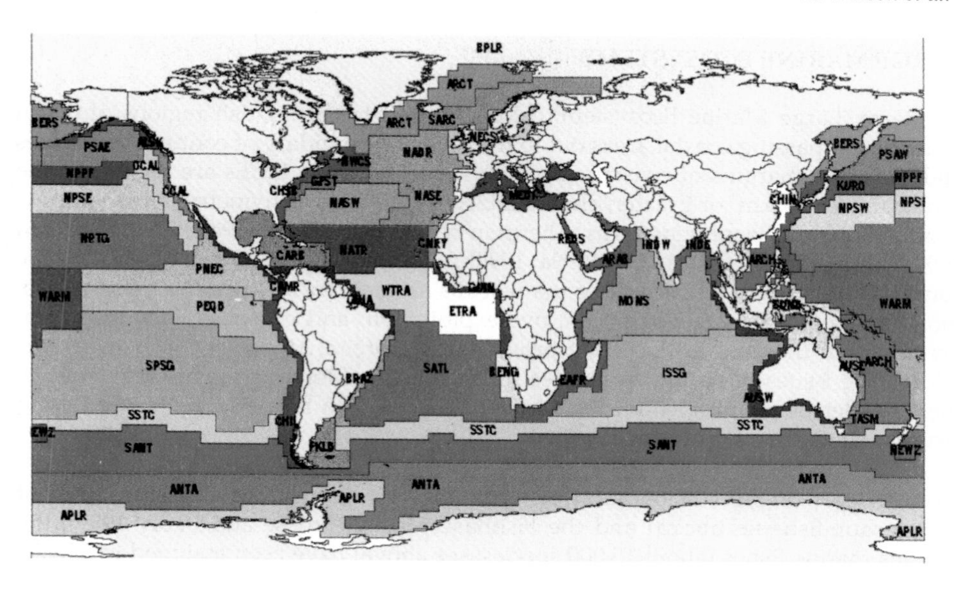

Figure 16-2. Map of the world ocean's 57 biogeochemical provinces, the second level in our proposed classification of the world oceans. (The borders of a few disjunct provinces, notably ARCH, were simplified; detailed file available from http://saup.fisheries.ubc.ca/lme/lme.asp).

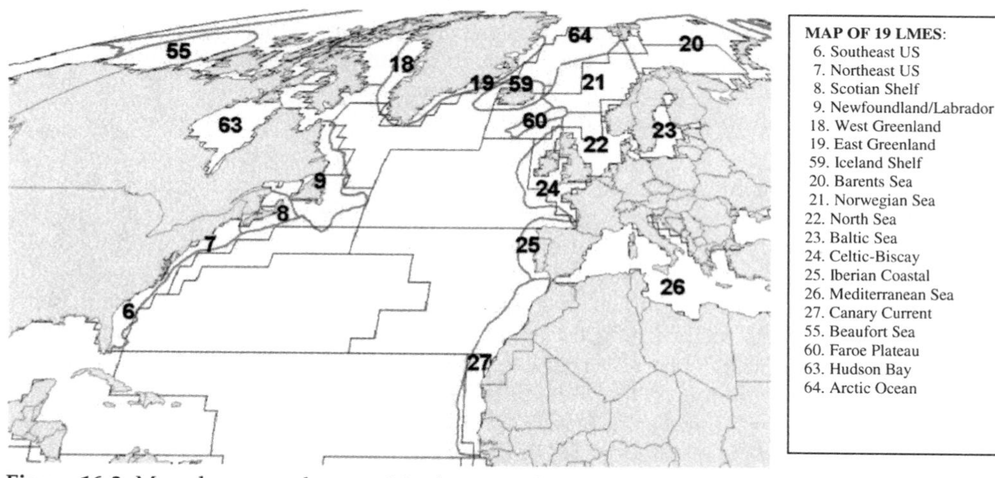

Figure 16-3. Map shows areal congruities between Coastal Biome biogeochemical provinces (BGCPs designated by straight lines) and the numbered Large Marine Ecosystems outlined in red (LMEs numbers are from the LME map, next page). Map by Adrian Kitchingman, UBC.

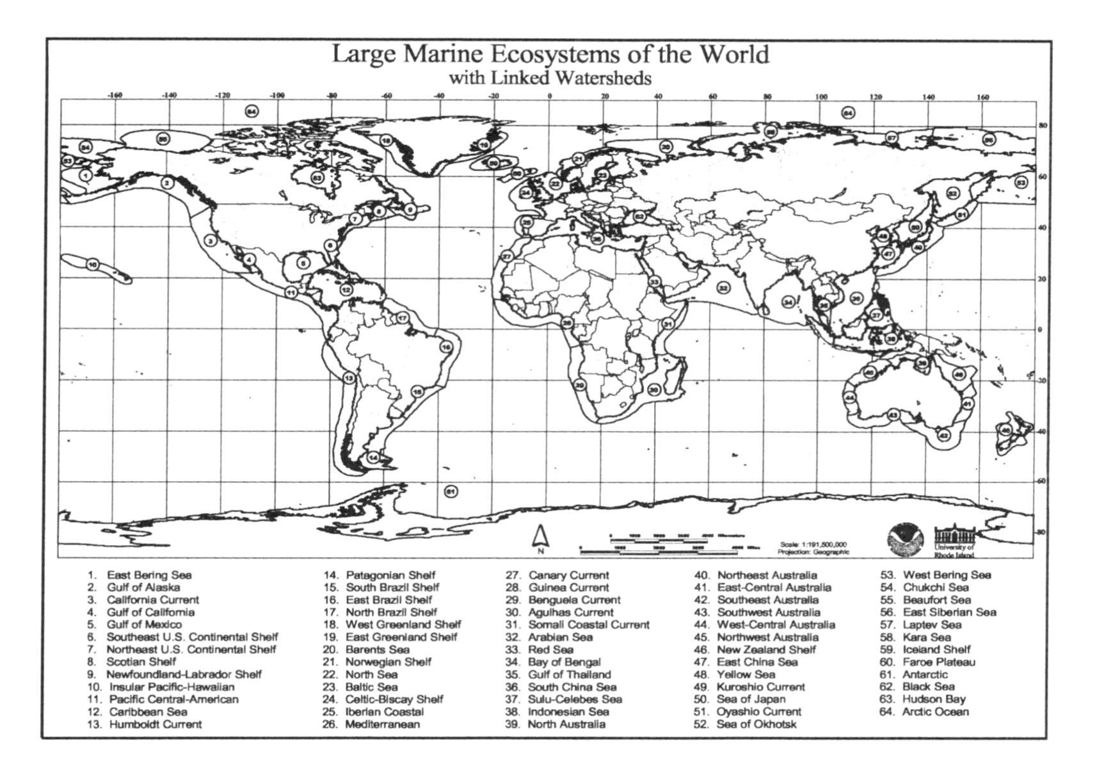

Figure 16-4. Map of the 64 Large Marine Ecosystems of the World

LMEs and the BGCPs of the same regions. A map for the global oceans, depicting the boundaries of all 64 LMEs is shown in Figure 16-4 and at www.lme.noaa.gov and www.edc.uri.edu/lme.

This mapping provides, we believe, the elements that had been lacking within each of the systems thus rendered compatible. Coastal BGCPs often overlap with LMEs that are, by definition, science-based units for fisheries and coastal area assessment and management. The LMEs obtain, via their incorporation into the scheme of biomes and BCGPs as discussed above, borders (here implemented in steps of half-degree cells), that allow GIS-based computation of system properties, such as mean depth, temperature, primary production (Figure 16-5), and other ecosystem attributes.

Another consideration is that our scheme for including access to LMEs together with BGCPs and biome level assessments can be used as an ecological complement to the coarse stratification scheme used by the Food and Agriculture Organization of the United Nations (FAO) to present global marine fisheries data, and which relies on 18 FAO statistical areas (7 for the Atlantic Ocean, 3 for the Indian Ocean and 8 for the Pacific Ocean). Table 16-2 lists examples of ecosystem data products available or soon to be available online.

Table 16-2. Ecosystem data products available or soon to be available on the University of British Columbia (UBC) (www.saup.fisheries.ubc.ca) and University of Rhode Island (URI) (www.edc.uri.edu/lme) websites:

- The bathymetry of the LME areas;
- The percentage of the world's ocean space for each LME
- The percentage of the world's coral reef area within the LME;
- The percentage of the world's gazetted seamounts within the LME;
- Productivity in $gC/M^2/yr$ as SeaWifs derived median values;
- Hot link to lists of fish found in the LME as recorded in FishBase;
- Hot link to Lindeman pyramid (trophic levels) of species in LMEs;
- Access to graphs showing multidecadal trends in catch composition (currently 12 groups: anchovies, herring-like, perch-like, tuna & billfishes, cod-like, salmon/smelt, flatfishes, Scorpionfishes, sharks & rays, crustaceans, molluscs, and 'others') for LMEs from 1950 to 2000;
- Images of selected oceanographic features within LMEs including SST and temperature profiles.
- The 64 LMEs, along with selected summary data, are depicted at websites www.lme.noaa.gov and at www.edc.uri.edu/lme.

To facilitate comparisons between catch data stratified by these two schemes, we have split the five circumpolar BCGPs into ocean-specific provinces. This procedure enables 'closure' of the Atlantic, Indian and Pacific oceans and thus allows direct comparisons, at least at ocean-level scale, between catch data stratified within the scheme proposed here, and that used by FAO for its global catch database. In this context, we have assigned the catches in the global FAO data set to BGCPs and LMEs using locally-derived data sets. Among other things, this allows for rapidly arraying fisheries catches and related ecological data for comparative analyses.

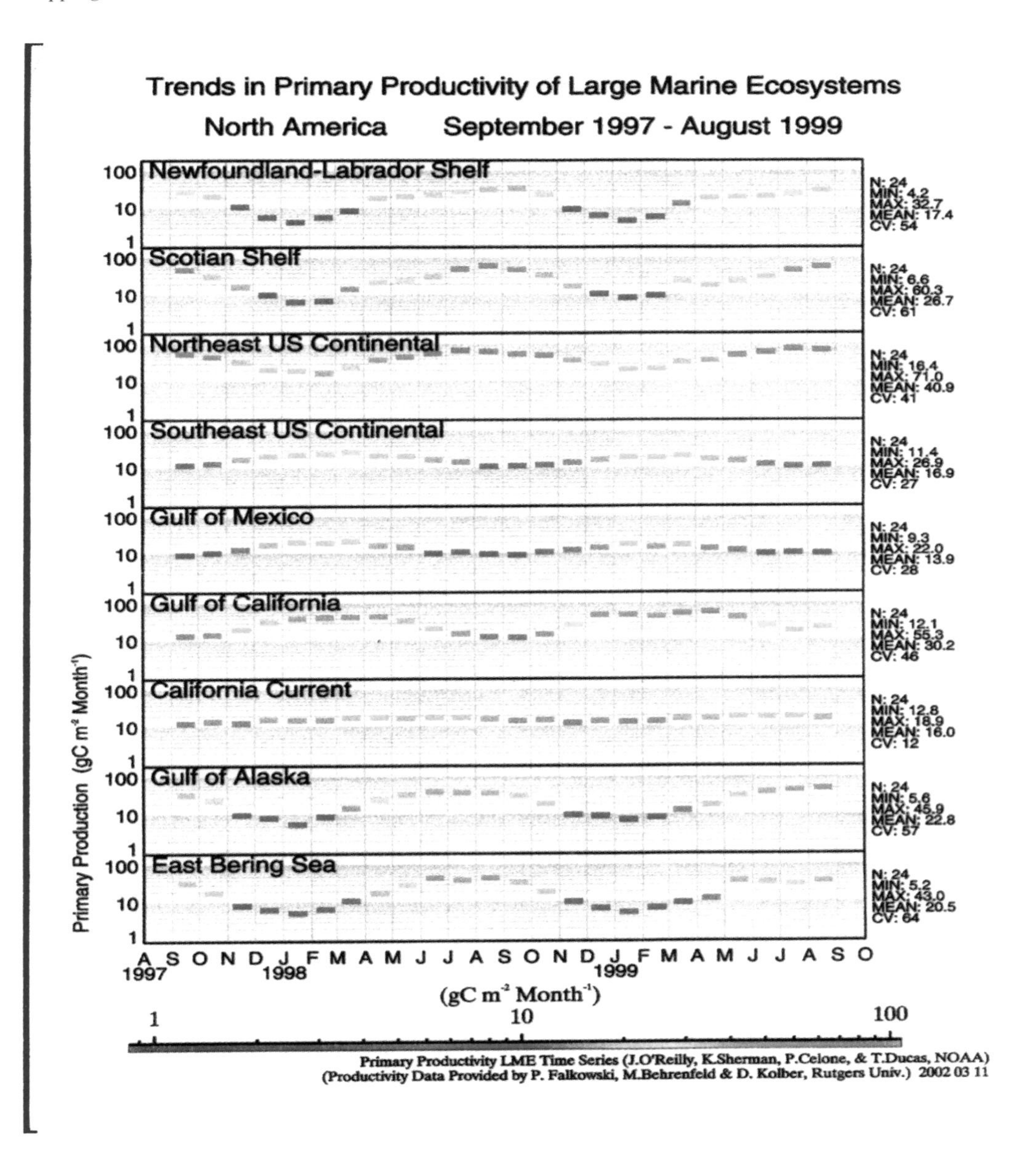

Figure 16-5. Trends in Primary Productivity of the Large Marine Ecosystems of North America, September 1997 - August 1999. Productivity data are based on SeaWiFS data and the Behrenfeld & Falkowski 1997 model. Primary productivity estimates were provided by P. Falkowski, M. Behrenfeld and D. Kolber, Rutgers University.

Table 16-3. Hierarchical relationships among the global ocean Biomes and Provinces, and the 64 LMEs that are merged with the biogeochemical coastal provinces.

<table>
<tr><th colspan="6">----POLAR BIOME----</th></tr>
<tr><td colspan="2">BOREAL PROVINCES
Hudson Bay LME
Beaufort Sea LME
Barents Sea LME
Kara Sea LME
Laptev Sea LME
East Siberian Sea LME
Chukchi Sea LME
Arctic Ocean LME</td><td colspan="2">PACIFIC POLAR PROVINCES
Eastern Bering Sea LME
Western Bering Sea LME</td><td colspan="2">ANTARCTIC POLAR PROVINCES
Antarctic LME</td></tr>
<tr><th colspan="6">----WESTERLIES BIOME----</th></tr>
<tr><th colspan="3">ATLANTIC PROVINCES</th><th colspan="3">PACIFIC PROVINCES</th></tr>
<tr><td colspan="3">Mediterranean Sea LME
Canary Current LME
Guinea Current LME
Benguela Current LME</td><td colspan="3">Kuroshio Current LME
Oyashio Current LME
Gulf of Alaska LME</td></tr>
<tr><th colspan="6">----TRADE WINDS BIOME----</th></tr>
<tr><th colspan="3">ATLANTIC PROVINCES</th><th colspan="3">PACIFIC PROVINCES</th></tr>
<tr><td colspan="3">Caribbean Sea LME
Gulf of Mexico LME</td><td colspan="3">Insular Pacific Hawaiian LME</td></tr>
<tr><th colspan="6">----COASTAL BIOME----</th></tr>
<tr><th colspan="6">NW ATLANTIC SHELVES PROVINCES</th></tr>
<tr><td colspan="3">Scotian Shelf LME
Newfoundland-Labrador Shelf LME</td><td colspan="3">NE US Continental Shelf LME
SE US Continental Shelf LME</td></tr>
<tr><th colspan="6">NE ATLANTIC SHELVES PROVINCES</th></tr>
<tr><td colspan="3">West Greenland Shelf LME
East Greenland Shelf LME
Iceland Shelf LME
Faroe Plateau LME
Black Sea LME</td><td colspan="3">North Sea LME
Norwegian Sea LME
Celtic-Biscay LME
Baltic Sea LME
Iberian Coastal LME</td></tr>
<tr><th colspan="6">SW ATLANTIC SHELVES PROVINCES</th></tr>
<tr><td colspan="3">East Brazil Shelf LME
Patagonian Shelf LME</td><td colspan="3">North Brazil Shelf LME
South Brazil Shelf LME</td></tr>
<tr><th colspan="6">AUSTRALIAN SHELVES PROVINCES</th></tr>
<tr><td colspan="3">West-Central Australian Shelf
Northwest Australian Shelf
Northeast Australian Shelf
North Australian Shelf LME</td><td colspan="3">East Central Australian Shelf
Southeast Australian Shelf
Southwest Australian Shelf</td></tr>
<tr><th colspan="6">PACIFIC COASTAL PROVINCES</th></tr>
<tr><td colspan="3">Gulf of California LME
California Current LME
Humboldt Current LME
Indonesian Seas LME
Sea of Japan LME
New Zealand LME</td><td colspan="3">Sulu-Celebes Sea LME
Pacific Central American Coastal LME
Sea of Okhotsk LME
Yellow Sea LME
East China Sea LME
South China Sea LME
Gulf of Thailand LME</td></tr>
<tr><th colspan="6">INDIAN OCEAN PROVINCES</th></tr>
<tr><td colspan="3">Bay of Bengal LME
Arabian Sea LME</td><td colspan="3">Somali Current LME
Agulhas Current LME
Red Sea LME</td></tr>
</table>

LMEs lend themselves to Ecopath modeling

The ECOPATH with ECOSIM and ECOSPACE modeling approach has been reviewed in several contributions (Christensen and Pauly 1992; Walters *et al.* 1997, 1999; Pauly Christensen and Walters 2000), and there is no need here to present its working or outputs. ECOPATH models exist for numerous parts of the world (details in www.ecopath.org), including the North Atlantic. Currently, well over 100 models have been published, and more than 1800 people in nearly 100 countries have registered as users of the ECOPATH software system. However, the ecosystem model coverage of various ocean basins is still spotty at best, hence precluding simple raising of flows and rates from ecosystem to basin scales. Thus, a stratification scheme is required, based on the geographic structure outlined above, which can be used to scale models from the sampling area of the field data used to parameterize the models to the wider area that is assumed represented by these same models. The strata for the Atlantic, Pacific, Indian Ocean and Polar regions are presented in Table 16-3.

LMEs are seen here as providing the key level for ecosystem model construction. For each LME, an Ecopath model can be constructed to describe the ecosystem resources and their utilization, and to ensure that the total fisheries catch of each LME is used as output constraint (just as their primary production will be used as input constraint). In addition, our stratification scheme can accommodate any number of additional ECOPATH models for each LME. This can be done so as to simultaneously address the issue of parameter uncertainty, as briefly described below.

The LME ECOPATH models require information on abundance, production and consumption rates and diets for all ecosystem groupings. Such information can be obtained from the following sources:

- Abundance, production and consumption rates, and diets of marine mammals are available from the Sea Around Us database for all (117) species of marine mammals (see also Pauly *et al.* 1998b, Trites and Pauly 1998);
- Fishery catches: available from the spatially structured catch database generated as described above, and covering all species groups;
- Occurrence, biology and ecology of marine fishes: available from FishBase (www.fishbase.org), and available at the LME-level. The relevant FishBase search routine was designed for optimizing extraction of Ecopath-relevant information, and is a result of the ongoing cooperation between FishBase and Sea Around Us projects;
- For marine invertebrates: only limited information (beyond the catches in the FAO database) is available from electronic databases, but a variety of

publications provide extensive information. Production rates can be estimated from the well-founded empirical relationships of Brey (1999), now included in Ecopath;

- Primary production estimates: establishment of a global database aimed at supplying fine grid-level satellite-based estimates of primary production is presently underway through a cooperation between the Space Applications Institute, EC Joint Research Centre, Ispra, Italy, and several authors of the present contribution.

The origin of each set of data (5 rate or state variables for each of the often 20-40 functional groups in a model, plus a diet matrix) can be described and a related confidence interval assigned to each of the input parameters. Confidence intervals can also be estimated, as 'posterior distributions' for the output parameters of models. In addition a module of Ecopath is designed to describe the 'pedigree' of Ecopath models, i.e. the degree to which the models are rooted in locally sampled and reliable data (described in more detail by Christensen and Walters 2000). This module estimates, based on the pedigree of its input data, an overall quality index for each model, which in turn can serve as a weighting factor, as required when dealing with discrepancies (e.g. between local vs. LME-wide catches), i.e., when raising one or several model(s) to the LME level. The LME-level Ecopath models will make up the backbone for our approach for addressing province, basin and global issues related to abundance, productivity, interactions and impacts for ecosystem resources e.g., by trophic levels. Being based on the best available estimates of productivity and utilization of the upper trophic levels, and on productivity for the primary producers, the models are constrained from the top as well as from below.

Where possible the LME-level models will be supplemented with additional models. The procedure for this is:

- New models are assigned to strata, based on the proportion of area covered falling within each of the depth strata < 10 m, 10-50 m, 50-200 m, 200-1000 m, and > 1000 m;
- For each new model, the confidence intervals of input and output parameters are estimated along with the pedigree index of the model;
- The LME-level model is assigned to depth strata using weights based on the relative primary productivity in each of the depth strata;
- Within each of the depth strata productivity, abundance, etc., are raised to the LME level using the quality index of the models as weighting factors for the associated confidence intervals.

With this structure in place, it will be easy to add new models as they become available, and it is feasible to assign confidence intervals to all estimates derived from the analysis.

EXCLUSIVE ECONOMIC ZONES

Exclusive Economic Zones (EEZs) are not science-based and are the most political of the units for which the interrelational database could supply information. Allocating freshwater species and their catches to countries is straightforward, as the international borders of countries are usually well defined. This is more difficult in the marine realm, where the fishes and invertebrates caught off the coast of a given country may be caught outside its territorial waters. The International Law of the Sea provides, at least in principle, a solution to this, in form of Exclusive Economic Zones, usually reaching 200 nautical miles into the open ocean, and linking countries with much of the productive areas, i.e., the shelves, adjacent to their coasts. However, not all countries have an EEZ accepted by all their neighbors, and in certain areas, such as the South China Sea, the same rocky outcrops are claimed by up to half a dozen countries (McManus 1992). It cannot be expected that this and similar situations in other parts of the world will be resolved soon and we cannot expect therefore, that official maps of the EEZ will appear that could be used for assigning fisheries catches to the countries of the world.

Nevertheless, various scholars and institutions have published EEZ maps of various parts of the world (see e.g. Mahon 1987 for the Caribbean), based on the rules for definition of EEZ established by the Law of the Sea Convention (Charney and Alexander 1993). We propose that such maps can be used to derive a coherent single map for the EEZ of the world, especially if care is taken to incorporate into such a map the delimitations so far agreed through bilateral or multilateral treaties (as compiled, e.g., in Charney and Alexander 1993). The advantage of such a map is that, unlike the map of LMEs and provinces mentioned above, it will enable the assignment of fish and other species, and of fisheries catch statistics to countries. This will enable comparisons of various features of the use and productivity of various countries' EEZ, with enough degrees of freedom for multivariate analyses, as are now routinely performed for the land-based resources. It is clear, of course, that such a designation will be unofficial and for scientific purposes only, and that it will have no bearing, implicit or explicit, on the status of any EEZ disputes between sovereign states.

Global distribution of coral reef systems can be quantified

Coral reefs, though presently under threat throughout much of their range (Buddemeier and Smith 1999), support important fisheries wherever they occur (Munro 1996). However, quantifying these catches in reliable fashion has proven particularly difficult. One reason is that most countries with coral reefs had administrative infrastructures that precluded detailed monitoring of their fisheries. As suggested by Smith (1978), who performed the first analysis of this type, global assessment of present and potential fisheries yields from coral reefs would be much improved by comparative studies wherein the coral reef fish and invertebrate catches from various EEZs would be matched against the surface area of coral reefs within these same EEZs.

However, while it is possible to assign to coral reefs, at least roughly, a fraction of the catches of each country with reefs in the global FAO fisheries catch database, a matching set of coral reef area per country is not available, despite various global reviews of coral reef distribution (see e.g. Wells 1988; Polunin and Roberts 1996).

The model of Kleypas et al. (1999) can be used, however, to estimate expected coral reef area for any part of the world ocean with a well defined depth, temperature and light regime, and thus can be used to predict coral reef areas within each of the EEZ defined above. We anticipate, once this model becomes widely available, that plots of coral reef fish and/or invertebrate catches vs. reef area will allow us to identify countries with problematic catch data, and/or estimated coral reef areas, and thus to gradually improve the underlying databases and models. Much progress is, however, being made toward making global maps of coral reef distribution (and that of other critical habitats) available by the World Conservation Monitoring Centre (http://www.unep-wcmc.org/marine/data/coral_mangrove/marine.maps.main.html).

Spatial expression of fisheries catch data

Fisheries catches are usually not reported on a per-area basis (e.g. as $t \cdot km^2 \cdot year^{-1}$), though the areas from which they are derived are often specified. Maps of catch per area are rare, and indeed exist only for local studies, often pertaining to single-species fisheries. Thus, one additional reason for the hierarchical system proposed above is that it would allow, and make worthwhile, consistent, basin-scale and ocean-wide mapping of catches onto the ecosystems from which they originate. We have initiated the emergence of such global maps through a procedure described in Watson *et al.* (2002) in which fisheries data reported by countries to FAO for taxa of differing levels of identification (ranging from species to 'miscellaneous marine fishes') for the large FAO statistical areas could be assigned to spatial cells measuring one half degree of latitude by one half degree of longitude
(see http://saup.fisheries.ubc.ca/lme/CatchAllocate.htm)

A database of the global distribution of commercial fisheries species was developed using information from a variety of sources including the FAO, FishBase, and experts on various resource species or groups. Some distributions were specific; others provided depth or latitudinal limits, or simple presence/absence data. A rule-based spatial disaggregation process was used which determined the intersection set of spatial cells within the broad statistical area for which the statistics were provided to FAO, the global distribution of the reported species, and the cells to which the reporting nation had access through fishing agreements. The reported catch tonnage was then proportioned within this set of cells.

CONCLUSIONS

The ecosystem classification proposed here is not meant as a panacea that will solve all our biogeographical problems, or all spatial problems of fisheries. It should not be necessary to stress this; however, it is likely that some readers will think we believe it. We don't. However, we know that no telephone registry would ever emerge, if regular debates were held as to the optimal way to arrange the letters in the alphabet. The ecosystem classification proposed here has been implemented globally by FishBase, which assigns all marine fish species so far described to their LME(s). It will also be used to give a geographic structure to an unofficial, 'spatialized' version of the FAO database of global fisheries catches (see above), thus complementing the atlas of tuna catches compiled by Fonteneau (1998), and allowing both to be related to estimates of primary production for example, mapped in similar fashion by Longhurst *et al.* (1995) and by O'Reilly and Zetlin (1998). Moreover, this classification is fully compatible with the LME approach of Sherman and co-workers, which has led to an extensive documentation of management issues at the LME scale (see references in Sherman and Duda 1999), and a number of field projects designed to address these issues, funded by various international granting agencies (Table 16-1). The merger of offshore biogeochemical biomes and provinces with the more coastal and ecologically defined LMEs provides a hierarchical framework for moving up from LMEs to global scale changes in ecosystem states, and scaling down from open-water pelagic seas to coastal LMEs. It is now possible with the GIS framework to better account for impacts on marine ecosystems of water mass and current perturbations, movements of highly migratory species (whales, tunas, billfish, turtles), changes in coastal pelagic and demersal species biodiversity and biomass yields, the spatial advances in eutrophication, and the frequency and extent of pollution events. Thus, we invite colleagues to join us in expressing their results using the classification and definitions proposed here. To support this collaboration, we will supply, via the Internet, tables presenting the details of the classification by half-degree cells.

ACKNOWLEDGMENTS

The members of the Sea Around Us Project thank the Pew Charitable Trusts for their support of the work leading to this contribution. Also, author Daniel Pauly acknowledges support of a grant from the Canadian National Science and Engineering Research Council.

REFERENCES

Anon. 1991. Bartolomew Illustrated World Atlas. HarperCollin, Edinburgh.

Behrenfeld, M.J and P.G. Falkowski. 1997. Photosynthetic Rates Derived from Satellite-based Chlorophyll Concentration. Limnol. Oceanogr. 42(1):1-20.

Bonfil, R., G. Munro, U.R. Sumaila, H. Valtysson, M. Wright, T. Pitcher, D. Preikshot, N. Haggan and D. Pauly. 1998. Impacts of distant water fleets: an ecological, economic and social assessment. In: The Footprint of Distant Water Fleet on World Fisheries. Endangered Seas Campaign, WWF International, Godalming, Surrey. 11-111. 111p.

Brey, T. 1999. A collection of empirical relations for use in ecological modelling. Naga, the ICLARM Quarterly 22(4): 24-28.

Buddemeier, R.W. and S.V. Smith. 1999. Coral adaptation and acclimatization: a most ingenious paradox. Amer. Zool. 39:1-9.

Caddy, J.F., F. Carocci and S. Coppola. 1998. Have peak fishery production levels been passed in continental shelf areas? Some perspectives arising for historical trends in production per shelf area. J. North Atl. Fish. Sci. 23:191-219.

Charney, J.I. and L.M.A. Alexander, eds. 1993. International Maritime Boundaries. The American Society of International Law. Kluwer, Dordrecht.

Christensen, V. and D. Pauly. 1992. The ECOPATH II - a software for balancing steady-state ecosystem models and calculating network characteristics. Ecological Modelling 61:169-185.

Christensen, V. and C.J. Walters 2000. Ecopath with Ecosim: Workings, capabilities and limitations. [Available from www.ecopath.org].

Duda, A. and K. Sherman. 2002. A new imperative for improving management of large marine ecosystems. Ocean & Coastal Management 45:797-833.

Ekman, S. 1967. Zoogeography of the Sea. Sidgwick and Jackson, London. 417p.

Fonteneau, A. (1998) Atlas of tropical tuna fisheries. Edition ORSTOM, Paris.

Froese, R. and Pauly, D., eds. FishBase 1998: Concepts, Design and Data Sources. ICLARM, Manila. [see www.fishbase.org].

Golley, F.B., 1993. A history of the ecosystem concept in ecology: more than the sum of its parts. Yale University Press, New Haven and London. 254p.

Grove, J.S. 1985. Influence of the 1982-1983 El Niño event upon the ichthyofauna of the Galápagos archipelago. In: G. Robinson and E.M. del Pino, eds. El Niño in the Galápagos Islands: the 1982-1983 event. Publication of the Charles Darwin Foundation for the Galápagos Islands, Quito. 191-198.

Grove, J.S. and R.J. Lavenberg. 1997. The Fishes of the Galápagos Islands. Stanford University Press, Stanford. 863p.

Kleypas, J.A., J.W. McManus and L.A.B. Meñez. 1999. Environmental limits to coral reef development: where do we draw the line? Amer. Zool. 39:146-159.

Levin, S.A. 1990. Physical and biological scales and the modelling of predator-prey interactions in Large Marine Ecosystems. In: K. Sherman, L.M. Alexander, and B.D. Gold, eds. Large Marine Ecosystems: Patterns, Processes and Yields. Amer. Ass. Adv. Sci. Washington, D.C. 179-187.

Longhurst, A.R. 1995. Seasonal cycles of pelagic production and consumption. Progress in Oceanography 36:77-167.

Longhurst, A.R. 1998. Ecological Geography of the Sea. Academic Press, San Diego. 398p.

Longhurst, AR., S. Sathyendranath, T. Platt, and C.M. Caverhill. 1995. An estimate of global primary production in the ocean from satellite radiometer data. Journal of Plankton Research 17:1245-1271.

Mahon, R. (Editor). 1987. Report and Proceedings of the Expert Consultation on Shared Fishery Resources in the Lesser Antilles. FAO Fish. Tech. Pap. 383. 278p.

McManus, J. 1992. The Spratley Islands: a marine park alternative. NAGA, the ICLARM Quarterly 15(3):4-8.

Munro, J.L. 1996. The scope of coral reef fisheries and their management. In: N.V.C. Polunin and C.M. Roberts, eds. Reef Fisheries. Chapman & Hall, London. 1-14.

NRC 1999. Sustaining Marine Fisheries. National Research Council. National Academy Press, Washington, D.C. 164p.

O'Reilly, J.E. and C. Zetlin. 1998. Monograph on the Seasonal, Horizontal, and Vertical Distribution of Phytoplankton Chlorophyll *a* in the Northeast U.S. Continental Shelf Ecosystem. NOAA Technical Report NMFS 139, Fishery Bulletin. 120p.

Pauly, D. 1998. Tropical fishes: patterns and propensities. In T.E. Langford, J. Langford and J.E. Thorpe, eds. Tropical Fish Biology. J. Fish Biol. 53 (Suppl. A): 1-17.

Pauly, D. 1999. Review of A. Longhurst's "Ecological Geography of the Sea." Trends in Ecology and Evolution. 14(3):118.

Pauly, D., V. Christensen, J. Dalsgaard, R. Froese and F.C. Torres Jr. 1998a. Fishing down marine food webs. Science 279:860-863.

Pauly, A. Trites, E. Capuli and V. Christensen 1998b. Diet composition and trophic levels of marine mammals. ICES Journal of Marine Science. 55:467-481.

Pauly, D., V. Christensen and C.J. Walters. 2000. Ecopath, Ecosim and Ecospace as tools for evaluating the ecosystem impact of fisheries. ICES J. mar. Sci. 57:697-706.

Pauly, D., V. Christensen, R. Froese, A. Longhurst, T. Platt, S. Sathyendranath, K. Sherman and R. Watson. 2000. Mapping fisheries onto marine ecosystems: a proposal for a consensus approach for regional, oceanic and global integration. In: D. Pauly and T.J. Pitcher, eds. Methods for Evaluating the Impacts of Fisheries on North Atlantic Ecosystems. Fisheries Centre Research Reports 8(2) [Originally presented at the ICES 2000 Annual Science Conference. Theme Session on Classification and Mapping of Marine Habitats CM 2000/T 16p.; available online at http://saup.fisheries.ubc.ca/report/method/pauly02.pdf]

Polunin N.V.C. and C.M. Roberts, eds. 1995. Reef Fisheries. Chapman & Hall, London. 477p.
Platt, T., C.M. Caverhill and S. Sathyendranath. 1991. Basin-scale estimates of oceanic primary production by remote sensing: the North Atlantic. Journal of Geophysical Research 96:15147-15159.
Platt, T. and S. Sathyendranath. 1988. Oceanic primary production: estimation by remote sensing at local and regional scales. Science 241:1613-1620.
Platt, T. and S. Sathyendranath. 1999. Spatial structure of pelagic ecosystem processes in the global ocean. Ecosystems 2:384-394.
Platt, T., S. Sathyendranath, O. Ulloa, W.G. Harrison, N. Hoepffner and J. Goes 1992. Nutrient control of phytoplankton photosynthesis in the Western North Atlantic. Nature 356:229-231.
Sala, O.E., F. S. Chapin, J. J. Armesto, E. Berlow, J. Bloomfield, R. Dirzo, E. Huber-Sanwald, L.F. Huenneke, R.B. Jackson, A. Kinzig, R. Leemans, D.M. Lodge, H.A. Mooney, M. Oesterheld, N.L. Poff, M.T. Sykes, B.H. Walker, M. Walker, and D.H. Wall 2000. Global Biodiversity Scenarios for the Year 2100. Science 287:1770-1774.
Sathyendranath, S. and T. Platt. 1993. Remote sensing of water-column primary production. ICES mar. Sci. Symp. 197:236-243.
Sathyendranath, S., T. Platt, C.M. Caverhill, R.E. Warnock and M.R. Lewis. 1989. Remote sensing of primary production: computations using a spectral model. Deep-Sea Research 36:431-453.
Sherman, K., L.M. Alexander and B.D. Gold, eds. 1990. Large Marine Ecosystems: Patterns, Processes and Yields. Amer. Assoc. Adv. Sci. Washington D.C. 242p.
Sherman, K., L.M. Alexander and B.D. Gold, eds. 1993. Stress, Mitigation and Sustainability of Large Marine Ecosystems. Amer. Assoc. Advan. Sci., Washington D.C. 376p
Sherman, K. and A.M. Duda. 1999. An ecosystem approach to global assessment and management of coastal waters. Marine Ecology Progress Series. 190:271-287.
Smith, S.V. 1978. Coral reef area and the contribution of coral reefs to processes and resources of the world's ocean. Nature. 273:225-226.
Sverdrup, H.U. 1953. On the conditions for vernal blooming of the phytoplankton. J. Cons. perm. int. Explor. Mer. 18:287-295.
Trites, A. and D. Pauly. 1998. Estimates of mean body weight of marine mammals from measurements of maximum body length. Can. J. Zool. 76:886-896.
Walters, C. V. Christensen and D. Pauly. 1997. Structuring dynamic models of exploited ecosystems from trophic mass-balance assessments. Rev. Fish Biol. Fish. 7(2):139-172.
Walters, C. D. Pauly and V. Christensen 1999. Ecospace: prediction of mesoscale spatial patterns in trophic relationships of exploited ecosystems, with emphasis on the impacts of marine protected areas. Ecosystems 2: 539-554.
Watson, R., Gelchu, A. and D. Pauly 2002. In: Mapping Fisheries Landings With Emphasis on the North Atlantic, Zeller, D., R. Watson and D. Pauly (eds)

Fisheries Impacts on North Atlantic Ecosystems: Catch, Effort and National/ Regional Data Sets. Fisheries Centre Research Reports, University of British Columbia, Vancouver, Canada. 9(3):11 (http://saup.fisheries.ubc.ca/report/datasets/CatchMaps_Watson1.pdf)

Wells, S.M., ed. 1988. Coral Reefs of the World. International Union for the Conservation of Nature, Gland, and United Nations Environment Programme, Nairobi. 3 Volumes.

V
Synopsis

Large Marine Ecosystems of the World
G. Hempel and K. Sherman (Editors)

17

Large Marine Ecosystems of the World: Synoptical notes

Gotthilf Hempel

The concept of Large Marine ecosystems (LMEs) was established thirty years ago as a reaction to the observed regime shifts in complex fisheries in the North Sea and the U.S. Northeast Continental Shelf. Over those three decades great changes have taken place in civil societies' relation to the Ocean and its resources.

CHANGES IN OCEAN AFFAIRS

UNCLOS put most of the productive parts of the Ocean under national ownership and custody. Most coastal states developed their national fisheries and excluded foreign fleets. National management, however, was not always more sustainable than the former international fisheries regulations. The precautionary approach to fishery management, multi-species regulations and the introduction of marine protected areas are attempts to reduce the consequences of the publicly subsidized over-capacity of fisheries worldwide (Hempel and Pauly 2002).

The political changes in the Eastern Bloc in the early 1990s resulted in the withdrawal of most of its distant water fleets. At the same time, the trade in fisheries products and also the fishing fleets themselves, became more globalized. Hence, national control became weaker. The quality of catch and effort statistics has deteriorated in many parts of the world, including those of the countries of the European Union. Although no breathtaking breakthroughs in fishing techniques have taken place, attempts were successful in selectively turning fishing towards targeted species and to make a more extensive use of remote sensing for monitoring of ocean conditions and of fishing operations.

Mariculture has expanded greatly, largely at the expense of natural coastal habitats like mangroves and by using wild fish as feed. The growing output of mariculture camouflaged the drop in world marine catch. The extraction of oil and gas from the sea bed has expanded into much greater depths down the continental slopes. A new generation of transoceanic cables has been placed on the ocean bottom and offshore wind parks are planned for many shallow water areas. The extraction of the vast deposits of gas hydrates is under discussion. Tourism and recreational activities have expanded in many parts of the world, as has the urbanization of coastal zones. All these activities have or will have considerable ecological implications.

New technologies in data collection, processing and communication have enhanced oceanography including its biological branches through improved *in situ* observations from research vessels, moored and floating sensor packages and ships of opportunity; by remote sensing from statellites and air craft; and by advanced modelling, including complex ecosystem models (Summerhayes and Rayner 2002).

Emphasis in oceanographic research has moved towards global climate change, i.e. the role of the ocean as a sink for carbon dioxide, as a source of dimethylsulphide and methane and as an archive of paleoclimate data. A new interest in the present and past circulation of the world oceans has led to large-scale international programs. Great advances in marine microbiology provided new insights into the biogeochemical functioning of the marine systems. Molecular genetics became a powerful tool in the study of populations of marine organisms including the effect of heavy exploitation on the gene pool. Ecosystem health and restoration, including the conservation of marine biodiversity on the population, species and community level became new paradigms. They require transdisciplinary thinking and call for national and international interdisciplinary cooperation as well as for international partnership in the formation of scientific and managerial capacity world wide (Field *et al.* 2002).

In the wake of the World Summits of Rio (1992) and Johannesburg (2002), not much positive trend can be stated for public awareness of ocean affairs and of the need for wise ocean management. In recent years, environmental concerns were pushed back by economic recessions and political conflicts. Only overexploitation of fisheries resources and oil spills receive wide public attention.

CHANGES IN THE LME CONCEPT

Within this framework the LME concept has developed over the years from a fisheries oriented ecological approach to a tool for the assessment, monitoring and management of coastal ocean ecosystems and their environments, and the living resources therein. Right from the start, the interaction between fisheries and fish stocks was seen in the broad context of trophodynamics and of abiotic impacts on recruitment and food ecology. Later, the role of pollution and of the invasion of alien species came into focus. Periodic and aperiodic shifts in the systems were observed and teleconnections were postulated on a global scale. Until recently the socioeconomic and governance dimensions of ecosystem management have been studied less intensively than the physical and biological dynamics. The present and planned LME projects are aimed at safeguarding and restoring the ecosystem and ensuring sustained use of living resources and habitats in balance with socio-economic needs and limitations.

The LME concept promotes multidisciplinary research, assessment, and transdisciplinary thinking and leads to the development of networks of iinternational cooperation among countries, institutions and individual scientists and resource managers. Under the framework of LME projects, building up of scientific and administrative capacity are enhanced.

SYNOPSIS OF VARIOUS GROUPS OF LMEs

Summary and chapter highlights with emphasis on driving forces

The present book provides case studies of LMEs from a global perspective beginning with the polar and boreal LMEs, continuing with the highly productive upwelling LMEs, followed by studies of tropical LMEs. Surveyed are: the semi-enclosed Black and Baltic Seas; the four large eastern boundary upwelling systems; tropical/subtropical LMEs including Brazilian waters, Gulf of Thailand, Yellow Sea and Great Barrier Reef; and boreal and polar LMEs off the Northeast Coast of the US, the Nova Scotian shelf, Bering Sea, Barents Sea and Weddell Sea. Some obvious candidates are missing like the North Sea, Gulf of Guinea, Bay of Bengal and Gulf of Mexico partly because each has been featured in a recent volume or volume chapter in the LME series. The final chapter presents a global hierarchical scaling of ecosystems, from biomes to national EEZs that merges the Longhurst biogeochemical provinces, especially the coastal BGCPs, with the coastal ocean and (semi)-enclosed LMEs.

Taken together, a picture emerges of the global extent of LMEs, their variation at different time scales, from years to centuries, caused by inherent natural variability and by their vulnerability to human impacts. The resulting classification allows the reader to recognize similarities among LMEs in different parts of the world and affords the reader the opportunity to detect ocean-wide and global connections.

Enclosed Seas LMEs

The Baltic Sea LME and Black Sea LME are of about the same size (0.4 million square-kilometer) but the Black Sea is much deeper, older and more haline than the Baltic, which is virtually a post-glacial brackish fjord. The exchange of water masses with the Mediterranean and the North Sea respectively is very limited, resulting in long residence times. The layers below the sill depth of the water masses and their pollutants are virtually stagnant and anoxic in the Black Sea, while in the Baltic stochastic influxes of highly saline North Sea water result in a periodic aeration of bottom water.

Both seas have very large catchment basins with great rivers passing through densely populated, highly industrialized areas. Over-fertilization was typical for the agriculture in much of the riparian countries belonging to the Eastern Bloc. Consequently, pollution and eutrophication are quite substantial. Fisheries have a long tradition in both seas and overfishing is common in spite of international and national regulations. Marine sciences are well established and LME projects are in progress for both the Seas.

In the Baltic and in the Black Sea, great changes have been observed over the past decades. In the Baltic, key driving forces are both man-made and natural, i.e. due to human impacts like nutrient inputs and overfishing as well as to climatic variations controlling the inflow of North Sea water. In the Black Sea, the changes are primarily caused by human activities, including land-based pollution, unintended introduction of alien species and heavy overfishing.

B.O. Jansson calls the young **Baltic Sea** a "sea of surprise" which is far from having reached a nearly steady state. Over its past history of 20,000 years it has undergone dramatic changes, switching to and fro, from fully marine to freshwater water conditions. In future, global warming with more rainfall over central and northern Europe will make the Baltic even more brackish and more stratified than today. In the past fifty years human impact on the Baltic Sea became very noticeable. Nutrient content has increased greatly, turning the Baltic from the oligotrophic into a eutrophic stage. The high input of toxic organic compounds and heavy metals in the 1970s has meanwhile been halted, largely thanks to the international regulations and monitoring activities of the Helsinki Commission (HELCOM). The fishing yields have increased from 200,000 t in the 1940s to 840,000 t in the 1980s. Since then, the yields have dropped. Eutrophication and the expansion of the anoxic zones due to increased input of nutrients and freshwater and reduced inflow of North Sea water combined with heavy fishing input, are the major causes for the observed general shift from cod and herring to sprat. In spite of the fact, that its nine riparian countries and 85 million inhabitants of its drainage basin differ considerably in their interest in the Baltic Sea and its resources and amenities, the Baltic Sea has the potential to become a well managed LME in terms of control of pollution and fisheries. Much of the legislation is already at hand, but enforcement is still inadequate.

Daskalov describes the **Black Sea LME**, particularly in its shallow northwestern part as the perfect example of the complex interaction between various human impacts resulting in the extinction of much of the bottom fauna and in a shift in the plankton communities from diatoms to flagellates, and from crustacea to jelly fish (Zaitsev and Mamaev 1997). All these changes affect the fish stocks in a drastic way, both bottom up via food and top down by predation on ichthyoplankton. In his chapter on the Black Sea, Daskalov included in his models the additional effects of overfishing. It started with the increased catch of the large predators like dolphins and swordfish, then turned to mackerel and ended up with small planktivorous fish. His ecosystem models provide insights for future management. He suggests that conserving and restoring natural stocks of fish and marine mammals can contribute greatly to sustaining viable marine ecosystems."

Baltic and Black Sea coastal countries are receiving substantial assistance from the Global Environment Facility to accelerate their movement toward ecosystem-based natural resources assessment and management. In both cases the countries are following a five module strategy to monitor the changing states of the LMEs; included are assessments of productivity, fish and fisheries, ecosystem health, socioeconomics and governance practices. Additional information is available in Duda and Sherman (2002).

Upwelling Current LMEs

All are dealt with in the present volume. What have the four eastern boundary upwelling current LMEs (Humboldt, Benguela, Canary and California Currents) in common? The surface currents transport cold water equator-wards except for the Benguela Current which is fed by warm waters from the Indian Ocean. Upwelling and its inter-annual variations are more pronounced in the southern compared to the northern hemisphere systems. The global atmospheric circulation and its interaction with the ocean have the greatest effect in the Humboldt LME, the home of El Niño (ENSO), but are noticeable in the other systems too (Bakun 1996). The term teleconnections was originally attributed to effects of ENSO in other parts of the world, particularly the California Current ecosystem. Now the term is in much wider use, tracing effects of ENSO in the other systems too but also the far reaching effects of other phenomena like the North Atlantic Oscillation (NOA).

The interannual variability of the upwelling and drastic changes in horizontal and vertical circulation and water mass distribution are reflected in major changes in water temperature and oxygen content. The marine flora and fauna are adjusted to those changes and undergo strong natural fluctuations. The high primary productivity and high concentrations of food permit short pelagic food chains with large steps in body size from one trophic level to the next and with birds and mammals as top predators. High sedimentation of organic matter results in oxygen deficiencies linked with a rich microbial life, particularly in the centres of heavy upwelling.

The physical and biological phenomena of the four currents and their upwelling zones have long attracted the curiosity of seafarers and oceanographers. Oceanographic work was greatly intensified with the onset of the exploitation of the pelagic fish resources, mostly for fish meal and oil in the 1950s. The California Cooperative Fisheries Investigations (CalCoFI) commenced in 1949 and Wooster published the first comprehensive account of El Niño in 1960. The US, followed by European countries (partly under the umbrella of FAO) put much effort into the establishment of research capacity in Peru and Chile. Some years later a similar development took place in the Canary Current region where ORSTOM of France took the lead. Several "western" and "eastern" countries of Europe together with USA joined in the CINECA programme of FAO and ICES in the 1970s (Hempel 1982). The Benguela Current had been studied intensively but not regularly by various European expeditions. Finally, systematic investigations on a regular basis were initiated in the 1980s by South Africa in its large scale Benguela Ecology Programme, which encompasses the entire ecosystem in a very comprehensive way.

Monitoring of the physical conditions and near surface phytoplankton abundance of the eastern boundary currents is nowadays much facilitated by remote sensing. The sea-going sampling programmes for monitoring of chemical and biological parameters is presently more extensive for the two Southern Hemisphere systems than for the northern ones. The CalCoFI grid, once the best in the world was thinned out between 1970 and 1985 and has been diminished in latitudinal extent since then. Monitoring of the Canary Current off NW-Africa seems presently the weakest of the four.

Well known, but still not fully understood, are the long-term variations in abundance and distribution of sardines and anchovies in all four systems which were reflected in the rise and fall of some of the largest fisheries in the world. The interaction of severe exploitation in a phase of natural decline proves critical for the stocks. Hake species and scombrids are further important resources under heavy stress. Economically sound and ecologically sustainable management of the systems will only be possible if the fishing fleet is much reduced, in Peru possibly to about a quarter of its present capacity.

In summary, in the upwelling current LMEs climatic factors rank first as cause for the changes in fishing yields, followed by the impacts of heavy fishing. There is little input of pollutants and the effect of anthropogenic eutrophication is negligible compared to that of the natural upwelling effects.

In an effort to advance toward ecosystem-based monitoring of the resources of Canary Current, Benguela Current and Humboldt Current LMEs, the GEF is providing substantial funding to the recipient coastal countries (Duda and Sherman 2002).

Except for the California Current LME, the dominant share of the pelagic fishing yields of the upwelling systems had been taken originally by foreign distant water fleets. They are now mostly replaced by local fleets supplying local processing firms that are greatly dependant on the global market. Management of the resources should be facilitated by the fact that the number of bordering coastal states is small for each of the upwelling current LMEs: two each in the Pacific (USA and Mexico, Peru and Chile), Portugal, Spain and four major African countries border the Canary Current LME, and Angola, Namibia and South-Africa for the Benguela Current. Coastal deserts like the Namibia and Atacama separate the coast from the rest of the country. In the majority of the coastal states, the sea is at their doorsteps. For Portugal, Morocco, Senegal, Chile and Peru, small scale fisheries are important and make use of the demersal fish and invertebrates. Japanese and other Asian vessels fish for commodities like squid in the regions.

The authors of the four chapters on eastern boundary current LMEs highlight very different aspects: M. Wolff *et al.* focus on ecological modelling of **the Humboldt Current LME** (HCLME) and its variability under El Niño. Although the variations in the pelagic stocks are clearly dominated by the El Niño cycles, no quantitative predictions of future yields can be made bottom up from oceanographic models linked with monospecies population dynamics. Strong top down effects of grazing and predation necessitate the application of multispecies approaches (ECOPATH/ ECOSIM). This is particularly the case for the demersal resources, which largely benefit from El Niño. Hake is drastically overfished and "There are signs of substantial changes in the structure of the food web in the HCLME that have occurred over the last decades of intensive fishing…." While the fish meal species, mostly anchovy off Peru and nowadays horse mackerel off Chile are economically most important for the fishing industry, a far greater part of the society depends directly on the great number of mostly demersal species for human consumption.

V. Shannon and M. O'Toole describe the **Benguela Current** LME as a "convex system" at the crossroads between the Atlantic, Indian and Southern Oceans: bounded in the North and South by warm water systems and with the most active upwelling cell off Lüderitz in the middle. The Benguela ecosystem has its own El Niño intruding from the North. It is not necessarily in phase with the Pacific ENSO, but a regional response to changes in the global atmosphere-ocean system. From the South, warm Agulhas intrusions and cold sub-Antarctic intrusions contribute to the great inter-annual variability in the physical environment, its communities and fishery resources. Decadal changes in the wind stress and a long term rise of SST have been recorded since the 1920s. Even more dramatic is the increase in zooplankton over the second half of the century. Species switching and regime shifts in the sardines and anchovies are out of phase with those of the Pacific stocks but in synchrony with shifts in the Canary Current LME.

International cooperation in research, monitoring and in the initial stages of joint management is most advanced in the Benguela Current LME. It is the only eastern boundary current system with both an on-going LME project (BCLME) and a well established regional research and training programme (BENEFIT). The authors conclude their chapter proudly with the statement on the "determination of three countries, which have been subjected to centuries of oppression and exploitation, to take joint action to correct the wrongs of the past and demonstrate to the rest of the world how a fragile and variable marine ecosystem can be managed sustainably."

Roy and Cury, in their contribution on the **Canary Current LME**, focus on the physical sea surface temperature (SST) distribution and its variations at different time scales. In contrast to the Southern Hemisphere upwelling systems, El Niño type events are less pronounced than decadal changes. A major cooling in the early and mid 1970s was followed by a general warming until around 1990 when a new cooling started. The changes in SST derived from the COADS database of merchant ship observations reflect changes in upwelling intensity. Teleconnections between the Northern and central Atlantic are obvious and links between NAO index and El Niño events are suggested. They result in global co-variations in the abundance of key pelagic species.

Fish stock abundance is driven by long term variations, by short term fluctuations and by the changes in fisheries after the withdrawal of much of the foreign fleets around 1990. Altogether, the fishing yields of the Canary Current LME are smaller than those of the other eastern boundary upwelling current LMEs. The Canary Current LME and its fish populations are rather open to the Gulf of Guinea. Occasional outbursts of some otherwise rare species reflect the great importance of predator control, particularly in demersal stocks.

A fresh look at the **California Current LME** is provided by D. Lluch-Belda *et al.* making use of an excellent data set. Long time series of observations exist. They cover the entire 20th century. That is more than for any of the other upwelling systems. Similar to the analysis by Roy and Cury for the Canary Current System, the authors describe the changes at different time scales: < 10, 10-120, > 50 years. In the California Current, very low frequency changes reflect regime shifts. Long trends were observed warming starting in the mid 1990s, cooling around 1941, warming in 1976/77 and again cooling in 1990. The analysis of the interannual changes demonstrates relations to ENSO and to the Pacific Decadal Oscillation (PDO). The impacts of the physical variations on primary production and on the pelagic stocks of sardines and anchovies are obvious but less so for zooplankton. The multitude of interaction between processes at the different time scales makes predictions very difficult. The authors suggest the existence of natural cycles in the system, "thus the system should be expected to return to previous states. The interaction of these scales of change with anthropogenic-induced variation has yet to be unveiled." That is particularly the case for man-induced global warming. Its potential present and future effect on the California Current LME is a matter of controversy and debate. GLOBEC studies now underway may provide the answer.

Temperate and cold water LMEs

The examples for temperate and cold LME's range from the Yellow Sea LME through the waters off the eastern coasts of southern Canada and northern US to the Bering Sea LME and Barents Sea LME and finally to the ice-covered Weddell Sea. Except for the Weddell Sea the seasonal signal in water temperature is high. Seasonality in light, stratification and hence nutrient supply result in pronounced periodicity in primary production and in the abundance of phyto- and zooplankton. Inter-annual differences in the matching of the reproduction phase in fish and with variations in abundance of food and predators of the fish larvae are considered major causes of year class variations. They are very pronounced in longlived fish where strong year classes appear at irregular and long intervals. In species like cod it is now considered important, to retain a sufficiently large number of old fish in order to bridge the often long gap between strong year classes.

The semi-enclosed **Yellow Sea LME** as described by Tang is characterized by exceptionally high seasonal variations with tropical surface water temperatures in summer and freezing in winter. Its productivity is variable and the living resources are highly diverse with warm-temperate species dominating. Multinational and multispecies fisheries are well developed and mariculture is of great importance to all bordering states. Extensive economic development in the coastal zones including mariculture causes considerable pollution and habitat destruction and hence loss in biodiversity and living resources. Phytoplankton and zooplankton biomass have reportedly decreased over the past 40 years, while red tides have become more frequent.

The Yellow Sea fish fauna is now in a state of severe over-exploitation with a pronounced trend in average landing size of fish and in species composition towards short lived, low value engraulids and other species. Apart from fishing, shifts in oceanographic conditions are also considered responsible for some of the changes. In order to restore the Yellow Sea LME in its role as one of the most important fishing areas in the World a mitigation action plan has been developed linking a GEF/LME-project to a regional GLOBEC programme and to monitoring under the international GOOS programme.

The shelf of the **U.S. Northeast Coast** from Cape Hatteras to the Gulf of Maine, is one of the most productive LMEs in the World with up to 500g $Cm^{-2}y^{-1}$. It has a centuries-long history of intensive exploitation by foreign and local fishers, particularly in the North, i.e. Gulf of Maine and Georges Bank. Continued ecological research dates back to Bigelow's surveys in the 1920's. Here, the LME concept was developed when Sherman and co-workers studied the impacts of the ichthyoplankton and zooplankton component on pelagic fish in the frame work of MARMAP 1977-1986. In the present volume Sherman *et al.* revisit the NE Shelf Ecosystem and describe major changes over three decades. The two northern subsystems show a noticeable change in the composition of the dominant copepods species. The overall zooplankton biomass of the LME decreased sharply in the 1970's but has recovered recently. "The species shifts and the inter-annual and decadal variability observed in zooplankton biomass appear to have allowed for sufficient residual sustainability in biodiversity and abundance to support the recovery of zooplanktiverous herring and mackerel stocks." The authors find evidence for system resilience and robustness.

Herring and mackerel had been very heavily exploited in the 1960s and early 1970s, but have recovered since 1980 to an unprecedented high level of five to ten times the former minimum stock biomass. While exploitation of pelagics decreased after the closure of the LME for European fleets, the exploitation rate of ground fish like cod, haddock and flounders continued. As a consequence, the ground fish biomass declined but the biomass of elasmobranchs went up until 1990 when increased fishing pressure on spiny dog fish and skates depleted the stocks. Progress has been noticeable in understanding the interaction between the major components of the pelagic system and its consequences for recruitment and growth in the various fish species. In addition to those achievements by quantitative science, the authors report on substantial management progress towards the recovery of demersal species. However, an adequate framework for governance for all components of the ecosystem is still lacking.

The Scotian Shelf is another LME with a long fishery tradition. Zwanenburg divides the LME in two distinct subsystems i.e. the eastern and western shelves which call for different fisheries management objectives. "Large interannual fluctuations in temperature and in the distribution of fishes, invertebrates and marine mammals have been documented. Exploitation, mainly in form of commercial fisheries, operates against this dynamic background." Shortcomings in the traditional regulatory management lead to the adoption of the precautionary approach of management. Substantial efforts are made to address the need for the maintenance of biodiversity and habitat productivity.

Watson *et al.* (this volume) stated, "There is a broad consensus in the scientific community that fisheries management should be ecosystem based but very little agreement as to what that means." Zwanenburg's traffic light approach goes a long way in this direction by the definition of status indicators both for fish species and for important elements of the ecosystem and its functions at various trophic levels and compartments (plankton, benthos). Social and economic indicators could be added making the traffic light scheme a powerful tool for sustainable management.

In the eastern **Bering Sea LME** as described by Schumacher *et al.* changes in the winter extent of sea ice have been partly attributed to El Nino-Southern Oscillation (ENSO) and decadal/multidecadal North Pacific Oscillation (PDO) but also to Global Warming. High heat flux in summer and low winter cooling put the near surface water temperature up by > 3^0 C in 1997 compared to the 1960s. Those hydrographic changes caused substantial shifts in primary production through extended spring blooms. Over the past 25 years fisheries harvest in the eastern Bering Sea was rather constant (ca. 1,6 mill $t.y^{-1}$ mostly targeting pollock). Nevertheless most stocks of fish underwent massive changes due to inter-annual and decadal variations in recruitment. Increased predation by the protected stocks of marine mammals might have been an additional cause for declines. Top down control linked to shifts in primary production seems responsible for decadal variations.

The authors suggest that the major changes in the ecosystem dynamics of the eastern Bering Sea observed over the past two decades are mainly due to climate change. The understanding of the underlying physical and ecological mechanisms has substantially advanced both by monitoring and by process-oriented studies. The authors speculate what might happen if the majority of the coming years will resemble the warm spring/summer conditions of 1997. They predict a cascade of changes in the ecosystem with a decrease in annual primary production but a prolonged spring bloom period, favouring the pelagic rather than the benthic compartment of the food web. Marine mammals will suffer from competition for food and will switch to alternative food resources. Altogether the eastern Bering Sea is a splendid area for studying the effects of climate forcing at different time scales in an ecosystem which is not over-exploited but is well monitored.

The comprehensive contribution by Matishov *et al.* on the **Barents Sea LME** is primarily based on the rich Russian literature and data sets and on ICES statistics. It provides a particularly good picture of the eastern part of the LME, while the Norwegian programmes like ProMare had focussed on the western part. The Barents Sea in its present configuration is only 10.000 years old. It is influenced by the North Atlantic and the Arctic Ocean meeting at the highly variable Polar Front. The formation of a stable stratification in summer and total mixing in winter and the seasonal ice cover are the two prominent features leading to a comparatively high primary production. The system undergoes changes at time scales from years to centuries. No local warming was observed at the end of the 20th century but it is predicted for the future. The pelagic food webs differ markedly: In the Atlantic waters large copepods and euphausiids support herring and capelin eaten by cod, birds and whales; in Arctic waters the carnivorous amphipod *Themisto libellula* resembles the antarctic krill *Euphausia superba* not only in size and longevity but also as staple food for birds and seals, partly via polar cod.

Intensive fishing by European distant water fleets and Russian vessels started hundred years ago. The decreasing trend in yields of herring and cod was camouflaged in the 1970s and early 1980s by a rich capelin fishery turning the Barents Sea LME again into one of the richest fishing regions in the world, but resulting in severe perturbations at the top levels of the food web. Bilateral fishery regulations agreed upon by Norway and Russia have gained in importance, although their enforcement is far from being perfect, particularly but not only in the Russian fishing industry.

In addition to climate variations affecting current and ice regimes, and in addition to the impact of fisheries, other anthropogenic impacts have to be mentioned. It took less than four decades for the Kamchatka king crab to occupy the entire shallow waters of the Barents Sea and invade even Lofoten waters with serious consequences for the benthic food webs. The Barents Sea is a sink of substantial atmospheric fall out, dumping and riverine run off of a wide array of pollutants. Furthermore exploration and possibly exploitation of oil and gas on the Barents Sea shelf has stimulated increasing awareness about the ensuing environmental risks of all these activities amongst the various stakeholders.

The Weddell Sea described by Hubold is a distinct sub-region of the Antarctic LME and an extreme representative of the Polar biome. It is virtually untouched by man except for possible former indirect effects of whaling in the northern Weddell Sea impacting the krill stocks to an unknown extent. At the northern fringes of the Weddell Sea, intensive fisheries started in the 1970s off the Antarctic Peninsula and the adjacent South Shetland and South Orkney Islands and quickly depleted the stocks of slowgrowing notothenoids. Afterwards CCAMLR introduced a regulation scheme for catches of fin fish and set a very conservative limit to krill exploitation in order not to seriously affect the food base for Antarctic mammals and birds. The Convention for the Antarctic Marine Living Resources was the first large scale fishery convention with an ecosystem concept right from the beginning. Hubold goes beyond considerations of potential indirect effects of fisheries and discusses the threats to the system by stratospheric ozone depletion and global warming. Any major increase in UV-radiance and particularly any loss in sea ice cover will affect the present food web as both phytoplankton production and the life history of krill are closely related to the extent and dynamics of the sea ice.

The chapter by Watson *et al.* on **mapping of fisheries** merges the approaches of academia and fisheries institutions in classification of marine ecosystems. The classification proposed by Platt, Sathyendranath and Longhurst comprises of four physically defined biomes and 57 biogeochemical provinces. The Coastal Boundary biome harbors most of the LMEs which are largely shelf-based, their boundaries often coincide with biogeochemical provinces. Combined mapping of integrated databases of different scope and scale makes features of productivity, exploitation and resilience comparable on regional, oceanwide and global level.

The fishery data bases of FAO and of regional and national fishery institutions as well as Fish Base and the marine mammal data base of Sea Around Us can now be knitted together with data bases of GIS-derived environmental properties collected by various assessment programs of UNEP (e.g. GIWA) and other organisations.

CONCLUDING REMARKS

What do the various case-studies of this volume tell us about exploitation, protection, and research in the shelf regions of the World Ocean?

Exploitation

The exploitation of fish and some invertebrates is heavy in most regions and many stocks have seriously declined over the past three decades even though the new Law of the Sea put most of the living resources under national jurisdiction. Fishing down the food web is a wide spread phenomenon. The EU with its high scientific and administrative capacity is still unable to agree on a sustainable fishery policy and to implement an adequate reduction in fishing effort. World fisheries are still very far from the goals set up by the Johannesburg Summit to introduce ecosystem based assessment and management practices by 2010 and to restore the world´s depleted stocks to maximum levels of sustainable yields by 2015. Recognition is growing for the need to diminish subsidies for the fishing fleets and for putting much of the regulatory management in the hands of local and provincial authorities and stakeholders. In many parts of the World there is a severe shortage of scientific and administrative infrastructure and know-how as well as of public awareness and political will for the longterm management based on socio-economic and ecosystem-oriented rules. In recognition of those needs various LME projects were initiated. They are primarily financed by the Global Environmental Facility (GEF) in cooperation with UN-organizations, regional and national fishery institutions as well as non-governmental organizations.

Protection

Protection of the marine environment has to combat five major risks: Overexploitation, pollution, habitat destruction, alien species, and global change. Most of the case studies do not go beyond the effects of **overfishing** on the marine ecosystems. Reduction of fishing effort, introduction of fishing practices with few "collateral effects", and the creation of marine protected areas are important, albeit not new, protective measures. **Eutrophication** and **toxic pollution** have been described as serious threats in enclosed and semi-enclosed seas. Over the past two decades inputs into coastal waters have been reduced in many but not all industrialized regions. Public sensitivity against oil spills is high. In other parts of the world, pollution and eutrophication by air borne transport, run-off and dumping from land, as well as point sources from industry, communes and fish and shrimp farms is still serious and even growing, causing *inter alia* harmful algal blooms and damage to coral reefs.

Little has been said in the present volume about **habitat destruction.** It is not only the disastrous damage of coral reefs which matters but also the destruction of other near shore habitats like mangroves, sea grass beds, salt marshes and estuaries. They all have far reaching ecological functions in the marine ecosystems. The role of mariculture in those habitats has to be considered in a holistic ecological as well as socioeconomic context. The establishment of oil rigs and other structures for the oil and gas exploration and exploitation as well as the creation of wind parks leads to the destruction of marine habitats but might result in areas protected against most fisheries. Extraction of sand and gravel and other seabed mining affects the benthos through mechanical damage and siltation. The expansion of fishing and of drilling for oil beyond the 1000 m isobath into the deep parts of the continental slopes threatens benthic deep sea communities of sponges and cold water corals and their rich associated fauna.

Two examples can be found for the effects of the unintended or intended introduction of **alien species**: The jellyfish *Mnemiopsis leydii* reached the Black Sea with ballast water, while the Kamchatka crab was implanted in the Barents Sea by Russian scientists. Both spread widely and had great consequences for the pelagic and benthic communities respectively. The increasing use of double hull vessels carrying large amounts of ballast water call for powerful protective measures against the introduction of alien species. Escapes from fish farms are a further threat to pristine marine communities.

In the long run, **global climate change** will be felt in all LMEs, although to a rather different extent. So far, shifts attributed to climate change have been observed in the geographical distribution of species and in the composition of the communities. The effects will be most serious for the polar and subpolar regions, particularly in the North, where hitherto permanent ice-covers melted in summer. Hence the seasonal sea ice zone extended northward while its southern edge retreats. Other seriously affected habitats are coral reefs, which are sensitive to long spells of high water temperatures. Shifts in the frequencies and intensity of El Niños will strongly affect the pelagic fish production in the eastern boundary upwelling regions and, via teleconnections in many other fishing areas. We cannot protect ecosystems against climate changes except by reducing their causes, i.e. the output of green house gases. But we can adjust our management to the effects of climate variations.

Global Change in a broader sense includes the increase in human population and its coastward migrations. Economic globalization, shifts in World markets, new technologies, increase in energy demands are further facets of "Global Change". Political and socio-economic changes in the past thirty years had often a greater impact on the utilization of the shelf ecosystems than natural trends. Vis-à-vis all those changes, sustainability cannot be seen as retaining a certain status of the past, but as a dynamic state with a sufficiently high resilience. "Maximum yield" as a prime goal will be increasingly difficult to reconcile with the goals of restoration and conservation of high biodiversity.

Research

Research in most LMEs has substantially advanced the understanding of physical processes leading to environmental changes at different time scales. Monitoring by remote sensing and by a fast-growing array of in-situ observation platforms as well as modelling have paved the way for predictions of the potential regional impacts of global climate change. Research in the variability and interaction of fish stocks has benefited from the growing length of time series. It suffered, however, from the deterioration of catch and effort statistics in various parts of the world. Ecosystem modelling like ECOPATH and large data bases like FishBase facilitate comparisons among LMEs. Pelagic life at the various trophic levels from phytoplankton to fish has been intensively studied. On the other hand, many pure and applied marine biologists seem to consider the sea as bottomless. At a time of growing concern about marine biodiversity, benthos should come into focus as part of the food web and as being affected by man-made perturbations such as demersal trawling, construction activities at the sea bed, extraction of non-living resources, and dumping. Benthos studies are more demanding in terms of sampling strategies and skilled manpower than routine plankton work; nevertheless, they are indispensable.

Scientists and administrators alike recognise the need to solve the conflict between short-term socioeconomic pressures and the long-term reduction of fishing activities and other human impacts on LMEs. Nevertheless, politicians will follow the easy shortsighted route as long as the scientists cannot provide them with adequate tools and arguments which help to come closer to sustainability without demanding too much from society and its lobby groups.

Multidisciplinary research of natural scientists and socio-economists has to be strengthened as well as their dialogue with stakeholders.

There is still much room for research to provide a scientifically sound basis for the sustainable exploitation of the marine living resources and for the protection of the marine biodiversity and other riches, goods and services of the world's LMEs.

REFERENCES

Bakun A. 1996. Patterns in the Ocean. California Sea Grant System. La paz, BCS, Mexico. 323p

Field, J.G., G. Hempel, C.P. Summerhayes. 2002. Oceans 2020. Science, trends, and the challenge of Sustainability. Island Press, Washington. 365p

Hempel, G., D. Pauly. 2002. Fisheries and Fisheries Science in their search for sustainability. In: J.G. Field, G. Hempel, C.P. Summerhayes, eds. Oceans 2020. Island Press, Washington. 109-135

Hempel, G. 1982. The Canary Current: studies of an upwelling system. Introduction. Rapp. P.-v. Réun. Cons. Int. Explor. Mer. 180:7-8

Sakshaug, E. ed. 1992. Økosystem Barentshavet Norges Allmennvitenskapelig ForsKningsråd Oslo. 304p

Summerhayes, C.P., R. Rayner. 2002. Operational Oceanography. In: J. Field, G. Hempel, C.P. Summerhayes, eds. Oceans 2020 Island Press, Washington. 109-135

Zaitzev, Yu., V. Mamaev. 1997. Marine biological diversity in the Black Sea: A study of change and decline. UN Publications, New York

Index